CHEMICAL CYCLES IN THE EVOLUTION OF THE EARTH

Edited by

C. Bryan Gregor
Robert M. Garrels
Fred T. Mackenzie
J. Barry Maynard

WILEY

A WILEY-INTERSCIENCE PUBLICATION

JOHN WILEY & SONS

New York / Chichester / Brisbane / Toronto / Singapore

For WILLEM NIEUWENKAMP
1903–1979
qui orbes reinvenit

Copyright © 1988 by John Wiley & Sons, Inc.

All rights reserved. Published simultaneously in Canada.

Reproduction or translation of any part of this work
beyond that permitted by Section 107 or 108 of the
1976 United States Copyright Act without the permission
of the copyright owner is unlawful. Requests for
permission or further information should be addressed to
the Permissions Department, John Wiley & Sons, Inc.

Library of Congress Cataloging in Publication Data:

Chemical cycles in the evolution of the earth / edited by C. Bryan Gregor . . . [et al.].
   p. cm.
  Based on discussions held in the forum of the Work Group on Geochemical Cycles.

"A Wiley-Interscience publication."
Bibliography: p.
Includes index.
ISBN 0-471-08911-7
  1. Geochemistry—Congresses. 2. Geology—Periodicity—Congresses.

I. Gregor. C. Bryan.
QE514.C47 1988
551.9—dc19   87-34941
                                                       CIP

Printed in the United States of America

10 9 8 7 6 5 4 3 2

CONTRIBUTORS

James I. Drever, Department of Geology and Geophysics, University of Wyoming, Laramie, Wyoming

C. Bryan Gregor, Department of Geological Sciences, Wright State University, Dayton, Ohio

William T. Holser, Department of Geology, University of Oregon, Eugene, Oregon

Yuan-Hui Li, Lamont-Doherty Geological Observatory, Columbia University, Palisades, New York

Fred T. Mackenzie, Department of Oceanography, University of Hawaii, Honolulu, Hawaii

J. Barry Maynard, Department of Geology, University of Cincinnati, Cincinnati, Ohio

Norman H. Sleep, Department of Geophysics, Stanford University, Stanford, California

Manfred Schidlowski, Max-Planck-Institut für Chemie, Mainz, Federal Republic of Germany

Ján Veizer, Lehrstuhl der Sedimentgeologie, Institut für Geologie, Ruhr Universität, Bochum, Federal Republic of Germany
and
Derry Laboratory, Department of Geology, University of Ottawa, Ottawa, Ontario, Canada

James C. G. Walker, Department of Atmospheric and Oceanic Science, University of Michigan, Ann Arbor, Michigan

Thomas J. Wolery, Lawrence Livermore Laboratory, Livermore, California

PREFACE

Atmosphere, ocean, and biosphere write their history in the rocks and move on, renewing themselves in a few years or millennia. Their record is preserved in those parts of the lithosphere, growing scarcer as time goes on, that escape the ravages of Zeus and Pluto. Much of this record is chemical; reading it requires the development of chemical hypotheses and models and the formulation of chemical questions to test them. This book deals with some of the broader questions being addressed to the chemical record today. The title implies the assumption that chemical mass transfer between global reservoirs is cyclic, that the intake of geologically permanent reservoirs is balanced in the long run by their output so that their size and composition remain, within rough limits, constant over long periods. Evidence for this quasi-steady state view is presented and discussed, together with data that can be used to trace fluctuations about the steady-state mean, and signs that indicate long-term secular change. There emerges a pattern of geochemical cycles going round and round on an earth that ages slowly in comparison, like the days in a human life.

These cycles are all linked at one point or another to the biosphere, whose importance in regulating some of them has long been recognized. There has even been envisaged a self-regulating system for the planet Earth controlled by biotic reactions and processes: a global "metabolism." Humans are currently intefering with this natural metabolism. Inputs to biogeochemical element cycles have been accelerated by man's industrial, agricultural, and fossil-fuel burning activities. The environment is being modified accordingly, and in a few cases (e.g., atmospheric carbon, sulphur, and nitrogen gases) the global chemical balance is being perceptibly disturbed. Furthermore, substances that do not exist naturally are now being released into the environment for the first time: thousands of synthetic compounds never before confronted by life in the arena of natural selection. The consequences cannot readily be anticipated nor remedies sought unless the theater affected can be formulated as a system of reservoirs and fluxes with controls whose mechanisms are (or are thought to be) understood; in other words, can be represented by a mathematical model. Because many systems have already been appreciably disturbed by human activities, a historical approach, such as the one taken here, is generally required to establish the range of "natural" (i.e., pre-human) conditions. This approach, in addition to providing an estimate of the background parameters of the system, can sometimes offer clues as to how it may have responded to natural disturbances during the geologic past.

Because geochemical processes appear cyclic, superimposed on the slow secular evolution of the earth, the modeling of environmental systems is the modeling of cyclic processes or of parts of them. There are by now many scattered papers dealing with or touching on geochemical cycles, most of them written for specialized readers. A few edited collections, as well as several books on the subject, have recently become available. But the idea is not familiar to scientists in general, or even to many geologists and still less to those individuals in the industrial and governmental sectors of science on whom falls the main burden of responsibility for our interaction with the environment. If a wider public were aware of cyclic processes in geology, there might be more fruitful interdisciplinary collaboration in studying these processes, which have important, sometimes dominant, nongeological aspects. It is hoped that this book, through its historical approach to the subject, will help to increase the scientific community's interest in geochemical cycles from both a theoretical and a practical point of view.

This work grew out of discussions in the forum of the Work Group on Geochemical Cycles over two decades that have seen considerable advances in both data collection and modeling capabilities. It reflects a broad consensus on most topics and occasional disagreement on some. After the prologue introducing the historical foundation of cyclic concepts, two rapidly cycling reservoirs of the exogenic system, the ocean and the atmosphere, are discussed in the first two chapters. The mass balance between igneous rocks and sediments is related to oceanic and atmospheric composition, with major- and minor-element sources and sinks for the ocean being identified in Chapter 1 before development of cycles of atmospheric gases in Chapter 2. Chapter 3 deals with interaction between exogenic element cycles and the mantle, and Chapter 4 is devoted to the cycles of carbon and sulphur, two elements that play a very active role in the exogenic system and whose reservoirs and fluxes are well enough known to justify detailed treatment. In Chapter 5 cycling concepts are integrated with signs of long-term secular change to arrive at a set of working hypotheses for the history of the exogenic cycle as a whole.

The major responsibility for each chapter is indicated in the chapter title, but it should be emphasized that all of the topics presented here were subjects of open discussion, much of it lively, among all concerned. Where there is disagreement, it stems for the most part from uncertainties in the continuously evolving data base of geochemical cycling research. The field is still in its infancy and changing rapidly. We hope this book presents something of the state of the art as it now exists, realizing that tomorrow will see many changes.

Our task has been lightened by help from many quarters. We owe special thanks to our friend, Konrad F. Springer, for giving us the idea of a book and then encouraging us to begin; to the National Science Foundation for supporting the Work Group on Geochemical Cycles (with Grants EAR77/21097, 78/20545, 79/18290, 79/18564, 81/18561 and 82/17337); to Denis M. Shaw of McMaster University, for reviewing the manuscript and making many helpful suggestions; to Emmaly Maynard for smoothing our rough English: *labor limae ac mora*; to Everett W. Smethurst (former editor at John Wiley & Sons), and to his successor, Stephen A. Kliment, for guiding us with enthusiasm and wisdom;

to Lisa Van Horn and Linda Shapiro and their able staff at Wiley (not forgetting the copy editor and typesetter); to the Department of Geology at the University of Cincinnati for drafting and word processing; and last but not least, to our fellow contributors, without whose forbearance and sustained goodwill the manuscript could never have reached the press. To all of these, and others not named here, we gratefully acknowledge a large debt for whatever success this book may enjoy.

Dayton, Ohio
February, 1988

C. BRYAN GREGOR
ROBERT M. GARRELS
FRED T. MACKENZIE
J. BARRY MAYNARD

CONTENTS

Willem Nieuwenkamp (1903–1979): A Biographical Note   1
C. B. Gregor

Prologue: Cyclic Processes in Geology, a Historical Sketch   5
C. B. Gregor

1. The Beginnings / 5
2. Neptunism and Plutonism / 7
3. Classical Magmatism / 11
4. The Sodium Cycle / 12
5. The Rock Cycle / 14
6. Biogeochemical Cycles / 15
7. The Future / 15

1 Geochemical Cycles: The Continental Crust and the Oceans   17
J. I. Drever, Y.-H. Li, and J. B. Maynard

1.1 Mass Balance Between Igneous Rocks and Sedimentary Rocks Plus Seawater / 19
1.2 Cycles of Major Elements in the Ocean / 24
1.3 Mass Balance Between River Inputs and Oceanic Sediment Outputs for Minor and Trace Elements / 37
1.4 Variations with Time / 42

2 Geochemical Cycles of Atmospheric Gases   55
J. C. G. Walker and J. I. Drever

2.1 The Modern Atmosphere / 55
2.2 Variations with Time / 68

3 Interactions of Geochemical Cycles with the Mantle 77
T. J. Wolery and N. H. Sleep

- 3.1 Plate Tectonics / 77
- 3.2 Experimental and Theoretical Investigations / 85
- 3.3 Elemental Fluxes / 86
- 3.4 Low-Temperature Alteration of Sea-Floor Basalts / 89
- 3.5 Summary / 103

4 Biogeochemical Cycles of Carbon and Sulfur 105
W. T. Holser, M. Schidlowski, F. T. Mackenzie, and J. B. Maynard

- 4.1 The Carbon Cycle / 107
- 4.2 The Sulfur Cycle / 118
- 4.3 Isotopic Fractionations in the Geochemical Cycles of Carbon and Sulfur / 129
- 4.4 Relations Among Isotope Age Curves / 143
- 4.5 Tectonics and Isotope Age Curves / 148
- 4.6 Long-Term Evolution of the Biogeochemical Cycles of Carbon and Sulfur / 161

5 The Evolving Exogenic Cycle 175
J. Veizer

- 5.1 Introduction / 175
- 5.2 The Earth System / 176
- 5.3 Recycling of Oceanic Crust / 182
- 5.4 Global Tectonic Realms and Their Recycling Rates / 184
- 5.5 Growth and Recycling of Continents / 189
- 5.6 The Global Sedimentary Mass and Its Recycling / 199
- 5.7 The Sedimentary System and Its Lithologic Evolution / 200
- 5.8 Chemistry of the Sedimentary System / 206
- 5.9 The Hydrosphere and Atmosphere / 211
- 5.10 Synopsis / 214
- 5.11 Conclusions / 218

References / 221

Author Index / 263
Subject Index / 270

WILLEM NIEUWENKAMP (1903–1979): A BIOGRAPHICAL NOTE

C. B. Gregor

Willem Nieuwenkamp was born in Lunteren (The Netherlands) on January 1, 1903 to Anna Wilbrink and W. O. J. Nieuwenkamp, the Dutch painter and

engraver. He died in Bilthoven on November 12, 1979. Notwithstanding an early interest in astronomy (which he never lost), on entering Utrecht University in 1919 Nieuwenkamp chose geology for his studies, seeing behind its then preponderantly descriptive and systematic character a future opportunity for exciting theoretical advances. His formal studies ended with the doctoraal examination in 1926, and he received the D.Sc. degree in 1932 for a dissertation on the crystal structure of lead bromide and lead fluorobromide. In 1947, he was appointed professor of crystallography and petrography (later of geochemistry and mineralogy) at Utrecht, a post he held until his retirement in 1968. In 1965, he was elected to the Royal Netherlands Academy of Sciences.

Nieuwenkamp's early experiences were varied. After two years in Patagonia with an oil company, he went in 1933 to Göttingen as assistant to V. M. Goldschmidt and later worked on gravity measurements, first with F. A. Vening Meinesz in submarines, then, with the onset of the war, on land with a bicycle. But his real interest lay in the search for a satisfactory theory of the origin of rocks. Quick to see the logical flaws in classical magmatism (at the time the orthodox view of petrogenesis), he traced it back to its roots in the neptunist–plutonist controversy of 150 years before, looking for where it had gone wrong.

The confining war years extended Nieuwenkamp's command of the literature and ripened his ideas. Exhaustive reading of Hutton and Playfair, and of their opponents in the Wernerian Society, and careful study of field observations, especially those of the great Scandinavian geologists Eskola and Sederholm, drew him to a neo-Huttonian viewpoint from which he would later create the *persedimentary theory* that came to be associated with his name. After World War II, he was able to make his own direct observations. Careful studies of granitic bodies and their field relations in Spain and in the Massif Central of France made it clear that many granites had originated from sedimentary rocks: an idea that is commonplace today but was highly controversial in 1950. These field studies formed the subject of a number of doctoral dissertations at Utrecht University, and the sedimentary heritage of granite became a pillar of Nieuwenkamp's developing persedimentary theory. Another was the cyclic behavior of sodium. The creation of granite from sediments appeared to require the addition of sodium, whereas the apparent steady state of the ocean required its removal from seawater. By a simple mass balance calculation, Nieuwenkamp showed that the excess in the sea corresponded roughly with the deficiency in the rocks. The sodium could be recycled, as could other mobile or volatile elements such as chlorine and boron.

For Nieuwenkamp, the earth was a cyclic system where every end was a new beginning. The gradual succession of events to complete the cycle had its counterpart in the composition of the rocks, each kind shading imperceptibly into the next in a *continuous rock series*, defying the artificial boundaries of petrography. In its extreme form, the persedimentary hypothesis pictured all rocks as being resurgent and as having passed through the sedimentary cycle at some time. The theory had to retreat somewhat from this position in the light of new discoveries about the oceanic crust and the origin of synorogenic magmas, and Nieuwenkamp later recognized the oceanic and continental cycles as separate but interconnected, both of them resurgent but only one persedimentary.

Nieuwenkamp has left no voluminous literature behind him. The main elements of his petrogenetic theory can be found in three papers: "Geochemistry of sodium" (1948), "Géochimie classique et transformiste" (1956), and "Oceanic and continental basalts in the geochemical cycle" (1968). Nor did his soft, conversational voice carry far in the great lecture halls of Europe's scientific societies. He never visted the United States. His considerable influence was felt instead at small colloquia, at roadside stops by his favorite outcrops, in quiet bars and restaurants, in his comfortable and elegant room at the old geological institute on the Oude Gracht and, by a lucky few, at his house in Bilthoven where he and his wife Sienie (also a geologist) kept a table justly reputed for its cuisine, wines, and conversation. Prominent among those influenced by Nieuwenkamp's ideas were Holmes, Raguin, Wegmann, Winkler, and perhaps above all Barth, who restated and augmented Nieuwenkamp's theories in his well-known textbook (*Theoretical Petrology*, 1962) and in a number of papers written in the early 1960s, one of which (Barth, 1961b) has practically the same title as one by Nieuwenkamp (Nieuwenkamp, 1965b). It has been mainly through the assimilation of his ideas into the writings of scientists like these (all of whom were on the most cordial terms with him) that Nieuwenkamp's cyclic views of geochemistry and petrology have achieved the still limited and largely anonymous currency that they enjoy today. In the last decade of his life, Nieuwenkamp turned his attention to scientific biography, writing articles on van Marum, von Buch, Krayenhoff, and Vening Meinesz. But his interest in the cycles never flagged. At the time of his death, he had begun a manuscript entitled (with characteristic modesty) "The new petrology of Tom F. W. Barth."

Nieuwenkamp's extraordinary erudition allowed him to view a subject from many vantage points, and a wide command of languages gave him access to a range of literature beyond the reach of most of his contemporaries. His underlying attitude was one of benign scepticism, and nothing amused him more than to discover in some dogma of the times an inconspicuous but pregnant inconsistency. On his retirement from the chair of geochemistry, his Utrecht colleagues and students gave him a beer mug engraved with the motto: "Who sows doubt shall reap insight" ("Wie twijfel zaait zal inzicht oogsten"). Always urbane and good-natured, he made many friends among scientific supporters and opponents alike. His wit was mordant but never cruel. He was a delightful companion, and traveling with him was a continual series of digressions rewarded by unusual discoveries, arresting views, unexpected meetings, and memorable meals. He had not much patience with stupidity but was never unkind to simple ignorance. His towering intellect was half hidden by an irrepressible sense of fun that made him approachable. On hearing that a student using pyknometers thought they must have something to do with pigs, he got the glassblower to make some actually shaped like pigs and surreptitiously substituted them for the real ones.

He loved the past, but always looked forward; and were he alive today, he would approve of the great new geological institute on the edge of Utrecht and gently chide us for wishing ourselves back in his familiar room on the Oude Gracht.

PROLOGUE: CYCLIC PROCESSES IN GEOLOGY, A HISTORICAL SKETCH

C. B. Gregor

*Historians make a
little stir, spinning
round those memorable
feats, while all
together drain down
the sink of time.*
  —W. Nieuwenkamp

1 THE BEGINNINGS

Influential scientific ideas have sometimes had to make more than one appearance before people were ready to accept them and see the new light they shed on the established order of things. This has happened more than once in the science of geology, which has recently been illuminated by the concept of continental drift, an idea advanced without success on no less than three distinct occasions since the mid-seventeenth century and predating, in a manner of speaking, the very existence of the science it was so profoundly to influence. Another example is the subject of this book. Rudimentary allusions to the cyclic aspect of geology can be found in writers of antiquity; definite formulations of it were presented in 1785 and in 1871, but both were ignored. Only in the last decades has it begun to achieve recognition; and even today, although textbooks pay lip service to the *rock cycle*, much (if not most) of the current literature in geology and geochemistry continues to ignore its broader implications.

Geology began to take shape as an organized, independent discipline in the eighteenth century, when extensive travel made widespread observations possible and rationalism provided a propitious intellectual climate for their interpretation. Until then, it was mostly shared between cosmology, where it took largely unsupported flights of fancy about the earth as a whole, and mining, in which its vision was generally though not always limited by the utilitarian needs of the craft it served. It is through the former of these two distinct heritages that we can trace the idea of cyclic processes in geology.

As far back as human memory can reach, cosmological stories have had a place in the folk tales, religion, or science of every culture. Many of these accounts of the world and its workings are built around one or another (sometimes both) of the following ideas:

1. The world has progressed in a secular manner from a starting condition to a final one, as in the six days of creation described in Genesis.
2. The world is essentially unchanging, maintained in a steady state by processes whose net effect is zero, such as those evoked in the Book of Ecclesiastes by Koheleth the Sage.

Thus, for the universe the *big bang* hypothesis (secular) is antithetical to the steady-state one (cyclic); for the earth, the idea that the land worn away by erosion is restored by uplift of new mountains made out of the resulting sediment (cyclic) has as its antithesis the idea that the lost material accumulates continually in the sea and is replaced by magma from inside the earth (secular).

The recognition of a process as secular or cyclic is partly a matter of time. The butterfly may not believe a rabbit who tells her that the seasons recur and that he is looking forward to next spring. Even if the rabbit shows the butterfly last year's dead leaves or the rings on a treestump, the butterfly may still be incredulous: "The world cannot be that old," the butterfly will say. "These things must have another meaning." In turn, the butterfly would have a hard time convincing a mayfly of the sunset. When our time is short compared to the period of the cycle, the process seems secular; the converse is also true. Seeing the rivers flow, we can reason that the water must return to the source to run again, else they would dry up. When time is long, we may be aware of processes that are in the long run secular while consisting of recurrent cycles of events that are short in comparison with the whole, so that successive cycles are barely distinguishable from one another. Such are the years in the life of a human (see Chapter 5) or epochs in the history of the earth.

According to the Book of Genesis, the earth's age can be reckoned at about 6000 years. The early Christian apologists calculated it; so did Martin Luther and, after him, Archbishop Ussher. On a time scale like this, the water cycle is obvious: if the river water ran always without coming back, the sea by now would be lapping the walls of the Kremlin. But it is hard to see how erosion could have sculptured mountains as some (like Agricola) believed. The great valleys of the Alps would have been lowered by less than the height of a human being. It is sometimes supposed that the Reformed Church prevented cosmologists of the sixteenth and seventeenth centuries from speculating freely about the age of the earth, by threats of punishment for heresy. But while it is true that the early Protestant Church generally approved the punishment and even execution of heretics, it does not seem to have gone out of its way to persecute scientists. Bruno and Galileo, the most celebrated mavericks of those times, were both victims of the Inquisition. With a few well-known exceptions (none of whom was punished in a protestant country), cosmologists of the

Reformation seem to have regarded the Mosaic time scale with equanimity. Having little idea of the rates of natural processes, they perhaps did not grasp the severity of its limitations.

2 NEPTUNISM AND PLUTONISM

By the eighteenth century, religious intolerance was unquestionably on the wane as far as scientists were concerned, although those who felt called on to disregard the biblical time scale might still have to risk a reproving frown or a slap. A few did. De Maillet (1748), supposing that the earth's crust had been laid down by a primeval ocean that had once covered its highest parts, calculated (from the retreat of the sea on the Nile delta) that 400,000 years must have elapsed since the beginning. Buffon more modestly asked 75,000 years for his account of the earth's evolution (Buffon, 1807). But the vastness of geologic time was not yet generally sensed.

The eighteenth century was an age of rationalism and of discovery and dissemination of knowledge. It was the age of Kant and Laplace, Cavendish and Lavoisier, Linnaeus and Buffon, Captain Cook, Voltaire, Diderot, and d'Alembert. De Luc, Voight, Schreiber, and other geologists in Europe were noticing (but not understanding) what are now called unconformities in rock outcrops where flat-lying strata rested on the eroded edges of older, upturned ones beneath them. James Hutton knew these for what they were: relics of an ancient world built on the ruins of a still older one and of materials derived from its destruction; witnesses of the deformation, uplift, and erosion of one set of strata, followed by subsidence and the sedimentation of another, then renewed uplift and erosion to lay the contact bare for human eyes. By 1788, Hutton had found the unconformity that divided the Systems later to be called Devonian and Silurian in Britain, noting the pebbles of Silurian *schistus* (graywacke) that commonly occur in the base of the Old Red Sandstone. He saw that rocks were recycled, new ones being formed from the erosional detritus of old.

Three years earlier, Hutton (1785, 1788) had presented to the Royal Society of Edinburgh his *Theory of the Earth*, whose main tenets were that the spoils of erosion accumulated on the sea floor in thick piles where they became hardened into new rock by the earth's internal heat and that by the agency of this same subterranean heat they were eventually raised up to form land in an apparently endless cycle of erosion, sedimentation, and uplift. Subterranean magma was widespread on account of the high temperature inside the crust, and it often surged up to invade the solid strata above. God's purpose in arranging things so was the continued renewal of the land to succour humankind. He worked with an unhurried hand. Presently observable causes, operating within the range of their current rates, could account for all that was to be seen in the rock record. The only other requirement was a hitherto unimagined expanse of time. The notorious neptunist–plutonist controversy broke out toward the end of the eighteenth century. The plutonists adhered more or less to Hutton's theory but emphasized its magmatic aspect at the expense of the cyclic one; the neptunists aligned themselves with the *geognosy* of A. G. Werner, professor of mineralogy at the Mining Academy of Freiberg in Saxony, according to whom the crust of the earth had been laid down (in the manner already proposed by de Maillet) as a series of worldwide formations by a primeval ocean that once reached above the highest mountains and then mysteriously fell to present-day sea level (Ospovat, 1971). Generally, granite, graywacke, and basalt were *primitive* chemical precipitates of this ancient ocean when at its height; later deposits were formed of material eroded from the primitive land as the sea retreated. For Werner, crustal evolution was a secular affair. His earth was cold and passive; volcanoes were no more than chance, local phenomena of recent times, caused by the accidental conflagration of underground coal seams, themselves of comparatively late origin. The face of the earth had been built and shaped, once and for all, by the agency of water.

The conflict was waged over much of Europe, and with peculiar bitterness in Britain where the Wernerian Society had its headquarters at Edinburgh. Werner himself, genial and averse to correspondence, did not participate. Hutton (1795; see also Geikie, 1899) was finally goaded, by neptunist attacks on his theory, into publishing an extended version of it just before his death, and this was amplified with lucid examples in another book by his friend John Playfair in 1802. Werner's model of the earth did not stretch the imagination of those accustomed to the Mosaic time scale. Hutton's model opened a vista of time receding into a past of inconceivable remoteness. Not all plutonists believed in a long time scale (rather the contrary), nor all neptunists in a short one. So there were (in this respect) four alternative ways of looking at things (exemplified below with four of the better-known names of that time):

	Plutonist	Neptunist
Short time scale	Hall*	Kirwan[†]
Long time scale	Hutton	de Maillet[‡]

Adherents of the short time scale would later be identified with the doctrine that came to be known as catastrophism; those who (like Hutton) believed in a long time scale were the forerunners of uniformitarianism.

The neptunists fell gradually silent in the early years of the nineteenth century. The evidence of subterranean lava with its intrusive contacts (the classic instance being the granite dikes in Glen Tilt discovered by Hutton in 1785) and the connections between layers of lava and the extinct volcanoes that had made them, already known to Guettard in 1752 in the Auvergne district of France and

---

*Sir James Hall of Dunglass, Bart. A friend of Hutton, generally regarded as the founder of experimental geology.
[†]An Irish chemist, the most vociferous of the neptunists.
[‡]Although his work *Telliamed* was not published until 1748, de Maillet really belongs to the previous century; but no other well-known neptunist seems to have believed in a long time scale.

rediscovered there by Desmarest in 1763 (Desmarest, 1774), were undeniable, and their meaning was clear to everyone. Granite and basalt were congealed melts, not chemical precipitates of the sea. Moreover, they could not be primitive when, as was often the case, they occurred in or (as lava flows) above strata belonging to the young formations of Werner's system.

To Hutton, subterranean lava was a by-product of the cyclic apparatus that kept the world in a steady state. To the later plutonists, it was a rising tide of fortune that carried them victorious into the nineteenth century. Hutton's cycles were forgotten. It is sometimes said that Hutton's writing is awkward, prolix, and incomprehensible; and up to a point this has to be admitted. (Playfair, his apologist, justified his own lucid exposition of the Huttonian theory on these grounds.) Yet it is difficult to believe that the central theme of the *Theory of the Earth* was condemned to obscurity by Hutton's own clumsiness. For one thing, the abstract of his two talks to the Royal Society of Edinburgh (1785) is clear and concise, superior in style to the full-length paper (1788) and the book (1795) ultimately developed from it; and for another, Playfair's *Illustrations* (1802), a model of clarity replete with carefully described examples, achieved a wide circulation and should have dispelled any doubts about Hutton's meaning from the minds of those who had not read the more narrowly distributed *Abstract*. The argument that people simply did not understand Hutton does not stand up to contemporary criticism: for instance, Kirwan (1793) and De Luc (1790–1791), whose misrepresentations of the Huttonian theory are plainly deliberate, Murray (1803), and Fitton's (1803) even-handed commentary in the *Edinburgh Review*. Basset's (1815) translations of Playfair's *Illustrations* and Murray's attempted rebuttal, published in a volume dedicated to the students of the École Normale in Paris (of which Basset was director), give some measure of the attention paid to Hutton's theory in Europe.

The igneous origin of granite and the recycling of sediments are both revealed by geometric relations (e.g., intrusive contact, unconformities) observed in rock outcrops and interpreted according to logically similar principles. In 1800, the difference between them was that the former had implications that need be followed only as far as its protagonists wished, whereas the latter plunged one into what John McPhee (1981) has called *deep time*, an experience for which the human intellect was not yet prepared. No doubt this accounts in large measure for Hutton's failure to get the cyclic idea to "take," but it is worthwhile to consider also the intellectual climate of the time, a climate whose development had been influenced by the earlier reaction against religious authority. In seventeenth-century England, the Restoration had weakened ecclesiastical power. Rationalism had gained force, and there was a growing respect for the inductive logic of Francis Bacon. Charles II (who disliked the church) founded the Royal Society for "the extension of natural knowledge as opposed to supernatural." This aim was to be accomplished by direct experiment, a Baconian stipulation that, even if not always honored to the letter, may have kept science in England, at least in appearance, nearer to the inductive ideal than it was in Germany or Scotland. The latter country, although reacting against the theological tyranny of the seventeenth century, did not altogether succeed in shedding its logical method. According to the historian of the Scotch intellect, Thomas Henry Buckle (1861), the Scottish school of philosophy that arose in the eighteenth century was a deductive one because "the theological spirit had taken such a hold on the Scotch mind, that it was impossible for the inductive method to gain a hearing." Buckle may have overstated his case, but his remark sheds light on the distaste with which the neptunist–plutonist controversy came to be regarded in England, where the Geological Society would shortly be founded with the object of "observing the condition of the earth but by no means generalizing the causes which [had] produced that condition." The rise of religious scepticism in France, culminating in the excesses of the Revolution and the emergence of Napoleon, produced a conservative reaction in England that brought back religious orthodoxy as a force that, if not to be feared, had at least to be reckoned with in scientific circles. Kirwan, attacking Hutton in 1793, observed "how fatal the suspicion of the high antiquity of the globe [had] been . . . to religion and morality."

Such was the English state of mind at the turn of the century when Hutton's unorthodox theory of cyclic change, with its deistic overtones and vision of unending time, made its appearance. Hutton was unconsciously deductive. He sincerely believed that he had built his theory of the earth inductively, reasoning from observation. In fact, he created it by a bold and lucky stroke of imagination and marshaled facts to confirm it. Its mainspring was Hutton's deism, the notion that the whole thing was designed for the benefit of humans.* His most significant observations (the unconformity beneath the Old Red Sandstone and the granite in Glen Tilt) were made after his presentation to the Royal Society of Edinburgh, not before. He divined the origin of marble not from Hall's famous experiment (which Hall delayed making until after Hutton's death) but because he learned from Black that limestone could be heated under pressure without losing its "fixed air." So it was too with his mistakes. He assumed that flint nodules and septarian concretions had been molten (even though the rocks containing them showed no signs of having been heated) because he was convinced of the importance of heat for changing loose sediment into hardened rock. Although amply provided with means and time for traveling, he never saw an active volcano. His repeated exhortations to "reason from our principles" and his well-known disdain for geologic experiments gave him away. His logical method could not pass unquestioned in the land of Francis Bacon and John Locke, where William Smith would toil 18 years assembling data to establish the principles of stratigraphy. No matter if Werner erred ten times more than Hutton in this respect: the greater wrong did not right the lesser one.

*It may be objected that this is not obvious in the *Abstract*, Hutton's first written sketch of the *Theory*, where the "argument from design" is relegated to a brief concluding statement; but in his later writings it reveals itself as his real inspiration. It has been suggested (Bailey, 1967) that Playfair really wrote the *Abstract* and kept the deistic argument toned down to avoid controversy.

Whatever the truth, Hutton's theory failed to transform geology as it might have done. The doctrine of present causes would come back as the cornerstone of geology's logical structure, and the vastness of geologic time would one day be demonstrated beyond question. Hutton would be acclaimed the father of geology, but the geologic cycle would not be seen again for many a year with the clarity of Hutton's original insight.

3 CLASSICAL MAGMATISM

Georges Cuvier led the catastrophists. In the Tertiary strata of the Paris basin he noticed that unconformities interrupted the succession of fossil forms and supposed they were the record of global calamities that had repeatedly exterminated life, which made a fresh start in each intervening period of calm. But it turned out that such hiatuses were only of local or at most regional extent. They were filled in elsewhere by strata where the record was continuous, so that they represented only losses resulting from erosion, not worldwide extinctions of life. In spite of a following in England under William Buckland, professor of mineralogy at Oxford, catastrophism yielded to uniformitarianism, the doctrine of present causes inherited from Hutton by Charles Lyell and advanced with the force of innumerable examples in his *Principles of Geology*. Uniformitarianism gained a strong ally in Charles Darwin, who in 1859 published *Origin of Species*. The idea of evolution by natural selection, like the forgotten cycles of Hutton, implied a time scale of unheard-of length. If no species had been seen to appear by human eyes, how much time would be needed for the evolution of all the living forms, for the rise and fall of all their extinct forebears? The long time scale became a subject for debate, even if still vigorously rejected in some quarters. The question of actually measuring its length had yet to be taken up seriously.

The tide of magma that ushered in the century had not abated, and a new school of geology was rising upon it. It started in Germany with Justus Roth and developed under the great petrographer Harry Rosenbusch into what is now called classical magmatism. The primitive crust of the earth was of magmatic rocks formed by congelation from the supposedly molten globe. These igneous rocks were weathered to produce sediments that accumulated at the surface, growing ever thicker as time went on, and their loss was made up by fresh eruptions from the still molten interior. When buried and subjected to heat or stress, sediments were converted into metamorphic rocks recognizable by their banded or foliated appearance. This could happen to igneous rocks too, so that sedimentary and igneous rocks of similar composition could become indistinguishable when metamorphosed. There was *orthogneiss* from granite and *paragneiss* from mudstone. But magma was paramount. Hundreds of different kinds of igneous rocks were discovered, examined, and classified with materialistic zeal.

The new petrology set the stage for renewed debate about the age of the earth, now a fiery ball of magma with an eggshell crust. William Thomson (later Lord

Kelvin) saw that agencies whose rate never flagged would amount to a *perpetuum mobile*, that the uniformitarians were defying the second law of thermodynamics (recently enunciated by himself and Clausius). Supposing the sun's heat to be derived from gravitational accretion, he quickly showed that, expended at the present rate, the sun could not continue to radiate more than a few tens of millions of years. Using Fourier's method for calculating the distribution of temperatures for a cooling sphere, he found it could not have taken much longer than that for the earth's surface to cool from a supposed initial temperature of 5000 K to its present one. The earth, though much older than the Bible implied, was certainly no more than 100 million years old. This limit was nicely confirmed by geologic reasoning from classical magmatism: the aggregate thickness of all the sedimentary formations, divided by their supposed rate of accumulation, gave about the same result (Reade, 1879).

Kelvin's estimate was unwelcome to those geologists who now believed in uniformitarianism and saw in addition the implications of Darwin's theory of evolution. But they had no quantitative arguments of their own, and Kelvin's reasoning seemed unassailable. It does not seem to have occurred to anybody that his premises might have been unsound: that the sun might be powered by something other than gravity or that the earth might be able to generate heat internally. Against the geologic argument that supported Kelvin, only one voice was raised—that of James Croll (1871), who pointed out that the quotient obtained by dividing the rate of sedimentation into the size of the sedimentary pile had nothing to do with the age of the pile because, instead of accumulating, the sediments were recycled. "As the materials composing our stratified beds," he wrote "may have passed through many cycles of destruction and reformation, the time required to have deposited at a given rate the present existing mass of sedimentary rocks may be but a fraction of the time required to have deposited at the same rate the total mass that has actually been formed." Croll's argument is the same as one might use against trying to find out the age of a person by dividing the amount of his blood by the rate of its movement through the heart. Yet it made no impression on his contemporaries. Neither T. M. Reade, who refined the sedimentation argument in 1893, nor John Joly, who in 1899 advanced a similar argument based on the supposed accumulation of salt in the sea (an idea that had occurred to Edmund Halley and, before him, to Leonardo da Vinci) so much as mentions it.

4 THE SODIUM CYCLE

In 1895, Becquerel discovered radioactivity, and it was soon apparent that here was a means not only of refuting Kelvin's assumption about cooling but of measuring the earth's age as well. By 1930, Rutherford, Boltwood, Strutt, Holmes, and their collaborators had shown the way to dating geologic materials radiometrically and had established the reliability of the new method. The long time scale was to prove longer than anyone had anticipated. There were sedimentary rocks billions of years old. The Cambrian Period had begun a mere 600 million years ago. Claire Patterson would later measure the age of the earth itself at 4.6 billion years. One might imagine that such a profusion of time would have embarrassed the magmatists. If magma kept the crust built up against the ravages of erosion and the waste products accumulated in the sea, at present rates of influx the ocean basins should long ago have been choked with sediment and salt. It would be difficult to argue that things had been very different in the past; ancient rocks are much like recent ones. Cross-bedding, filled channels, ripple marks, oolites, and mud cracks evoke the dunes, streams, beaches, waves, and mudflats of our own experience. The same processes were seemingly at work as now. To have kept things in balance they must have been cyclic. Submarine sediment must become land again; salt must somehow leave the ocean.

But the meaning of time was lost on the magmatists, whose torch passed to F. W. Clarke (1908), one of the founders of modern geochemistry, and then, still burning, to V. M. Goldschmidt (1933). Each viewed sedimentation as an essentially secular process in which igneous rocks were split by weathering into sediment and sodium, the latter accumulating in seawater; each attempted to calculate the total mass of sedimentary rocks by setting the difference between their supposed sodium content and that of an equivalent mass of parent igneous material equal to the amount of sodium in the sea. Neither grasped the fact, pointed out long before by Sterry Hunt (1875), Osmond Fisher (1900), and G. F. Becker (1910), that sea salt goes into marine sediments by entrapment in pore water, adsorption on clay, and formation of evaporites. In fact, the sediments store about as much sodium in this way as the sea itself, but this does not show up in analyses of samples taken from surface outcrops, because by then it has been leached out and is on its way back to the sea again.

A difficulty about metamorphism was that paragneiss contained about as much sodium as granite, 1% or so more than was found in outcrops of the sediments from which it had been formed. The classical solution of this problem was to suppose the existence of a mysterious fluid called "ichor" that ascended from the earth's interior to places where metamorphism was going on, bringing the necessary sodium with it. Like the magma that was supposed to keep replenishing the crust, the ichor was "juvenile," supplied de novo from the molten depths. S. J. Shand (1943) questioned the usefulness of such an ad hoc hypothesis, and the Dutch geochemist Nieuwenkamp (1948a) pointed out that the search for sodium to make paragneiss complemented the need to get rid of it from the sea. Nieuwenkamp developed (1956a,b) the idea that the chemical elements in the crust really move in closed cycles, and from this he elaborated a neo-Huttonian view of geology that achieved some recognition in Europe and in the British Isles through its influence on Barth, Holmes, Winkler, and others (see, for instance, Barth, 1961a,b; Holmes, 1965; Winkler and von Platen, 1961). In 1949, Harald Carstens, then a student of Barth and later director of the Geological Survey of Norway, made a mathematical model showing how the sea could have reached its present salinity early in the earth's history and remained essentially in a steady state ever since. He assumed that the rate of removal of sodium (by the mechanisms referred to above) would be proportional to its concentration in the water. In his presidential address to the Geochemical Society in 1971, H. D. Holland (1972) showed that the sedimentary record is compatible with a steady-state ocean and constrains the concentrations of the major dissolved species over Phanerozoic time to within a factor of 2 of their present values. The search for mechanisms to buffer the ocean's chemical makeup, begun by Sillen in 1961, is treated here in Chapter 1.

5 THE ROCK CYCLE

The ocean, a homogeneous solution, has no intrinsic property that reveals the cyclic behavior of its salts. This is inferred from their present concentrations and rates of supply. We cannot tell how long a particular sodium ion has inhabited the ocean any more than we can tell how long a particular banknote has lain in the bank. It is otherwise with the sedimentary rocks. These are marked by fossils at the time of their deposition, so that we know their ages. Alexander Ronov and his collaborators at the Soviet Academy of Sciences have devoted more than 20 years to the compilation of stratigraphic isopach maps of the world from which they have estimated, by planimetric measurements, the volumes of the different Phanerozoic Systems (Ronov, 1959; Ronov et al., 1980). The resulting mass–age distribution can be modeled mathematically to yield information about the past behavior of the sedimentary cycle (Gregor, 1968, 1980, 1985). Garrels and Mackenzie in their book *Evolution of Sedimentary Rocks* (1971a) deal comprehensively with the cycle of the sedimentary rocks from a geochemical viewpoint. The crystalline rocks are also recycled in the sense that the continental basement associated with geosynclines has its age reset when these undergo tectogenesis. Patrick Hurley and John Rand (1969) of the Massachusetts Institute of Technology have made a compilation of the mass–age distribution of basement rocks, and this can be treated in the same way as the distribution of the sediments (see Chapter 5). There is a two-way flux between the sedimentary and crystalline reservoirs: high-grade metamorphism converts some sediments into crystalline rocks, and erosion of crystalline rocks produces sediments (see Gregor, 1970). Oceanic crust is lost by subduction under island arcs and cordilleran continental margins and renewed at ocean ridges. The rate of this cycle can be calculated from rates of ocean-floor spreading as determined by the distribution of magnetic anomalies. Other connections between crust and mantle are the subduction of sediment riding on the ocean floor and the eruption of mantle-derived magma into or onto the continents. These fluxes are unknown at present but are unlikely to be large in comparison with those already mentioned. If they do not balance each other in the long run, the difference between them represents a small secular component in the evolution of the crust, perhaps a grain of truth in classical magmatism. There has been considerable recent progress in understanding the chemical links between the mantle and the exogenic cycle, which are the subject of Chapter 3. The rock cycle and its history

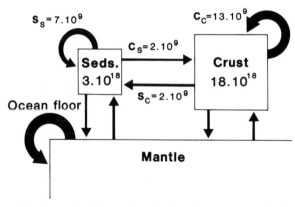

Figure 1 Generalized model of the rock cycle: S, sediments; C, crystalline (continental) crust; S_s, erosion of sediments; C_s, metamorphism; S_c, erosion of crystalline rocks; C_c, recycling of crystalline rocks (resetting of ages during tectogenesis). Units are tons; fluxes are in tons per year. The cycling of oceanic crust is represented by an unlabeled loop (lower left). Sedimentation rate = $S_s + S_c$, which here add to about 9×10^9 tons/yr.

are discussed at length in Chapter 5. A diagrammatic sketch of it, with rough estimates of the different fluxes and reservoirs, is given here in Fig. 1.

6 BIOGEOCHEMICAL CYCLES

The most active geochemical realm is the biosphere. With a mass only $1/(4 \times 10^6)$ that of the continental crust, it may well have fixed and metabolized in Phanerozoic time (about the last 600 million years) a mass of carbon greater than that of the continents. The major elements that participate in the biosphere's hurried cycles are carbon, oxygen, hydrogen, nitrogen, and sulfur, and the reactions that control their fluxes generally involve isotopic fractionation, so that the different reservoirs storing these elements in the rock record preserve isotopic information bearing on the activity of the biosphere through time and, via its linkages with the physical environment, on changes in atmospheric composition, global temperature, fluxes of nutrient elements, and so on. The best-known records in this respect are those of carbon and sulfur, and they are discussed in Chapter 4. The biosphere has played a crucial role in the evolution of the atmosphere and now controls its steady-state composition at least in regard to its major constituents. Atmospheric chemical cycles are the subject of Chapter 2.

7 THE FUTURE

In the two centuries since Hutton developed his *Theory of the Earth*, geologic thought has undergone many vicissitudes; doubtless it will experience more in the future. But in the establishment over the past two decades of a global tectonic theory, geology has provided itself with something it has not had until now—an integrated, well-tested model of the earth in terms of which to interpret field and laboratory observations of all the different subdisciplines of geophysics, geochemistry, paleontology, stratigraphy, petrography, mineralogy, and tectonics. A. N. Whitehead would have called it a general idea that illuminates the whole.

So while our understanding of the earth is sure to continue to evolve, its future changes are unlikely to be as dramatic as those of the past. As far as the geochemical cycle is concerned, we have to recognize that geochronology has reached back to the beginning of which Hutton could see no vestige—to a period of secular change whose record may lie at least partly within the reach of geochemistry. In fact, progress is already being made in the study of this secular component of the earth's history, and some of it is reviewed in Chapter 5. The future will no doubt make us increasingly familiar with the distant past, so that we shall one day view the present, with its controversies over secular and cyclic evolution, with equanimity.

ACKNOWLEDGMENTS

Thanks are due to Professor Cecil J. Schneer, of the University of New Hampshire, for a thorough review of this chapter that included important corrections and many helpful suggestions. Any errors that may remain are the author's own.

1 GEOCHEMICAL CYCLES: THE CONTINENTAL CRUST AND THE OCEANS

J. I. Drever, Y.-H. Li, and J. B. Maynard

People have long been curious about the source of the ocean's saltiness. In Russian folklore, there is a "salt mill" at the bottom of the sea that keeps it salty, but more modern thinking has been focused on what keeps it from becoming too salty. From at least the time of Edmund Halley (1646–1742), it has been realized that the source of the salts in the ocean is the small amount of material dissolved in rivers (Gregor, 1967, p. 2). It was also believed that these salts, at least NaCl, accumulated, producing a progressively saltier ocean. This concept was used by Joly (1899) to estimate the age of the oceans by comparing their sodium content with that of rivers. His result, 90×10^6 years, was eagerly embraced by geologists, many of whom had independently made similar estimates based on thickness of strata. It provided vindication for them in their continuing debate with physicists, led by Kelvin, who steadfastly maintained that the earth could be no more than 25×10^6 years old, based on the rate of heat flow from the interior (Burchfield, 1975, pp. 151–153). The emergence of radioactive dating over the next few decades was to show how hopelessly wrong both estimates were. It quickly became obvious that the earth is generating its own internal heat, not cooling steadily from its primordial state.

The composition of the ocean has proved more difficult to explain. Most geochemists continued to regard the oceans as simply accumulating salts, trying out various corrections to Joly's method. None of these proved satisfactory, and it became increasingly clear that the ocean, if it has existed for anything like the age now accepted, must be in some sort of steady-state balance.

The notion of steady-state composition goes back at least to the work of Osmond Fisher (1900), who pointed out that Silurian sediments contain as much salt as modern ones. Over the succeeding years, it became obvious that since the beginning of the Cambrian period, possibly even for the last 2×10^9 years, the compositions of the sedimentary rocks and the nature of the fossil record do not indicate any radical change in seawater composition (Barth, 1961b, Part V; Holland, 1972, 1984). The problem was then to explain how the steady-state is maintained and what is the likely extent of departures from it.

The next stage in thinking about seawater was the construction of equilibrium models. If the ocean has a relatively constant composition, then the simplest control mechanism is equilibrium with sedimentary mineral phases. This view was stimulated by the work of the physical chemist L. G. Sillén (1961), who was able to show that a constant seawater composition could be maintained by equilibration with the phases quartz, calcite, several clay minerals, and the atmosphere (CO_2 gas). This concept was then modified and extended by a number of workers (e.g., Helgeson et al., 1969; Helgeson and Mackenzie, 1970).

A consideration of the composition of river waters shows that some reaction between the solid phases and dissolved substances is necessary to convert river water into seawater (Conway, 1943; Mackenzie and Garrels, 1966a). That is, seawater is not simply evaporated river water but requires modification beyond the precipitation of evaporite minerals. The cations in river water are balanced mostly by HCO_3^-, whereas those in the oceans are balanced by Cl^-. Thus, it is necessary for HCO_3^- to be removed from the oceans, otherwise there would be an excessive buildup of alkalinity. The predominant process for removing HCO_3^- is precipitation of $CaCO_3$, but not enough Ca^{2+} is carried by rivers. Accordingly, some other reaction is required. As in Sillén's model, Mackenzie and Garrels (1966b) suggested that reactions of clay minerals were important.

These ideas prompted investigators to look for evidence of reaction of river-borne clays with seawater. By and large, the results were negative (Dasch, 1969; Russell, 1970; Drever, 1971; Maynard, 1976; see also reviews by Drever, 1974b; Wollast, 1974). As a consequence, other removal processes were sought and it became apparent that interactions of the oceans with newly formed basaltic crust may greatly influence seawater chemistry. This process was first identified by Deffeyes (1970), then elaborated by Hart (1973). There is now a large literature, discussed in Chapter 3.

The failure to find extensive clay mineral reactions and the identification of the interaction of seawater and basalt led workers to develop models in which kinetic as well as thermodynamic processes are considered (Broecker, 1971a; Pytkowicz, 1975; Maynard, 1976; Mackenzie and Wollast, 1977). Equilibrium processes now appear to be most important for elements strongly influenced by surface adsorption. Biological processes, many of which operate far from equilibrium, are particularly important for elements such as silicon and nitrogen. Also, reactions in parts of the ocean that are isolated from the bulk of seawater, either small basins or water in sediment or basalt, are the dominant control for many elements. The effect of these isolated processes on seawater depends on exchange rates with open seawater as well as on the thermodynamics of reaction.

This review has tried to show the evolution of thought about seawater chemistry, and why steady-state models have been so prominent. The body of this chapter will deal with the processes occurring in the modern ocean and atmosphere. Insofar as possible, we attempt to estimate all the fluxes directly, without resorting to the assumption of a steady state. Then the degree of balance of the fluxes can provide a test of the appropriateness of steady-state models for each element. The results should also set the stage for comparisons with the geologic record.

1.1 MASS BALANCE BETWEEN IGNEOUS ROCKS AND SEDIMENTARY ROCKS PLUS SEAWATER

The total mass of sedimentary rocks and dissolved ions in seawater today is the end result of the weathering of crustal igneous rocks by degassed magmatic volatiles (e.g., H_2O, CO_2, HCl, H_2S) throughout geologic time. Conservation of mass requires that the total mass of any nonvolatile element i released from weathering of the crustal igneous rocks should be equal to that found in sedimentary rocks plus seawater. For example,

$$M_{ig} C_{ig}^i = M_{sed} \sum A_j C_j^i + M_{sw} C_{sw}^i, \qquad (1.1)$$

where
M_{sed} = the total mass of sedimentary rocks today: the best recent estimate is $2.5 \pm 0.4 \times 10^{24}$ g (Li, 1972; Ronov and Yaroshevskiy, 1976);

M_{ig} = the total mass of crustal igneous rocks weathered to produce the sedimentary mass: M_{ig} is estimated to be 0.88 M_{sed}; M_{sed} includes volatile elements degassed from the interior of the earth (e.g., CO_2, H_2O) as well as elements derived from weathering of crustal igneous rocks (Li, 1972);

M_{sw} = the mass of seawater, 1.4×10^{24} g (Sverdrup et al., 1942);

A_j = the fraction of sedimentary rock type j (e.g., sandstone, limestone, shale): $\sum A_j = 1$;

C_{ig}^i, C_j^i, and C_{sw}^i = the average concentration of element i in the crustal igneous rocks, in sedimentary rock type j, and in seawater.

The estimates of the fractions of major sedimentary rock types vary appreciably, especially the fraction of carbonate rocks (for a summary, see Garrels and Mackenzie, 1971a). For illustrative purposes, we adopt the relative abundance of shale (including oceanic pelagic clay) : sandstones : carbonates (including oceanic carbonates) : evaporites = 74 : 11 : 15 : 2 (= 0.726 : 0.108 : 0.146 : 0.020 in fractions) as given by Garrels and Mackenzie (1971a). Because the oceanic pelagic clays and the oceanic carbonates are distinctive in their chemical compositions, it is worthwhile to treat them separately from continental shale and carbonates. According to results from JOIDES drillings and seismic profiles, the average thickness of oceanic sediments (layer 1) is about 0.5 km. With an average dry density of about 0.7 g/cm³, the total mass of oceanic sediments is only about 0.12×10^{24} g. If one assumes that roughly one-half of the oceanic pelagic sediment is pelagic clay and the other half oceanic carbonates (in reality, carbonates are probably slightly more abundant than pelagic clay), the fractions of the major sedimentary rock types are shale (sh) : sandstone (ss) : carbonate (c) : evaporites (ev) : oceanic pelagic clays (op) : oceanic carbonate (oc) = 0.702 : 0.108 : 0.122 : 0.020 : 0.024 : 0.024. Dividing Eq. (1.1) by M_{sed} and substituting estimated values for the various coefficients, we obtain

$$0.88 C_{ig}^i = 0.702 C_{sh}^i + 0.108 C_{ss}^i + 0.122 C_{cc}^i + 0.020 C_{ev}^i + 0.024 C_{op}^i$$
$$+ 0.024 C_{oc}^i + 0.56 C_{sw}^i. \quad (1.2)$$

For simplicity, the right-hand side of Eq. (1.2) can be designated by $\bar{C}_s^i$, the average concentration of element i in average sediment (including seawater).

Substituting concentration data for various rock types into Eq. (1.2), we can calculate $\bar{C}_s$ and the $\bar{C}_s/0.88 C_{ig}$ ratios for various elements (Table 1.1). The data are mainly from Turekian and Wedepohl (1961) and exceptions are indicated in the footnotes of Table 1.1. The C_{ig} values are estimated by $C_{ig} = 0.35 C_{basalt} + 0.65 C_{granite}$ according to Mead's (1914) suggestion. The average composition of granitic rocks is, in turn, an average of the high-calcium (granodiorites) and the low-calcium granites (Turekian and Wedepohl, 1961). Taylor (1964) obtained C_{ig} by averaging C_{basalt} and $C_{granite}$. Mead's C_{ig}, however, gives a better mass balance, although the difference may not be significant considering the many other uncertainties.

A $\bar{C}_s/0.88 C_{ig}$ ratio of 1.0 represents a perfect mass balance, but considering the uncertainties involved in the composition data themselves and in assigning the weighted fraction for each rock type, a $\bar{C}_s/0.88 C_{ig}$ ratio of 0.8–1.3 can be a satisfactory balance (group I elements in Table 1.1). A $\bar{C}_s/0.88 C_{ig}$ ratio of greater than 1.3 (group II elements in Table 1.1) may indicate extra sources (e.g., volcanic exhalations and hydrothermal solutions) other than weathering of igneous rocks.

As one would expect, the largest number of elements belongs to group I. Manganese is included in this group, in contrast to some earlier work. Within the uncertainty of the mass balance calculation, manganese is well balanced or, at best, only slightly in excess, whereas a large manganese excess was found by Horn and Adams (1966). Horn and Adams (1966) greatly overestimated the fraction of oceanic pelagic clay in the sediment mass by assuming that hemipelagic sediments have the same composition as pelagic clay. Because oceanic pelagic clay has a very high manganese content, the overestimation of the fraction of oceanic pelagic clay caused a large excess manganese value in their mass balance calculation. Boström (1976) reports that hemipelagic and shelf sediments are similar in composition, including manganese, to average shale.

The elements that cannot be balanced (group II) are chlorine, sulfur, mercury, bromine, boron, tellurium, selenium, bismuth, antimony, arsenic, iodine, tin, molybdenum, lithium, calcium, cesium, lead, and uranium in order of decreasing $\bar{C}_s/0.88 C_{ig}$ ratio. Cesium, lead, and uranium, however, are marginal cases ($\bar{C}_s/0.88 C_{ig}$ = 1.5–2.0) and may well belong to group I. Except for

1.1 MASS BALANCE BETWEEN IGNEOUS ROCKS AND SEDIMENTARY ROCKS

TABLE 1.1. Mass Balance Between Average Crustal Igneous Rocks (C_{ig}) and Average Hypothetical Sedimentary Rock (C_s) in ppm[a]

	Group I: $\bar{C}_s/0.88\,C_{ig} = 0.8\text{--}1.3$		
	$\bar{C}_s$	C_{ig}	$\bar{C}_s/0.88\,C_{ig}$
Be	2.19	1.98	1.26
F	638	585	1.24
Na	17.0×10^3	23.0×10^3	0.84
(Mg)[b]	17.7×10^3	19.7×10^3	1.02
Al	61.9×10^3	77.4×10^3	0.91
Si	24.3×10^3	29.3×10^3	0.94
(P)	624	879	0.81
K	20.9×10^3	24.7×10^3	0.96
(Sc)	9.86	13.8	0.81
(Ti)	3.57×10^3	4.65×10^3	0.87
(V)	99.2	130	0.88
(Cr)	70.7	68.0	1.18
(Mn)	927	827	1.27
(Fe)	36.4×10^3	44.3×10^3	0.94
(Co)	15.3	19.4	0.90
(Ni)	56.5	51.8	1.24
(Cu)	38.9	43.5	1.02
(Zn)	75.7	68.9	1.25
Ga	15.9	17.0	1.06
Ge	1.29	1.30	1.13
Rb	108	102	1.20
Sr	342	338	1.15
Y	29.4	31.7	1.05
Zr	142	151	1.07
Nb	14.1	20.0	0.81
(Pd)	$\sim 1.1 \times 10^{3c}$	$\sim 1.1 \times 10^3$	~ 1
(Ag)	0.054	0.067	0.92
(Cd)	0.23	0.20	1.29
In	0.08	0.10	0.91
Ba	469	525	1.02
La	33.5	37.8	1.01
Ce	71.9	73.0	1.13
Nd	32.6	29.8	1.24
Sm	6.80	7.97	0.97
Eu	1.39	1.26	1.25
Tb	1.11	1.26	1.00
Yb	3.10	3.17	1.11
Hf	2.53	2.82	1.02
Ta	1.57	1.81	0.99
W	1.54	1.38	1.26
(Au)	1.2×10^{-3}	1.5×10^{-3}	0.98
Tl	1.10	1.06	1.17
Th	9.16	9.96	1.05

TABLE 1.1. Continued

	Group II: $\bar{C}_s/0.88\,C_{ig} > 1.3$			
	$\bar{C}_s$	C_{ig}	$\bar{C}_s/0.88\,C_{ig}$	$\bar{C}_s - 0.88\,C_{ig}$
Li	50.1	26.8	2.12	27
B	83.3	7.93	11.9	76
S	4.36×10^3	300	16.5	4.1×10^3
Cl	17×10^3	128	151	17×10^3
(Ca)	67.4×10^3	36.5×10^3	2.1	35×10^3
As	9.69	1.81	6.1	8.1
Se	0.45	0.05	10.1	0.41
Br	44.2	3.96	12.7	40.7
Mo	2.62	1.27	2.34	1.5
Sn	4.27	1.99	2.44	2.5
Sb	1.11	0.20	6.28	0.93
Te	0.85	0.082	10.4	0.78
I	3.31	0.50	7.52	2.87
Cs	3.69	2.34	1.79	1.63
Hg	0.29	0.022	15.1	0.27
Pb	18.0	13.2	1.55	6.4
Bi	0.45	0.06	8.5	0.40
U	3.0	2.3	1.47	0.98

[a] Data are mainly Turekian and Wedepohl (1961). Exceptions are rare earths for marine sediments and shale (Piper, 1974); Th and U for pelagic clay (Ku, 1966); Si for shale (Wedepohl, 1960); Au and Pd for igneous rocks (Crocket and Kuo, 1979); Bi for igneous rocks, shale, and clay (Morowsky and Wedepohl, 1971); Te for igneous rocks, shale, sandstone, and limestone (Beaty and Manuel, 1973); Ta for igneous rocks, shale (Randle, 1974), and pelagic clay (Piper et al., 1979); and I for pelagic sediments (Chester and Aston, 1976).
[b] Parentheses indicate the elements with higher concentration in average basaltic rocks than in average granitic rocks.
[c] Assuming $\bar{C}_s$ equal to Arctic soil (Crocket et al., 1973).

calcium, the group II elements have higher concentrations in granite than in basalt, indicating a concentration of these elements in the later stage of magmatic differentiation.

The $\bar{C}_s/0.88\,C_{ig}$ ratio for calcium is only about 2, but because calcium is one of the major elements in various rock types, the excess calcium, defined here as equal to $\bar{C}_s - 0.88\,C_{ig}$, is the largest among group II elements, even larger than the excess chlorine (Table 1.1). Even if one assumes that the calcium content in the average granitic rock is close to that in a high-calcium granite (i.e., granodiorite), the $\bar{C}_s/0.88\,C_{ig}$ ratio for calcium is still about 1.5 and the excess calcium is as large as the excess chlorine.

Garrels and Mackenzie (1971a) suggested that the excess calcium was supplied by preferential leaching of calcium from volcanogenic sediments by carbonic acid during normal weathering processes. If this mechanism applies and if one assumes a complete leaching of calcium from volcanogenic sediments, the total mass of volcanogenic sediments should be equal to that of the normal

1.1 MASS BALANCE BETWEEN IGNEOUS ROCKS AND SEDIMENTARY ROCKS

sedimentary rocks. Partial leaching of calcium would require an even bigger mass of volcanogenic sediments. It is very unlikely, however, that the necessary volcanogenic sediments exist in normal stratigraphic sections (Ronov and Yaroshevskiy, 1969, 1976).

Recently, the mid-ocean ridges have been suggested as the sources of the excess calcium (e.g., Wolery and Sleep, 1976; Edmond et al., 1979a). According to laboratory experiments (e.g., Bischoff and Dickson, 1975; Seyfried and Bischoff, 1977; Mottl and Holland, 1978; Mottl et al., 1979) and field observations (e.g., Bjornsson et al., 1972; Edmond et al., 1979a; Von Damm et al., 1985), the high-temperature seawater–basalt interaction at mid-ocean ridges can be expressed schematically by the following major reactions:

$$6Mg^{2+} + 4CaAl_2Si_2O_8 + 4H_2O \longrightarrow$$
$$Mg_6Si_8O_{20}(OH)_4 + 4Al_2O_3 + 4Ca^{2+} + 4H^+. \quad (1.3a)$$

$$2Mg^{2+} + 2SO_4^{2-} + 11Fe_2SiO_4 + 2H_2O \longrightarrow$$
$$Mg_2Si_3O_6(OH)_4 + FeS_2 + 8SiO_2 + 7Fe_3O_4. \quad (1.3b)$$

$$2Na^+ + CaAl_2Si_2O_8 + 4SiO_2 \longrightarrow Ca^{2+} + 2NaAlSi_3O_8. \quad (1.3c)$$

$$Na^+ + KAlSi_3O_8 \longrightarrow K^+ + NaAlSi_3O_8. \quad (1.3d)$$

The Ca^{2+} released from these reactions is deposited as $CaCO_3$, releasing H^+:

$$Ca^{2+} + H_2CO_3 \longrightarrow CaCO_3 + 2H^+. \quad (1.3e)$$

The end results of the above reactions are loss of Mg^{2+}, SO_4^{2-} and, to a minor extent, Na^+ from seawater and a gain of H^+ and, to a minor extent, K^+. Therefore, the mid-ocean ridges may serve as important sites for *reverse weathering* (as defined by Mackenzie and Garrels, 1966a). It is also evident from the above reactions [(1.3a) and (1.3e)] that 1 mole of Mg^{2+} produces 2 moles of hydrogen and 2/3 mole of $CaCO_3$. Therefore, if the mid-ocean ridges are the main source for the so-called excess calcium, one would also expect about 1.5 times the equivalent depletion of magnesium in the magnesium mass balance calculation. Of course, this is not the case (Table 1.1) as pointed out earlier by Humphris and Thompson (1978a). Similarly, the exchange of seawater magnesium with calcium of basalts in layer 2 of the ocean floor at lower temperature (e.g., Lawrence et al., 1975) is an unlikely explanation for the excess calcium in the mass balance calculation.

Of course, if one reduces the fraction of carbonate rocks in the sedimentary mass by about one-half (i.e., from 15 to 7.5%), the excess calcium problem disappears without much effect on the mass balance calculation of other elements. However, the most recent estimate of the fraction of carbonate rocks in the average sedimentary column is even higher, about 30% (Ronov and Yaroshevskiy, 1976). Therefore, the excess calcium is still a geochemical dilemma.

Sibley and Vogel (1976) suggested two ways to solve the excess calcium dilemma. First, if one assumes that the average sediment consists of equal amounts of oceanic pelagic clay and average continental sediment (according to Ronov and Yaroshevskiy, 1976)—the average continental sediment consists of 49% shale, 23% sandstone, 27% carbonates, and 1% evaporites—the fraction of carbonate is reduced by at least one-half. Second, one could assume an increase in the average calcium content of the crustal igneous rocks by using an average of Canadian Shield and tholeiitic basalt. However, considering 50% of the total sedimentary mass to be pelagic clay is an overestimate by at least an order of magnitude, because it is based on a mean residence time of oceanic pelagic clay of 1.5×10^9 years instead of about $0.15 \pm 0.5 \times 10^9$ years (Li, 1972). Also, the assumption that half of the sedimentary mass is pelagic clay leads to an excess manganese content as well as an excess for other elements such as cobalt, nickel, barium, molybdenum, zinc, and copper in the mass balance calculations.

Because pelagic sediments do not seem to be accumulating at continental margins in significant volume (Ronov and Yaroshevskiy, 1976), they must be largely metamorphosed and remelted back into crystalline rocks at subducting plate boundaries. Before the Cretaceous, these "lost" sediments would have been low in $CaCO_3$, and so their absence from the mass balance may explain the calcium anomaly.

1.2 CYCLES OF MAJOR ELEMENTS IN THE OCEAN

In Section 1.1, we identified imbalances between igneous rocks and sediments plus seawater for some elements. Among major elements, chlorine, sulfur, and calcium are clearly imbalanced; whereas sodium, magnesium, potassium, and silicon are within the range for constant composition. However, these last four elements are present in such large amounts in the solids, compared with the amount in solution, that even a small change in the composition of the solids could have a strong effect on seawater. Therefore, in this section we turn to a detailed consideration of the amounts of these major elements carried to the ocean in the dissolved load of rivers as compared to the amounts in the ocean and known processes removing them. The trace elements are considered separately.

1.2.1 Contributions of Major Elements to the Ocean by the Dissolved Load of Rivers

The first half of our mass balance attempt is an estimate of the amounts of material delivered in solution by rivers to the ocean. Fortunately, such a calculation has been made by Meybeck (1979a) using data presented by Meybeck (1977) and Martin and Meybeck (1978, 1979). His results (Table 1.2) for the average concentrations of substances in river water, except for much lower K^+ and H_4SiO_4 and somewhat higher Na^+, do not differ by more than 10% from the earlier estimate of Livingstone (1963), on which most previous balance calculations have been based. Both Livingstone's and Meybeck's estimates give a slight excess of positive change. The net flux (last column in Table 1.2) has been adjusted to electrical neutrality by distributing this "error" over all the species.

This gross river flux to the oceans must now be corrected for cyclic salts and, if we wish to establish "normal" conditions, for pollution. Cyclic salts are sea salts blown from the sea surface onto the land. Because they constitute part of the oceanic reservoir of salts rather than being derived from weathering of the continents, they should not be included in the river flux for our calculation. Studies of the composition of rainwater in coastal areas are used in this sort of correction, and we have accepted Meybeck's calculations (Table 1.2). For comparison, his numbers give 20% of the river flux of sodium as cyclic. Other estimates are 46% (Livingstone, 1963), 52% (Gregor, 1967, Table 17), 35% (Garrels and Mackenzie, 1971, Table 4.8), and 19% (Holland, 1978, Table 4-15). Berner (1971) estimated that 6% of river sulfate was from cyclic salts, comparable to the 4% in Table 1.2, but that an additional 24% was derived from H_2S gas of marine origin.

Pollution contributions to the river flux are hard to estimate because we lack good records from the preindustrial era, but it is vital to obtain such estimates if we are to predict the possible effects of human activities or to understand the geologic past. Meybeck (1979a, Table 4) has estimated the pollution flux by examining the change of composition with time of certain large rivers since 1900. He used these data to establish per capita loadings of streams, which can then be multiplied by the earth's population to give total contribution of pollution to the river flux. Alternatively, loadings can be established based on energy consumption rather than population, but the results are much the same. The largest corrections are for Na^+ and Cl^-, the elements also most affected by the cyclic salt correction, and for SO_4^{2-}. Fossil fuel combustion probably accounts for most of this excess SO_4^{2-}, although much may also come from oxidation of pollution-derived H_2S (Turekian, 1971, p. 14). Table 1.2 shows 29% of the river load of SO_4^{2-} coming from pollution, whereas Holland (1978, Table 4-15) gives 39% and Berner (1971) gives 28%.

It is interesting to calculate what effect this sudden new SO_4^{2-} supply might have on the ocean. If we assume for the sake of argument that there is no additional new SO_4^{2-} removal and that concentration of Ca^{2+} in the ocean remains constant, then the extra 1.29×10^{12} moles/yr, added to a total in the ocean of 38.5×10^{18} moles, would lead to saturation of the ocean with respect to gypsum in about 300×10^6 years. Our ability to add new SO_4^{2-} from fossil fuels covers but a tiny fraction of this time. Thus, the ocean is really very resistant to changes in composition, if only because of its size. Athough this excess SO_4^{2-} is unlikely to affect seawater chemistry, it does pose problems in deciphering the isotope geochemistry of sulfur. This pollution component is unlikely to have the same isotopic composition as normal river-borne SO_4^{2-} and

TABLE 1.2. Flux of Major Elements from Rivers to the Oceans in Solution

	Concentration in rivers (mM) (Livingstone, 1963)	Concentration in rivers (mM) (Meybeck, 1979a)	Gross Flux[a] (10^{12} moles/yr)	Cyclic Salts[a]	Pollution[a]	Net Flux[b]
Na^+	0.247	0.313	11.7	2.39	3.39	5.91
K^+	0.059	0.036	1.36	0.05	0.13	1.17
Mg^{2+}	0.169	0.150	5.59	0.29	0.41	4.85
Ca^{2+}	0.375	0.368	13.7	0.05	1.18	12.36
Cl^-	0.220	0.233	8.69	2.82	2.63	3.27
SO_4^{2-}	0.117	0.120	4.49	0.16	1.29	3.07
HCO_3^-	0.957	0.894	33.4	—	1.64	32.09
H_4SiO_4	0.218	0.173	6.47	—	—	6.47
Totals:						
Mass	118 ppm	110 ppm	4.19×10^{15} g	0.18×10^{15} g	0.10×10^{15} g	3.91×10^{15} g
Charge	+0.010 meq/L	+0.018 meq/L	$+0.57 \times 10^{12}$ equiv			0

[a] Meybeck (1979a, Table 6). Based on a total river discharge of 37.4×10^{15} L/yr.
[b] Adjusted to electrical neutrality.

makes up such a large proportion of the river contribution that it is almost certain to alter the observed isotopic composition (Holser and Kaplan, 1966).

Another aspect of this problem is the proportion pollution forms of the *suspended* load carried by rivers. The present-day rate of supply of suspended solids is estimated to be 15.5×10^{15} g/yr (Meybeck, 1977, p. 30), much higher than seems to have been the case over most of geologic time (Garrels and Mackenzie, 1971a, p. 260). Some of the excess may be due to pollution and some to unusual geologic conditions in the present, but the exact proportions are unknown. Because some of the processes influencing seawater composition depend on the amount of suspended sediment, fluctuations in the proportion of suspended and dissolved load may conceivably have modified either the composition of seawater or the composition of sediments in the geologic past (Sedimentation Seminar, 1977).

We now have an estimate of the amounts of material, derived from weathering of the continents, that are being delivered to the ocean in solution. If the ocean is truly in a steady state, then equal removal processes must be operating within the ocean. An examination of the known possibilities follows; these are then summarized to provide an indication of whether there are likely to be others. This procedure enables us to evaluate critically the proposed magnitudes of removals. In some cases, far more material has been calculated to be leaving the ocean than is in fact entering.

1.2.2 Processes Removing Major Elements from the Ocean

1.2.2.1 Ion Exchange

River-borne clays exhibit a significant amount of cation exchange, particularly of Ca^{2+} for Na^+, when they encounter the new chemical environment of seawater. Sayles and Mangelsdorf (1977, 1979) have improved the analytic techniques for measuring exchangeable cations and found that more Na^+ and less Mg^{2+} are involved than in earlier estimates (Table 1.3). Ion exchange and related surface reactions appear to be relatively unimportant for major elements other than Na^+, but these processes are important for trace elements, as discussed in Section 1.3. Note that the ability of this process to buffer changes in the major element chemistry of seawater is limited by the availability of exchangeable Ca^{2+} on the river clays.

1.2.2.2 Burial of Pore Water

Because all sediments have a finite porosity, there is some seawater lost from the ocean each year by the permanent burial of pore water. The dissolved substances in this water, modified by reactions with the associated minerals, reappear in connate waters in ancient rocks. We have assumed (Table 1.8) for this removal that the pore water has the composition of normal seawater and an annual sediment mass (suspended plus dissolved load) of 20×10^{15} g/yr (Martin and Meybeck, 1979, p. 174) with an average final porosity of 30% (Holland, 1978,

TABLE 1.3. Additions to the River Flux from Ion Exchange Between River-Borne Clays and Seawater[a]

	Experimental[b]	Amazon[c]	Average	Percentage of River Flux[d]
Na^+	−1.58	−1.47	−1.53 ± 0.06	26
K^+	−0.12	−0.27	−0.20 ± 0.08	17
Mg^{2+}	−0.14	−0.49	−0.32 ± 0.08	7
Ca^{2+}	0.86	1.05	0.96 ± 0.10	8
H^+	0.26	0.58	0.42 ± 0.16	—

[a] Units are 10^{12} moles/yr; minus sign indicates a removal from seawater.
[b] Sayles and Mangelsdorf (1977).
[c] Sayles and Mangelsdorf (1979). Both assume a total particulate load of 15.5×10^{15} g (Meybeck, 1977, p. 20) with CEC of 25 meq/100 g (Sayles and Mangelsdorf, 1977, p. 958; Sayles and Mangelsdorf, 1979, p. 777).
[d] Corrected for cyclic salts and pollution.

Fig. 5-28). This removal was probably less over most of geologic time because of the lower sediment load. The greatest uncertainty in this estimate is the porosity. Drever's estimate was 50% (1974b, p. 399), which we have used here as a comparison to establish the uncertainty limits (Table 1.8).

1.2.2.3 Interactions Between Sediments and Seawater

A large number of reactions involving transformations of detrital grains or precipitation of new minerals can potentially affect seawater (e.g., Holland, 1978, Table 5-6). Most such reactions that involve the growth of silicate minerals cannot proceed in direct contact with seawater because its silica concentration is too low (Wollast and DeBroeu, 1971). Instead, silicate authigenesis involves pore waters of sediments in which the dissolved silica concentration is much higher because of the dissolution of siliceous organisms. Contact between these pore waters and seawater is by diffusion, and the resulting flux of material to or from seawater can be modeled if the appropriate diffusion coefficients and concentration gradients are known. In this way, the effect of the formation of authigenic silicates on seawater can be estimated without knowing the details of the individual reactions.

Sayles (1979) made detailed measurements of pore water composition, using an improved sampling system, and calculated fluxes between the ocean and its bottom sediments (Table 1.4). Two earlier estimates, based on more limited data, are also shown for comparison. The fluxes usually conform to charge balance constraints at individual sites, but the averages do not. Accordingly, the fluxes in column 3 of the table have been adjusted for electrical neutrality. The values are high when compared with the net river flux. For instance, note that the proposed diffusive flux would remove far more than the total available K^+ and all of the Mg^{2+}, whereas we believe that some Mg^{2+} is used up in hydrothermal reactions at mid-ocean ridges. Using a different method, Drever (1974b,

1.2 CYCLES OF MAJOR ELEMENTS IN THE OCEAN

TABLE 1.4. Diffusional Fluxes from Seawater into the Pore Water of Marine Sediments[a]

	Previous Estimates		Sayles[b] (1979)	Percentage of River Input
	Maynard (1976)	Manheim (1976)		
Na^+	0.83	?	5.0	85
K^+	−0.43	−0.72	−1.8	150
Mg^{2+}	−0.65	−2.50	−4.6	105
Ca^{2+}	−0.65	0.75	7.2	58
HCO_3^-	—	—	8.3	26

[a] Units are 10^{12} moles/yr; a minus sign indicates removal from seawater.
[b] After Sayles (1979, Table 7, column 6), adjusted for charge balance.

TABLE 1.5. Composition of Modern Sediments from Near-Shore Marine Environments[a]

	Na	K	Mg	Ca	Si	Al	Fe
Gulf of Paria[b]	0.27	0.15	0.17	0.02	2.82	1.00	0.28
Buzzard's Bay[b]	0.41	0.27	0.12	0.09	—	1.00	0.28
Oregon shelf[b]	0.37	0.16	0.29	0.13	4.24	1.00	0.21
Barants shelf[b]	—	0.20	0.21	0.01	2.17	1.00	0.32
Godavari delta[c]	0.08	0.11	0.18	0.06	—	1.00	0.38
California borderland[d]	0.27	0.20	0.24	0.12	2.22	1.00	0.36
Bristol Channel[e]	0.09	0.23	0.27	0.10	3.91	1.00	0.29
Bearing shelf[f]	0.46	0.15	0.20	0.10	5.56	1.00	0.22
Gulf of Alaska[f]	0.49	0.11	0.23	0.12	3.68	1.00	0.30
Sarawak shelf[g]	0.02	0.16	0.12	0.01	1.65	1.00	0.18
Orinoco shelf[g]	0.03	0.12	0.10	0.02	1.81	1.00	0.23
Niger delta[g]	0.01	0.06	0.07	0.02	1.47	1.00	0.25
Average	0.23	0.16	0.18	0.11	2.95	1.00	0.27

[a] In molar ratio to aluminum; corrected for $CaCO_3$.
[b] Calvert (1976, Tables 33.1, 33.2).
[c] Kaleska et al. (1980, Table 3), clay fraction only.
[d] Emery (1960, Tables 15 and 17).
[e] Hamilton et al. (1979, Table 3).
[f] Sharma (1979, p. 450).
[g] Porrenga (1967, Tables 2, 5, 7), clay fraction only, $CaCO_3$-free basis.

p. 350) calculated that a maximum of 1.35×10^{12} moles/yr of Mg^{2+} is removed by diagenetic processes, only one-third that amount calculated from the pore water profiles.

We can gain some idea of the maximum values of these fluxes by comparing the composition of clastic marine sediments with that of the suspended solids in rivers (Sayles, 1979, p. 542). Unfortunately, such data are limited and show considerable variation from place to place (Table 1.5). For instance, note that

TABLE 1.6. Annual Flux from Rivers to Oceans in Suspension, Compared with Composition of Marine Sediments

	River Particulates			Marine Sediments		
	Total[a] (ppt)	Net[b] (ppt)	X/Al[c]	Shallow[d] X/Al	Deep Water[a] X/Al	Change[e] (10^{12} moles/yr)
Na	7.1	9.2	0.11	0.23	0.25	−6.6
K	20.0	20.7	0.15	0.16	0.20	−0.76
Mg	11.8	12.6	0.14	0.18	0.21	−2.3
Ca	21.5	18.8	0.14	0.11	0.07	+2.0
Si	28.5	28.5	2.92	2.95	2.87	−1.2
Al	94	94	1.00	1.00	1.00	0
Fe	48	48	0.25	0.27	0.30	−1.2

[a] Martin and Meybeck (1979, Table 3).
[b] Corrected for ion exchange using Table 1.3.
[c] X/Al is the molar ratio to aluminum.
[d] Average from Table 1.5.
[e] Assumes 90% of sediment is shallow water and an annual river load of 15.5×10^{15} g. Also assumes aluminum is conserved during diagenesis.

sediments in tropical areas are depleted in Mg^{2+} and K^+ relative to those in subarctic and temperate regions. The tectonics of the source area are also important. The two samples from the ocean off Alaska are both sediments derived from a tectonically active source area, but the Bering Shelf sediments were deposited on the back (landward) side of the island arc whereas the Gulf of Alaska sediments lie in the fore-arc region. The Na/K ratio in the first case is 3.1, but it is 4.5 in the second case. Also, Mg^{2+} and Fe^{2+} are higher in the fore-arc sediments. These considerations show that each type of sediment needs to be weighted according to its proportion in the total sediment mass, but our database is at present inadequate for such a calculation. Accordingly, the compositions have simply been averaged to give some idea of the possible magnitudes of any changes from the river particulates (Table 1.6). Comparison with the river solids, assuming that aluminum, because of its very small solubility, is conserved, shows changes for K^+ and Mg^{2+} that are about one-half those calculated from the pore water data (Table 1.4). For Ca^{2+}, the calculated return to seawater is substantially less than in Table 1.4, but this is probably due to the inclusion of some $CaCO_3$ in the analyses of Table 1.5. The Na^+ is also very different, showing a large release to seawater in Table 1.4 but a large uptake into the sediments in Table 1.6. Many of the analyses in Table 1.5 seem to include NaCl from sea salts that were not removed by washing the sediment. Accordingly, the Na^+ flux in Table 1.6 is almost certainly due to the burial of pore water, as in the calculation presented earlier, and not to diagenetic reactions. In conclusion, it seems likely that the relative changes in Table 1.4 are correct but the absolute magnitudes are too high by a factor of at least 2. Error limits have been assigned by using the earlier estimates as lower limits.

1.2.2.4 Chemical Precipitates

$CaCO_3$. Certain nonsilicate minerals are near or above saturation in seawater or in seawater in restricted contact with the main part of the ocean such as the Black Sea. Such minerals can form within the water column or at the sediment–water interface and are not included in the diffusion calculations of Section 1.2.2.3. Principal among these are the carbonates. Only calcite and aragonite (both $CaCO_3$) seem to be forming in any abundance, although thermodynamic calculations show that dolomite [$CaMg(CO_3)_2$] should be precipitating everywhere. Its absence is one of the best-known instances of reaction rates, as opposed to thermodynamics, controlling the composition of seawater. A typical rate for $CaCO_3$ deposition is that for the Pacific Ocean, 0.39×10^{-3} g/cm² · yr (Worsley and Davis, 1979). Extending this rate to the whole ocean floor, which has an area of 3.6×10^{18} cm² (Menard and Smith, 1966), gives a total removal of 1.40×10^{15} g/yr or 14.0×10^{12} moles/yr. This amount may be slightly low because $CaCO_3$ deposition is more extensive in the Atlantic than in the Pacific Ocean (Berger, 1976, Table 29.4). Another estimate is available from the work of Broecker (1971b). Using his approach, modified to account for aragonite deposition and redissolution (Berner, 1977), the annual deposition rates for calcite and aragonite are 0.83×10^{15} and 0.016×10^{15} g/yr, for a total of 0.85×10^{15} g/yr. If deep-sea sediments receive 80% of carbonate deposition (Maynard, 1976, p. 1524), the worldwide rate would be 1.06×10^{15} g/yr or 10.6×10^{12} moles of Ca^{2+} per year. For comparison, Lerman (1979, Table 7.1) calculated a $CaCO_3$ removal rate of 3.5×10^{-6} moles/cm² · yr. For the whole ocean, the flux would then be $12.6 \pm 1.8 \times 10^{12}$ moles/yr. The mean of these three estimates is $12.4 \pm 1.8 \times 10^{12}$ moles/yr. The corresponding removal rate for HCO_3^- is twice this amount or 24.8×10^{12} moles/yr:

$$Ca^{2+} + 2HCO_3^- \longrightarrow CaCO_3 + CO_2 + H_2O.$$

In addition to Ca^{2+}, some Mg^{2+} is removed with the carbonates. This Mg^{2+} enters into a limited solid–solution form in the calcite lattice rather than forming dolomite. The average amount is about 5 mol% in shallow-water carbonates (Garrels and Mackenzie, 1971a, Fig. 8.6) but only 0.5 mole% in the deep-water oozes. Accordingly, about 0.17×10^{12} moles of Mg^{2+} go into carbonates each year.

The cycles of calcium and the carbonate species and the processes regulating their composition in seawater are closely interrelated. Seawater is *approximately* in equilibrium with calcite (and apparently has been since the early Precambrian), so that the solubility of calcite is one important control. For a solution in equilibrium with calcite, if any two of the seven quantities—calcium concentration, pH, carbon dioxide concentration or partial pressure, bicarbonate concentration, carbonate concentration, alkalinity ($= m_{HCO_3^-} + 2m_{CO_3^{2-}}$), and total carbonate species ($= m_{H_2CO_3} + m_{HCO_3^-} + m_{CO_3^{2-}}$)—are specified, all the others are uniquely determined. Thus, if we identify the processes controlling any two of these quantities, we have identified the processes controlling *all* of them.

Broecker (1971a) suggested that the two controls are partial pressure of CO_2 and CO_3^{2-} concentration. The CO_2 pressure in the atmosphere (and hence the CO_2 or H_2CO_3 concentration of the ocean) is determined by the requirement that the rate of CO_2 consumption by weathering must equal the rate of CO_2 return by metamorphism. The CO_3^{2-} concentration is set by the requirement that the removal of oxidized carbon (as $CaCO_3$) must exactly balance the input. Calcium concentration, pH, and alkalinity would then be consequences of these two controls.

Several other controls are possible. Any rise in the pH of seawater would cause precipitation of $Mg(OH)_2$ interlayers in clay minerals (Deffeyes, 1965, quoted in Russell, 1970), and so we might hypothesize that pH is buffered at the level where precipitation of interlayer hydroxides begins. This hypothesis can be expanded to involve controls on the cycle of magnesium. Another hypothesis would be that circulation through submarine basalts removes all alkalinity as vein-filling calcite, and so the rate of such circulation would control the concentration of carbonate species in seawater. This last hypothesis seems less probable in view of the apparent small extent of chemical interaction between seawater and oceanic crust at spreading centers (see Chapter 3).

Evaporites. Evaporites, either NaCl (halite) or $CaSO_4$ (anhydrite or gypsum) form the other common type of chemical deposit. Although such minerals can be precipitated very rapidly from solution, only under certain rather rare geologic conditions do solutions form that are sufficiently concentrated to cause them to precipitate. The essential requirement is a large basin with a limited connection to open seawater in an area with a very high ratio of evaporation to rainfall. Small basins of this type are known today, but the mass of Cl^- and SO_4^{2-} deposited in terms is trivial compared with the river flux. Because no other process, except the burial of pore water that we already considered, removes Cl^- from the ocean, periodic episodes of large evaporite deposition must account for most of the removal of Cl^- and perhaps of SO_4^{2-} (see section 1.4.2.3).

Because sulfur is removed as sulfide minerals in shales as well as in the sulfates in evaporites, there is no simple control over SO_4^{2-} concentration in seawater set by the deposition of evaporites, although past variations with respect to other solutes are limited by the preserved record (see Section 1.4.1). Some limits can be placed on Cl^-, however. The salinity of the oceans is largely determined by the concentrations of dissolved sodium and chloride, and so Na^+, Cl^-, and salinity can be discussed together. For reasons of charge balance, the same processes that remove chlorine (evaporites and burial of interstitial water) must also be the major removal mechanisms for sodium, although sodium is also removed by ion exchange, as discussed above, and possibly by interaction with hot basalt during hydrothermal circulation at spreading centers (see Chapter 3, Section 3.1.1). The processes of burial of interstitial water and ion exchange have probably varied little with time (although both will vary with the total input of terrigenous sediment and hence with the mean elevation of the continents), but the formation of halite in evaporites is highly episodic. The last major deposition was in the Mediterranean Basin about 6×10^6 years ago (Hsü et al., 1977). The effect of this episodic removal process is moderated by the long residence time of chloride in the ocean (more than 100×10^6 years).

The total mass of chloride in sedimentary rocks and pore waters is about equal to that in the ocean (Garrels and Mackenzie, 1971b), so that even if chloride removal from the oceans ceased completely, the maximum level to which the salinity of the oceans could rise is twice the present value. This is a strong upper limit, and it is unlikely that the salinity of the oceans has been more than 50% above its present value since the early Precambrian. The chloride concentration in the ocean in the first 10^9 years of earth history depended largely on the relative rates of release of water and chloride from the earth's interior and on how soon significant chloride was removed from seawater as halite. The simplest assumption is that the relative rates of release were the same, in which case the chloride concentration in the early ocean would have been double that in the present ocean and would have gradually decreased as halite became buried in evaporites. An *approximate* steady state would be reached when the rate of erosion of old evaporites was equal to the rate of formation of new evaporites. The steady state would be approximate because both the rate of erosion of old evaporites and the rate of formation of new evaporites must have varied appreciably with time.

There is no simple lower limit to the chloride concentration in the ocean. However, the rate of removal of chloride by both burial of interstitial water and evaporite formation would decrease as chloride concentration decreased, providing a feedback mechanism that would prevent the salinity from dropping to very low values. Holland (1972) concluded that the salinity of the oceans was probably never less than half its present value.

Sulfides. Sulfur is removed from the oceans not only as SO_4^{2+} in evaporites but also as S^{2-} in sulfide minerals, mostly pyrite. The balance between these two modes influences the isotopic ratio of sulfur in seawater and also has important implications for the cycles of carbon and oxygen (see Chapter 4). In Chapter 4, Section 4.2, a total S^{2-} removal of 0.60×10^{12} moles/yr is calculated. The stoichiometry of sulfate reduction and pyrite precipitation requires that an equal amount of HCO_3^- must be returned to the ocean (Maynard, 1976, p. 1524). This quantity of sulfur is quite small in comparison with the river flux (3.07×10^{12} moles/yr) and suggests that SO_4^{2-} must presently be accumulating in seawater. The excess, after accounting for burial of pore waters, is about 2.4×10^{12} moles/yr.

Biogenic Silica. Much silica is also deposited in deep-sea sediments in the form of biogenic opal. As mentioned, the silica concentration of seawater is considerably below that required for the precipitation of almost all silicate minerals so that silica is controlled biologically, much like phosphate and nitrate. The chief organisms responsible are diatoms, and to a lesser extent radiolaria. (Earlier studies that seemed to show inorganic reactions of silica in estuaries have not been confirmed: Fanning and Pilson, 1973; Boyle et al., 1974). Most of the silica deposited on the sea floor when these organisms die is redissolved (Berger, 1976, p. 323), but some is buried in the sediments. Maynard (1976) calculated a burial rate of 0.34×10^{15} g/yr or 5.7×10^{12} moles/yr based on the lithologic descriptions of JOIDES cores. This amount would then appear as chert in the geologic record. Lerman (1979, Table 7.1) gives 3.6×10^{12} moles/yr. Biogenic silica is also removed as a constituent of near-shore clastic sediments. Milliman and Boyle (1975) found that diatom production in the Amazon estuary totals 15×10^{12} g/yr of SiO_2. It is not known how much of this amount is redissolved, however. Lerman (1979, Table 7.1) suggests that only 4% of the SiO_2 precipitated by organisms in the open ocean survives to be incorporated in deep-sea sediments. Because of much shorter residence times in the water column and on the sea floor, the percentage that survives dissolution in shallow water must be higher. As an upper limit, let us assume that 50% of the SiO_2 would survive. A lower limit is probably about twice the deep-sea amount, or 8%. Accordingly, between 1.2 and 7.5×10^{12} g/yr of biogenic silica are added to the sediment mass created by the Amazon River. From the suspended sediment discharge of this river of 880×10^{12} g/yr (Meade et al., 1979), this amount of SiO_2 is equivalent to between 0.14 and 0.85% of the suspended load. If the average is typical of other rivers, the worldwide suspended load of 15.5×10^{12} g/yr would produce each year a clastic sediment mass incorporating $0.077 \pm 0.053 \times 10^{15}$ g of SiO_2, most of which will ultimately go into authigenic silicates. DeMaster (1981) similarly estimated a maximum removal of 0.08×10^{15} g/yr. The total biogenic silica deposition is then about $0.42 \pm 0.05 \times 10^{15}$ g/yr, or $7.0 \pm 0.9 \times 10^{12}$ moles/yr.

This large flux of biologic material to the sea floor suggests that perhaps significant amounts of other elements are also removed biologically. Work on suspended matter of the Amazon estuary by Sholkovitz and Price (1980) has provided a means of estimating this flux. They were able to calculate concentrations of each element that can be attributed to the inorganic particulate load. The difference between this amount and the observed concentration is the amount associated with organic matter (Table 1.7). Most of this excess must be

TABLE 1.7. Removals of Major Elements from the Oceans by Siliceous Microorganisms[a]

	Excess Concentration		Removal
	(10^{-6} moles/L)	X/Si	(10^{12} moles/yr)
Si	6.21	1.00	7.0
Mg	0.0833	0.0134	0.09
Ca	0.024	0.00403	0.03
K	0.031	0.00499	0.03

[a] Data from Sholkovitz and Price (1980).

related to diatoms, as shown by the high silica concentration. The resulting fluxes are quite small, suggesting that other organisms such as foraminifera also have little effect on the transfer of dissolved substances into sediments, except perhaps for the incorporation of Mg^{2+} into calcite, which was included in our previous calculation.

The possible range of H_4SiO_4 variations in seawater in the geologic past can be set fairly closely. The concentration in modern surface seawater, where diatom growth strongly depletes silica, seldom falls below 10^{-5} moles/L. The upper limit is set not by the precipitation of the stable phase—quartz, which should occur at about $10^{-4.0}$ moles/L—but by the precipitation of amorphous SiO_2 at about $10^{-2.7}$ moles/L.

1.2.3 Comparison of River Flux and Removal Processes

The processes discussed above cover all the known significant removals from seawater that are strictly within the exogenic cycle. Interaction of seawater with basalt at both high and low temperatures is also a major influence on the composition of seawater. Such interaction represents a coupling of the exogenic and endogenic cycles and is discussed in detail in Chapter 3 and illustrated schematically in Fig. 1.1. An idea of the quantitative importance of basalt–seawater interaction can be gained by comparing the other known input and output fluxes of solutes in the ocean. Such a comparison, including an estimate of the uncertainty (Table 1.8), shows that a few substances are probably balanced, others might be, and still others are badly out of balance. H_4SiO_4 and K^+ balance well; accordingly, there should be little net exchange of these two substances during interaction of internal and external cycles. Cl^- and SO_4^{2-} are

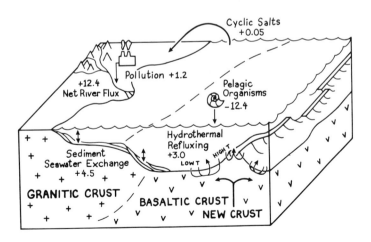

Figure 1.1 Cartoon of major calcium fluxes into and out of seawater.

TABLE 1.8. Mass Balance Between Rivers and the Ocean for Major Elements[a]

	Net River Input[b]	Ion Exchange[c]	Diffusion to Sediments[d]	Burial of Pore Water	Chemical and Biologic	Exogenic Balance[e]	Revised Exogenic Balance[f]
Na^+	5.91	-1.53 ± 0.06	2.5 ± 1.7	-0.96 ± 0.64		5.9 ± 2.4	5.5 ± 2.0
K^+	1.17	-0.20 ± 0.08	-0.9 ± 0.5	-0.02 ± 0.01		0.05 ± 6	0.20 ± 0.44
Mg^{2+}	4.85	-0.32 ± 0.08	-2.3 ± 1.6	-0.09 ± 0.04	-0.26 ± 0.02	1.9 ± 1.7	3.2 ± 0.4
Ca^{2+}	12.36	0.96 ± 0.10	3.6 ± 3.0	-0.02 ± 0.01	-12.5 ± 1.8	4.5 ± 4.9	2.4
Cl^-	3.27			-1.07 ± 0.71		2.2 ± 0.7	2.2 ± 0.7
SO_4^{2-}	3.07			-0.06 ± 0.04		2.4	2.4
HCO_3^-	32.09	-0.42 ± 0.16	4.7 ± 4.0		-24.2 ± 3.6	11.6 ± 7.8	11.4 ± 7.0
H_4SiO_4	6.47				-7.0 ± 0.09	-0.5 ± 0.09	-0.5 ± 0.09

[a] Units are 10^{12} moles/yr; a minus sign indicates removal from seawater.
[b] Table 1.2, column 6.
[c] Table 1.3, column 3.
[d] Table 1.4, one-half of column 3.
[e] Sum of previous columns.
[f] Assumes balanced calcium.

both apparently accumulating in the ocean until the next episode of evaporite deposition. Remember that chlorine, sulfur, and calcium are identified as imbalanced in our comparison of igneous rocks versus sediments plus seawater.

Of the remaining substances, Na^+ and HCO_3^- show large imbalances, while Mg^{2+} and Ca^{2+} show similar imbalances but within the uncertainty limits. Part of the Na^+ excess is balanced by the Cl^- accumulation, and part of the Ca^{2+} by SO_4^{2-}. Even so, the magnitudes of the remaining imbalances suggest either important interactions with oceanic crust or an unidentified error in the estimates of the other processes.

Ca^{2+} is particularly instructive in this regard. There is an excess of $4.5 - 2.4 = 2.1 \times 10^{12}$ moles/yr of Ca^{2+}, after setting aside Ca^{2+} to balance the SO_4^{2-} accumulation. Interaction of seawater with basalts, however, *releases* Ca^{2+} to seawater, so that the imbalance can only get worse. It must be that we have seriously overestimated the flux of Ca^{2+} from one of the exogenic processes, probably diffusion out of sediments. A revised balance can be calculated by reducing this component sufficiently to balance the Ca^{2+} (Table 1.8, column 7). In the process, all the other fluxes are reduced proportionately. Note that there is still no allowance for Ca^{2+} transferred from basalt to the ocean; any Ca^{2+} released by basalt–seawater interaction would have to be precipitated within the basaltic part of the oceanic crust. The abundant calcite veins in altered sea-floor basalts suggests that this contraint is not unreasonable. If this view is correct, seawater–basalt interaction cannot be used to explain the discrepancy in Ca^{2+} content between sediments and igneous rocks (Section 1.1). In addition to Ca^{2+}, this procedure also gives us the minimum consistent Mg^{2+} imbalance: any addition of Ca^{2+} to seawater by interaction with basalt requires a further reduction of the diffusion flux, and hence a worsening of the Mg^{2+} balance.

Thus, we see a reasonable exogenic mass balance for some major elements but for others a suggestion of important interactions with oceanic crust. Let us now compare the important processes removing major elements with those responsible for controlling trace elements in seawater.

1.3 MASS BALANCE BETWEEN RIVER INPUTS AND OCEANIC SEDIMENT OUTPUTS FOR MINOR AND TRACE ELEMENTS

In Section 1.1, we examined the difference in composition between igneous rocks and sediments plus seawater to see whether any elements could not be accounted for. Among major elements, calcium, chlorine, and sulfur are obvious problems. By examining the balance between river water and seawater, we concluded in Section 1.2.3 that chlorine and sulfur are presently accumulating in the ocean; the reasons for the calcium imbalance are unclear. Among minor elements, many show an excess in sediments over what would be predicted from weathering of igneous rocks (group II of Table 1.1). Let us examine the mass balance between rivers and seawater for minor and trace elements to see if we can understand the reasons for these imbalances.

For a first approximation, one can assume that the total river loadings of element i, including dissolved and particulate forms, are mainly deposited in estuaries and on continental shelves with typical shale composition and on the deep-sea floor with oceanic pelagic clay composition (for simplicity, the oceanic carbonates are ignored since their minor element concentrations are generally low). The mass balance then requires that

$$M_T C_p^i + F C_R^i = (M_T + F \sum C_R^i - M) C_{sh}^i + M C_{op}^i,$$

where

M_T = the suspended particulate flux from rivers to the ocean = $18 \pm 3 \times 10^{15}$ g/yr (Holeman, 1968; Meybeck, 1979a);

M = the sedimentation rate of oceanic pelagic clay = $1.1 \pm 0.4 \times 10^{15}$ g/yr (Ku et al., 1968);

F = the river water flux = $33 \pm 3 \times 10^{18}$ g/yr (Garrels and Mackenzie, 1971a);

$C_p^i, C_R^i, C_{sh}^i, C_{op}^i$ = the concentrations of element i in river-suspended particles, filtered river water, shale, and oceanic pelagic clay;

$F \sum C_R^i$ = the summation of the dissolved river loadings excluding Ca^{2+} and HCO_3^-, which are deposited separately in the ocean as oceanic carbonates, and the recycled sea salt components assuming 0.9×10^{15} g/yr of Cl^- is recycled through the atmosphere (Livingstone, 1963);

$(M_T + F \sum C_R^i - M)$ = sedimentation rate of nearshore and hemipelagic muds.

Dividing by M_T and inserting known values, one obtains

$$C_p^i + 1.8 \times 10^3 C_R^i = 0.99 C_{sh}^i + 0.06 C_{op}^i,$$

where C^i values are all in parts per million. Since C_p^i, C_R^i, C_{sh}^i, and C_{op}^i are also known, one can compare the balance between the left-hand side (= the total riverine input, C_{in}^i) and the right-hand side (= the total sediment output, C_{out}^i) for various elements, as summarized in Tables 1.9 and 1.10. The data for riverine-suspended particles (C_R^i) are from Martin and Meybeck (1979) except for silver (Gordeyev and Lisitsyn, 1978). The river water data (C_R^i) are from Goldberg et al. (1971) except for rare earths, aluminum, iron (Martin and Meybeck, 1979), uranium (Mangini et al., 1979), thorium (Moore, 1967), and cadmium (Boyle et al., 1976).

The elements in Table 1.9 are grouped according to the following criteria:

Group A	$C_{out}/C_{in} > 0.7$		Group B	$C_{out}/C_{in} < 0.7$
Subgroup A_1	$C_{op}/C_{sh} = 0.6–1.5$		Subgroup B_1	$C_{op}/C_{sh} > 1.4$
Subgroup A_2	$C_{op}/C_{sh} > 1.5$		Subgroup B_2	$C_{op}/C_{sh} < 0.7$

1.3 MASS BALANCE BETWEEN RIVER INPUTS AND OCEANIC SEDIMENT OUTPUTS

TABLE 1.9. Mass Balance Between River Inputs Both Dissolved ($1800 C_R$) and Particulate (C_p) and Marine Sediments, Using Shale ($0.99 C_{sh}$) and Pelagic Clay ($0.06 C_{op}$)[a]

	$0.99 C_{sh}$	$0.06 C_{op}$	$1800 C_R$	C_p	C_{out}/C_{in}^b	C_{op}/C_{sh}
	Group A: $C_{out}/C_{in} > 0.7$.		Subgroup A_1: $C_{op}/C_{sh} = 0.6$–1.5			
Sc	12.9	1.14	0.007	18	0.78	1.46
V	128	7.2	1.6	170	0.79	0.92
Ga	18.9	1.2	0.2	25	0.80	1.05
Mg	14,800	1,200	7.380	11,800	0.83	1.40
Ti	4,550	280	5	5,600	0.86	1.00
Al	79,000	5,000	100	94,000	0.89	1.05
Cs	5.0	0.4	0.04	6	0.89	1.20
Th	11.9	0.8	0.08	14	0.90	1.11
Cr	89.1	5.4	1.8	100	0.93	1.00
Ce	82.1	6.0	0.14	95	0.93	1.22
La	40.5	2.7	0.1	45	0.96	1.10
Si	273,000	15,000	12,000	285,000	0.97	0.91
U	363	0.16	0.7	3	1.02	0.70
Ca	21,900	27,700[c]	27,000	21,500	1.02	1.31
Fe	46,800	3,900	70	48,000	1.05	1.38
Sr	300	170	130	150	1.07	0.60
Sm	7.43	0.50	0.002	7	1.13	1.12
Eu	1.59	0.11	0.002	1.5	1.13	1.18
K	26,000	1,500	4,140	20,000	1.14	0.94
Nd	37.6	2.6	0.07	35	1.14	1.13
Lu	0.60	0.04	0.002	0.5	1.27	1.17
Rb	139	7	2	100	1.43	0.79
As	12.8	0.8	3.6	5	1.58	1.00
Li	65.3	3.4	5.4	25	2.26	0.86
	Group A: $C_{out}/C_{in} > 0.7$		Subgroup A_2: $C_{op}/C_{sh} > 1.5$			
Ni	67	14	0.5	90	0.89	3.5
Co	19	4.4	0.2	20	1.2	3.9
Mn	840	400	13	1,050	1.2	7.9
Ba	570	140	36	600	1.1	4.0
B	99	14	18	70	1.3	2.3
Mo	2.6	1.6	1.1	3	1.0	10
F	730	78	180	—	—	1.8
	Group B: $C_{out}/C_{in} < 0.7$.		Subgroup B_1: $C_{op}/C_{sh} > 1.4$			
P	690	90	36	1,150	0.66	2.1
Cu	45	15	13	100	0.53	5.6
Pb	20	4.8	5.4	150	0.16	4.0
Zn	94	9.9	36	350	0.27	1.7
Cd	0.29	0.03	0.12	1	0.29	1.7
Au	0.00099	0.00016	0.0036	0.04	0.026	2.6
Ag	0.069	0.0066	0.54	0.13	0.11	1.6
Na	9,500	2,400	11,340	7,100	0.65	$4.2(3.0)^b$
Cl	179	1,260	14,000	—	0.10	$116(0)^b$
Br	4.0	4.2	36	5	0.20	$18(\sim 0)^b$
I	2.2	1.7	13	—	0.30	13

TABLE 1.9. Continued

	$0.99C_{sh}$	$0.06C_{op}$	$1800C_R$	C_p	C_{out}/C_{in}^b	C_{op}/C_{sh}
	Group B: $C_{out}/C_{in} = 0.7$.		Subgroup B_2: $C_{op}/C_{sh} < 0.7$			
Hg	0.40	0.006	0.13	0.65	0.52	0.25
Sb	1.5	0.06	3.6	2.5	0.25	0.67
S	2,380	78	6,660	—	0.37	$0.54(0.13)^b$
Se	0.59	0.01	0.36	—	—	0.3
Sn	5.9	0.09	0.07	—	—	0.25
W	1.8	0.06	0.05	—	—	0.56

^{a}All in parts per million; $C_{in}^i = 1800C_R^i + C_p^i$, $C_{out}^i = 0.99C_{sh}^i + 0.06C_{op}^i$.
^{b}Corrected for sea salts.
^{c}Including oceanic carbonates $(= 0.08C_{oc})$.

TABLE 1.10. The Excess River Inputs $(C_{in}-C_{out})$ of Group B Elements and the Marine Aerosol Contributions to the Excess River Inputs $(C_{ma}^i/C_{ma}^{Cl})(C_{in}^{Cl} - C_{out}^{Cl})$ All in ppm

	$(C_{in}^{Cl}-C_{out}^{Cl})(C_{ma}^i/C_{ma}^{Cl})$	$(C_{in}-C_{out})$
	Subgroup B_1: $C_{op}/C_{sh} > 1.4$	
P	400	5
Cu	53	7
Pb	130	40
Zn	280	50
Cd	0.8	1
Au	0.042	0.003
Ag	3.5	0.12
Na	6,540	9,000
Cl	12,560	12,560
Br	33	28
I	$\geqslant 9$	12
	Subgroup B_2: $C_{op}/C_{sh} < 0.7$	
Hg	0.37	0.3
Sb	4.6	1
S	$> 4,200$	2,400
Se	—	1
Sn	—	0.3
W	—	0.05

The subgroup A_1 elements are, in general, well balanced ($C_{out}/C_{in} = 0.7–1.4$) except lithium and arsenic. The high C_{out}/C_{in} ratios for lithium and arsenic (2.26 and 1.58 in Table 1.9) are either caused by too low C_p^i (concentration in river-suspended particles) values given by Martin and Meybeck (1979) or extra inputs other than rivers, for example, mid-ocean ridges (Edmond et al., 1979a). For the subgroup A_2 elements (Ni, Co, Mn, Ba, Mo, B, and probably F), the balance between inputs and outputs is again satisfactory ($C_{out}/C_{in} = 0.89–1.3$); therefore, there is no need to invoke large sources other than rivers to explain

1.3 MASS BALANCE BETWEEN RIVER INPUTS AND OCEANIC SEDIMENT OUTPUTS   41

their high concentrations in oceanic pelagic clay (C_{op}/C_{sh} = 2–10). The desorption or remobilization of manganese, cobalt, nickel, barium and, to a lesser extent, molybdenum from the riverine-suspended particles during or after deposition in estuaries and on the continental shelf ($C_p^i > C_{sh}^i$) can supply the needed amounts for deep-sea sediments. By careful study of geochemical cycles in the Gulf of St. Lawrence, Yeats et al. (1979) demonstrated that this is the case for manganese. But Edmond et al. (1979b) estimated that the mid-ocean ridges alone can supply from one to three times the total manganese deposition observed in deep-sea sediments today. Apparently, their fluxes calculated for manganese as well as other elements are overestimated (see also Chapter 3, Section 3.1.1). In fact, molybdenum is not enriched in Reykjanes hydrothermal solution (Bjornsson et al., 1972) and nickel is even depleted in the hydrothermal solution of the mid-ocean ridges (Edmond et al., 1979b) compared to seawater. Therefore, one can independently show that the mid-ocean ridges are not important sources for molybdenum and nickel in deep-sea sediments.

The excess river inputs of the group B elements (Table 1.10), probably can be supplied by (1) volcanic exhalation (Mros and Zoller, 1975) and volatilization of metals from land vegetation (Beauford et al., 1977) or surface rocks (Goldberg, 1976), (2) recycling from the surface ocean through the atmosphere as marine aerosols and back to rivers by rain, and (3) pollution input. Alternatively, the excess could be only apparent, caused by bad data. If the excess river inputs are solely caused by the first processes, then in a steady-state ocean, the excess river inputs to the ocean should be balanced by sinks somewhere other than the ocean sediment reservoirs (e.g., the mid-ocean ridges). The flow rate of seawater through mid-ocean ridge systems is estimated to be at most about 0.14×10^{18} g/yr and more likely about one-sixth this value (see Chapter 3, Section 3.1.1), as compared to the annual river inflow rate of 3.3×10^{18} g/yr (Garrels and Mackenzie, 1971a). Therefore, even assuming that all the dissolved phosphorus, copper, lead, zinc, cadmium, gold, silver, mercury, antimony, selenium, tin and tungsten were taken up by mid-ocean ridges when seawater flows through hydrothermal systems, the ridges can take up most only a few percent to a fraction of 1% of the excess river inputs.

Alternatively, it is plausible that the excess river inputs to the ocean by the first set of processes could be returned via marine aerosols from the ocean's surface to the continents. Therefore, the next question is: How efficient is the marine recycling of aerosols? If one assumes, as a maximum, that the excess river input of chlorine ($= C_{in}^{Cl} - C_{out}^{Cl}$) is derived entirely from the ocean through marine aerosols, then the marine aerosol contribution to the excess river inputs for the group B elements can be estimated roughly by

$$(C_{ma}^i/C_{ma}^{Cl})(C_{in}^{Cl} - C_{out}^{Cl}),$$

where C_{ma}^i/C_{ma}^{Cl} is the concentration ratio of element i and chlorine in the marine aerosols. The results are given in Table 1.10. The North Atlantic aerosols given by Buat-Menard (1979) are assumed to be representative of the northern hemisphere aerosols. Additional data for tungsten, phosphorus, and tin are from Rahn (1976), for sulfur from Cuong et al. (1974), and for iodine from Moyers and Duce (1972). The marine aerosol contributions can almost entirely explain the excess river inputs of bromine, iodine, sodium, cadmium, and mercury (within a factor of 1.5, Table 1.10). It is well documented that seawater under ultraviolet irradiation, especially combined with bubbling, can release iodine (e.g., Miyake and Tsunogai, 1963; Seto and Duce, 1972). Miyake and Tsunogai (1963) also demonstrated that the evaporation rate of iodine from the ocean surfaces can account for the observed excess river input of iodine. Note that this calculation would also apply to the case where much of the chlorine excess is allowed to accumulate in the ocean, to eventually be removed as evaporites, as was assumed in Section 1.2.3. Sodium, sulfur, bromine, and iodine do in fact accumulate in evaporites along with chlorine.

As shown by Li (1981), mercury and cadmium are elements that have natural sources for marine aerosols. If the enriched mercury and cadmium in marine aerosols are primarily of marine origin, one can conclude that the excess river inputs of mercury and cadmium into the oceans are supplied by volcanic exhalation and volatilization from land vegetation or surface rocks and are mostly recycled back to the continents from the sea interface through bubbling and/or biologic activities. On the other hand, if mercury and cadmium in marine aerosols are supplied from the continents through the atmosphere, the excess river inputs of mercury and cadmium into the oceans have no place to go (neither the mid-ocean ridges nor atmosphere). Therefore, they should be mostly pollution inputs (or simply caused by bad data). The studies by Slemr et al. (1981) and Fitzgerald (1982) show that the gaseous mercury in marine air and particulate mercury in marine aerosols are mostly of continental origin, which supports the latter explanation.

The excess river inputs of sulfur are about two times greater, and of phosphorus, lead, copper, zinc, antimony, silver, and gold are at least four times greater, than the marine aerosol contributions, which, in turn, contain larger pollution components (Li, 1981). Therefore, one can conclude that the excess river inputs of phosphorus, lead, copper, zinc, antimony, silver, gold, sulfur, mercury and cadmium are largely caused by pollution or bad data.

1.4 VARIATIONS WITH TIME

To conclude our discussion of cycles in the chemistry of the ocean, let us turn to an examination of these cycles in the geologic past. In this section, we first discuss constraints imposed by the rock record. Then we turn to some more or less speculative considerations of possible changes induced by particular geologic or biologic events. Some of these are known events, some are more in the nature of "what if" thought experiments, but all are designed to illustrate the principles developed in this chapter.

1.4.1 Constraints Imposed by the Rock Record

Most sedimentary rocks were originally deposited in the oceans and should thus contain information about the composition of the ancient oceans. There are, however, several problems. Since the time of deposition, most rocks have experienced elevated temperatures and pressures associated with deep burial, and most have been exposed to circulating groundwater. Thus, for example, systematic differences between the clay minerals found in ancient rocks and those in modern sediments can be attributed to differences in the chemistry of the ancient oceans (Weaver, 1967) or simply to the effect of burial (Garrels and Mackenzie, 1974; Hower et al., 1976). Also, "the oceans" are not a single, uniform chemical solution. The modern oceans contain temperature gradients, hypersaline lagoons, brackish estuaries, and anaerobic basins where sulfate is being reduced to hydrogen sulfide. It is conceivable that the ancient oceans were divided into separate basins of different composition, and it is possible that a particular suite of sediment samples chosen for study were deposited in an atypical environment. Another problem is that the further back one goes in time the less complete is the sedimentary record (see Chapter 5). The older sediments tend either to have been destroyed by erosion or to have undergone metamorphism, obscuring their original properties. The sedimentary record is essentially continuous in the Phanerozoic (the last 600×10^6 years); that is, sediment corresponding to every stage exists somewhere in the world. The record is incomplete although still usable for many geochemical purposes in the Proterozoic (2500 to 600×10^6 years) but fragmentary in the Archean (older than 2500×10^6 years). No sedimentary rocks older than 3800×10^6 years have been discovered, and so sedimentary rocks provide no direct evidence for conditions that existed in the earth's surface during the first 900×10^6 years of earth history. The sedimentary rock record indicates that liquid water has been present at the surface of the earth for the last 3.8×10^9 years. Traces of living organisms have been found in rocks 3.5×10^9 years old (Schopf et al., 1983). The composition of water at the earth's surface has thus been tolerable for aquatic organisms since the early Precambrian.

1.4.1.1 Carbonates

Calcite ($CaCO_3$) is present in rocks of all ages, indicating that for the last 3.8×10^9 years the oceans have been *approximately* in equilibrium with calcite. This implies that an oceanic pH below 6 is improbable. At the present level of atmospheric carbon dioxide ($10^{-3.5}$ atm), the calcium concentration in the solution in equilibrium with calcite at pH 6 would be more than 10 molal, which is completely unreasonable (other calcium-bearing minerals would precipitate). At higher P_{CO_2} values, pH 6 becomes less unreasonable: at a P_{CO_2} of 10^{-1} atm, the equilibrium calcium concentration would be about 60 times the present value and at a P_{CO_2} of 1 atm it would be about double the present value. Thus, unless the carbon dioxide pressure in the ancient atmosphere were 1 atm or greater, the presence of widespread calcite precludes a pH value below 6.

Furthermore, the abundance of calcite and the absence of sodium carbonate minerals in marine sediments makes a pH value above 9 improbable. At the present atmospheric CO_2 pressure, the calcium concentration in water at pH 9.0 in equilibrium with calcite would be about 5 ppm (compared to the present calcium concentration of 400 ppm and pH of about 8). Any higher P_{CO_2} would result in a correspondingly lower equilibrium concentration of calcium. If the calcium concentration in seawater were less than 5 ppm, calcite would most likely not be as widespread and abundant in sedimentary rocks as it is, and sodium carbonate minerals would be much more common.

Dolomite [$CaMg(CO_3)_2$] is rare in modern sediments but is abundant in ancient (particularly Paleozoic) sedimentary rocks. Modern synthetic dolomite forms from calcite or aragonite when some process raises the molar Mg/Ca ratio in solution to a value greater than about 10 (the value in seawater is about 5). Thus, the abundance of dolomite in ancient rocks has been interpreted as indicating a higher Mg/Ca ratio in the ancient oceans. On the other hand, much of the dolomite could equally well have formed from calcite as a consequence of burial (Garrels and Mackenzie, 1974). The abundance of dolomite in ancient rocks may say nothing about the Mg/Ca ratio in ancient seawater.

Sandberg (1975) made the contrary suggestion that the Mg/Ca ratio was *lower* in the Paleozoic oceans than in the modern oceans. His argument was that ooids (inorganically precipitated spheroids of calcium carbonate) formed in the Paleozoic oceans appear to have been originally calcite, whereas modern marine ooids are all deposited as aragonite (a less stable polymorph of $CaCO_3$). The reason that modern ooids are aragonite is that the high magnesium concentration in seawater inhibits calcite precipitation (Berner, 1975), and so the formation of calcite ooids would imply a lower Mg/Ca ratio in solution. Pigott and Mackenzie (1979) and Sandberg (1983), however, attributed the formation of calcite ooids to a higher pressure of carbon dioxide in the atmosphere rather than to a change in the Mg/Ca ratio. A lower Mg/Ca ratio may also favor diagenetic dolomitization for kinetic reasons (Badiozamani, 1973; Folk and Land, 1975).

Lowenstam (1961) showed that the Mg/Ca, Sr/Ca, and $^{18}O/^{16}O$ ratios in brachiopod shells as far back as the Mississippian (340×10^6 years) are the same as those in modern brachiopod shells and concluded that these ratios in seawater have not changed since the Mississippian. There is always a question, however, of whether the ratios have really been preserved unchanged in the shells. In summary, there appear to be systematic changes in the carbonate minerals of sediments through time, but what these changes imply for the chemistry of the oceans is unclear because of postdepositional changes.

1.4.1.2 Absent Minerals

Certain minerals that could, in principle, form from seawater are absent or rare in normal marine sediments, and their absence places some constraints on the possible composition of ancient seawater (Holland, 1972). Sedimentary rocks show no evidence that seawater in the open oceans has ever become saturated with respect to gypsum (CaSO$_4 \cdot$ 2H$_2$O). This places an upper limit on the activity product $a_{Ca^{2+}} \cdot a_{SO_4^{2-}}$ during the last 2500 × 10^6 years or so. The sedimentary record in the early Precambrian is so incomplete that saturation with respect to gypsum cannot be ruled out. In any event, precipitation of gypsum would prevent the activity product from rising much above the solubility product of gypsum. The absence of brucite [Mg(OH)$_2$] places an upper limit on the activity product $a_{Mg^{2+}} \cdot (a_{OH^-})^2$ that corresponds to a pH between 9.0 and 9.5 in oceans with the present-day Mg^{2+} concentration. Sepiolite [approximately Mg$_4$Si$_6$O$_{15}$(OH)$_2 \cdot$ 6H$_2$O] is by no means absent from ancient marine sediments, but it appears to be confined to rather unusual environments. Its solubility can be described by the equation

$$Mg_4Si_6O_{15}(OH)_2 \cdot 6H_2O + 8H^+ + H_2O = 4Mg^{2+} + 6H_4SiO_4.$$

The corresponding solubility product expression is

$$K_{sp} = (a_{Mg^{2+}})^4 (a_{H_4SiO_4})^6 (a_{H^+})^{-8}.$$

At the Mg^{2+} concentration of the present oceans, the solubility of sepiolite places an upper limit on oceanic pH of about 8.6 at a silica activity of 10^{-4} (a reasonable lower limit for the average silica concentration in seawater); higher silica concentrations would correspond to lower pH values. On the other hand, we have no data that constrain the Mg^{2+} concentration in ancient seawater, except that the widespread occurrence of penecontemporaneous dolomite in ancient sediments suggest that the Mg/Ca ratio in seawater has not been much below 1 during the time interval represented by well-preserved sedimentary rocks (Holland, 1978).

1.4.1.3 Marine Evaporites

Marine evaporites represent salts that were deposited when ancient seawater underwent evaporation and are thus good potential indicators of the composition of the oceans in the past. The amounts of different salts in evaporites do not, however, correspond closely to the amounts of the corresponding ions in seawater, because there is usually some back-mixing of saline brine into the oceans. Dissolved ions that form relatively insoluble salts (notably Ca^{2+} and SO$_4^{2-}$) are overrepresented, and ions that form highly soluble salts only (notably K$^+$ and Mg^{2+}) are greatly underrepresented in the solid evaporites.

Although *amounts* of different salts present in evaporites cannot be used to infer the composition of ancient seawater, the *order* in which the different salts were precipitated can be used to set limits on possible compositional variations (Holland, 1972). According to Holland, the order (CaCO$_3$ or dolomite–gypsum–halite–chlorides and sulfates of K$^+$ and Mg^{2+}) is the same in all ancient evaporites (although there are few well-preserved evaporite deposits older than the latest Precambrian). This order imposes the following constraints on seawater composition:

1. The activity product $a_{Ca^{2+}} \cdot a_{SO_4^{2-}}$ can never have been sufficiently high for seawater to become saturated with respect to gypsum or so low that halite precipitated before gypsum.
2. The concentration of calcium was always less than the equivalent sum of the sulfate and bicarbonate:

$$m_{Ca^{2+}} < m_{SO_4^{2-}} + \tfrac{1}{2} m_{HCO_3^-}$$

If this were not the case, all sulfate would be removed by gypsum precipitation, and the soluble salts precipitated late in the evaporation sequence would contain calcium but no sulfate, the reverse of what is observed.

3. The concentration of calcium was always greater than that of bicarbonate (in equivalents):

$$m_{Ca^{2+}} > \tfrac{1}{2} m_{HCO_3^-}$$

If this were not the case, all calcium would be precipitated as a carbonate, no gypsum would form, and sodium carbonate minerals would probably be precipitated from the final brine.

These constraints, coupled with equilibrium with respect to calcite and an upper pH limit of 9.0, are shown graphically in Fig. 1.2. Holland discussed the effect of variations in the concentrations of sulfate, magnesium, sodium, and chloride (Fig. 1.3) and concluded that, in the last 700×10^6 years, it is unlikely that the concentration of any of the major ions in seawater varied by a factor of more than 2.

1.4.1.4 Chert

Bedded inorganic cherts are common in the Precambrian but are rare or absent in Recent sediments. Probably, silica was precipitated inorganically in the Precambrian eons, whereas in the Phanerozoic eon it was extracted from seawater by diatoms and radiolaria. The concentration of silica in seawater thus probably decreased from a value of about $10^{-2.7}$ M (120 ppm) in the Precambrian to the modern value of about 10^{-4} M (6 ppm).

1.4.1.5 Isotopic Evidence

The isotopic ratios $^{86}Sr/^{87}Sr$, $^{34}S/^{32}S$, $^{18}O/^{16}O$, and $^{13}C/^{12}C$ in the substances precipitated from seawater have all varied systematically through geologic time (Claypool et al., 1980; Veizer et al., 1980). The implications of these

1.4 VARIATIONS WITH TIME 47

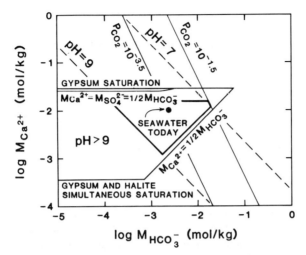

Figure 1.2 The possible range in the Ca^{2+} and HCO_3^- concentrations in present-day seawater that would be consistent with the observed mineralogy of marine evaporites. Note that in a solution in equilibrium with calcite pH and P_{CO_2} are uniquely determined if $m_{Ca^{2+}}$ and $m_{HCO_3^-}$ are specified. After Holland (1972).

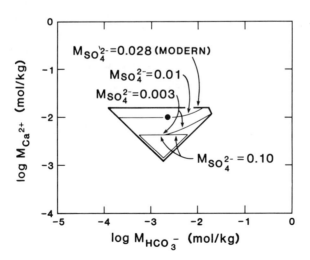

Figure 1.3 Illustration of the effect of changing sulfate concentrations on Fig. 1.2. Any change from the present-day sulfate concentration causes a decrease in the possible range of Ca^{2+} and HCO_3^- concentrations. After Holland (1972).

variations for related geochemical cycles are discussed in Chapter 4 and are not discussed again here. The isotopic data show that the form in which sulfur is removed from the ocean (pyrite versus gypsum) has varied greatly, but the data do not really constrain the composition of ancient seawater.

1.4.1.6 Summary

Limits to the possible variations in the composition of seawater in the geologic past can be established by study of ancient sediments deposited in seawater and by consideration of the mechanisms controlling the concentration of each solute. The most striking fact is that sediments deposited over the last 2×10^9 years look remarkably like modern sediments; the only changes in seawater composition for which there is good evidence are a decrease in silica concentration as siliceous organisms developed in the Precambrian and the appearance of free oxygen in the atmosphere and ocean about 2×10^9 years ago (See Section 2.2.1.1). The question then becomes: What variations in seawater composition could have occurred without the effects being detectable in the sedimentary record? The tightest constraints are imposed by the chemistry of ancient evaporites; in the last 600×10^6 years it is unlikely that the concentration of any major constituent of seawater was different by more than a factor of 2 or 3 from its present value. There are few well-preserved evaporites older than 800×10^6 years, and so the composition of seawater prior to that time is not rigorously constrained.

On a more general note, we know that the composition of seawater is a consequence of the overall cycles of the individual elements, which are coupled and are subject to various types of feedback. We would thus expect the composition of seawater to change in response to changes in global parameters such as the rate of tectonic uplift and the rate of sea-floor spreading, but we really do not know yet how much these parameters have varied in the past or the magnitude of the ocean's chemical response. The mean concentrations of solutes involved in biologic cycles are also sensitive to the rate of circulation of water within the oceans (Broecker, 1971a). The apparent constancy of seawater composition through the geologic record may be a consequence of the inherent stability of coupled cycles (Pytkowicz, 1975; Lasaga, 1980). It does not appear to be a consequence of control by simple thermodynamic equilibrium.

1.4.2 Some Perturbations

These then are the constraints imposed by the rock record and by theory. Let us now see how some plausible disturbances might alter the ocean within these constraints.

1.4.2.1 Tectonics

In Section 1.2.1, we discussed the processes controlling the concentrations of the major elements in seawater; in this section, we outline briefly how these concentrations might change in response to major changes in the tectonic regime. Such changes take the form of increased topographic relief of the continents (the relief at present is anomalously high), of increased volume of the mid-ocean ridges, or of certain geometries of the continents that give rise to evaporite basins, glaciation, and so on. Some aspects that relate to carbon and sulfur are treated in detail in Chapter 4, Sections 4.5 and 4.6. Here we explore briefly a few general considerations.

Continental Relief. An increase in the rate of uplift of the continents should produce increased relief and a greatly increased rate of physical erosion, but only a moderately increased rate of chemical erosion (Holland, 1978). A further effect of uplift might be to reduce atmospheric P_{CO_2} as a consequence of increased weathering rates. This would have the effect (see Chapter 2, Section 2.1.3) of decreasing atmospheric temperatures, decreasing rainfall, and hence reducing the net increase in chemical erosion. The net increase in the input of solutes to the ocean as a result of tectonic uplift should be relatively small; the main effect would be on the input of suspended sediment.

The effect on solute concentrations in seawater would depend on the removal mechanism for the individual solute. The seawater concentrations of elements that are removed primarily by incorporation into detrital sediments should decrease as a result of the relative increase in sediment flux, whereas the concentrations of elements removed by other processes (to the extent that the other processes are independent of tectonic uplift) should increase or remain constant. For example, increased uplift should have no effect on an element such as silicon, whose concentration is determined by biologic processes and circulation rates within the oceans. On the other hand, elements that are removed in large part by ion exchange on mineral sufaces might decrease in concentration, presuming that the mineralogy of the river load does not change appreciably. Among the major elements, Na^+ is the most likely to be affected (Table 1.3). For minor and trace elements, iron–manganese oxides precipitated within the ocean basins seem to be far more important than detrital clays (Section 1.3), and so any change in the amount of detrital input should not cause a significant change in their concentrations. Thus, the only likely effect of increased sediment load on seawater chemistry is a decrease in sodium concentration, but the long residence time for this element (on the order of 100×10^6 years) suggests that increased uplift would have to persist for a very long time for any significant change to occur.

Rate of Sea-Floor Spreading. An increase in spreading rate would result in a rise of sea level on the continents, as a result of the increased volume of the mid-ocean ridges, and in an increase in the rate of convective circulation of seawater through spreading centers, as a result of steeper thermal gradients. The influence of a change in the rate of convective circulation on the composition of seawater depends on how important hydrothermal circulation is as a source or sink for the major elements in seawater. It can be argued (see Chapter 3,

Section 3.2) that, for the post-Archean, the fluxes into and out of seawater caused by hydrothermal processes have been relatively minor, in which case the effect of changes in the circulation rate would also be minor. We recognize, however, that others (e.g., Holland, 1978; Edmond et al., 1979a,b; Von Damm et al., 1985) disagree with this conclusion and consider that hydrothermal circulation is the main process removing Mg^{2+} from seawater and that it has a major influence on the cycles of other elements. In view of this uncertainty, it is reasonable to discuss the *direction* of the possible changes in seawater composition but to recognize that the magnitude of the changes may be trivial, at least in the Phanerozoic.

Hydrothermal circulation at the ridge axis (see Chapter 3, Section 3.3) removes magnesium, sulfate, and alkalinity (bicarbonate) from seawater and adds calcium, potassium, and silica. If the rate of hydrothermal circulation were to increase, we would expect the concentrations of magnesium and sulfate in seawater to decrease, the alkalinity and pH to decrease, and the concentration of calcium to increase. Silica concentration would not change because the additional input would be precipitated by organisms; potassium would tend to increase, but uptake by clays or basalt at low temperature should prevent any major rise in concentration, leaving calcium, magnesium, alkalinity, and pH as varying—although note that all these are buffered by carbonate reactions.

Although the overall significance of hydrothermal circulation for seawater chemistry in the Phanerozoic is subject to debate, such circulation was almost certainly important in the Archean (see Chapter 5). The pH and the concentrations of magnesium and sulfate should thus have been lower, and the concentration of calcium should have been higher. The supply of iron to the deep ocean in the Archean might also have had an important effect on seawater chemistry. In the modern oceans, iron supplied by hydrothermal solutions is precipitated immediately by oxidation as the solutions mix with seawater, whereas in an oxygen-free ocean, much of this iron might remain in solution as Fe^{2+}.

Perhaps the most important effect of an increase in spreading rate would be the general rise in sea level caused by the increase in the volume of the mid-ocean ridge system. One result would be a decrease in the input of both sediment and solutes from the continents, which would enhance the relative importance of hydrothermal circulation as an influence on seawater chemistry, as suggested by strontium isotope variations (see Chapter 5). Another result might be the expansion of anoxic environments and hence an increase in the removal of sulfate in the form of sulfide. These ideas are explored in detail in Chapter 4, Section 4.5.

1.4.2.2 Evaporitic and Anoxic Events

In Section 1.2.3, we noted that Cl^- and SO_4^{2-} appear to be accumulating in the ocean at appreciable rates. These substances must be removed in occasional bursts owing to the formation of restricted basins in which large volumes of evaporites or sulfides are deposited. These bursts can be referred to as evaporitic or anoxic events. Both are documented in the rock record, the Mediterranean Basin accumulated large volumes of evaporite minerals in the late Miocene, while the Atlantic Ocean accumulated abundant sulfide-containing sediment in several episodes during the Cretaceous.

Is the size of such occasional events great enough to account for the river flux of Cl^- and SO_4^{2-}? In the late Miocene, about 25×10^{18} moles of NaCl were deposited in the Mediterranean Sea, Red Sea, and nearby basins (Holser et al., 1980). This amount is about 4% of the present mass of NaCl in the oceans. The excess flux from rivers calculated in Table 1.8 is 2.2×10^{12} moles/yr. Thus, the depletion of NaCl by the Miocene event can be replaced in 11×10^6 years, compared with the approximately 5×10^6 years that have elapsed since the event. Therefore, such events appear to be well able to absorb the river excess over long periods of time, but their irregular occurrence suggests that the salinity of the oceans has varied substantially on a shorter time scale. Holser et al. (1980) suggest that salinity may in fact have declined in spurts from about 45‰ in the Paleozoic to 35‰ in the modern oceans.

We cannot as yet make an analogous calculation for anoxic events. Besides the Cretaceous Atlantic, there are widespread anoxic deposits in the Upper Devonian of North America and in the Ordovician of Newfoundland and Great Britain. The volume of sulfur deposited in each is unknown, but the irregular distribution in time of these deposits suggests again that removal may be episodic and therefore that the proportion of sulfur being deposited as sulfide may have varied significantly through geologic time (see Chapter 4, Section 4.4).

1.4.2.3 Biologic Evolution

The most obvious noncyclic influence on the ocean and atmosphere is biologic evolution. There are certain events in the history of life, known from paleontology or biochemistry, that seem to us capable of having influenced the composition of the ocean–atmosphere system sufficiently to be detectable in the sedimentary record. We discuss the well-known example of the appearance of oxygen-yielding photosynthesis in Chapter 2. Here we mention briefly some other events that are equally important biologically but that seem to have left a much more meager record.

Eukaryotes. The appearance of eukaryotes (by 1.4×10^9 years ago) is the most important evolutionary step after the appearance of life. It has left no trace in the composition of sediments, however that has been detected so far.

Metazoans. For sediments, the most important difference between metazoans (by 0.7×10^9 years ago) and protozoans is that the former produce bioturbation, a stirring of the sediments by their burrowing for food or for protection from predators. This stirring should enhance exchange between sediments and seawater and so may lead to decreased burial of organic carbon. The change from nonbioturbated to bioturbated sediments should be readily detectable in the rock record, but so far it has not been studied in enough localities. Bioturbation appears to be absent in 1.3×10^9 year-old sediments of the Belt Supergroup (Byers, 1976). In fact, trace fossils seem to be confined to strata a few tens to perhaps 100 m below the earliest body fossils (Frey and Seilacher, 1980), making them little older than 0.6×10^9 years.

$CaCO_3$ Skeletons. This event (0.6×10^9 years ago) is the real beginning of the fossil record. Its influence on seawater chemistry is doubtful, however. Although today most $CaCO_3$ removal from seawater is through the activities of shelled organisms, the ocean is so nearly saturated with $CaCO_3$ that its inorganic removal would provide the same constraints.

Land Plants. Land plants (about 0.4×10^9 years ago) make up the bulk of the earth's biomass; they also accelerate rock weathering by increasing the amount of CO_2 in the air in soils and by secreting organic acids. At the same time, they tend to retard erosion by binding the soil and inhibiting rain splash. Thus, it would be reasonable to suppose that their evolution would lead to a more highly weathered sediment mass delivered to the ocean plus an increase in the ratio of dissolved to suspended load. As yet, no unequivocal evidence for such a change has been reported. Perhaps tectonic (relief) and climatic (rainfall) factors predominate over the biologic ones. Moore (1983) has argued that the growth of a large plant biomass in the Carboniferous period had little effect on atmospheric CO_2, but Berner and Raiswell (1983) have suggested that C/S ratios in sediments were affected.

Planktonic Foraminifera. As mentioned earlier, the appearance (about 0.14×10^9 years ago) of pelagic $CaCO_3$-secreting organisms shifted the locus of maximum $CaCO_3$ deposition from shallow shelves to the deep sea. Seawater composition may have been affected in two ways: first, because dolomitization is apparently caused mostly by early diagenetic alteration of shelf carbonates, Mg^{2+} concentration may have been lower before this shift; second, the new position of $CaCO_3$ deposits on the deep-sea floor makes them far more susceptible to subduction. There is no clear evidence for the amount of sediment subducted (see Chapter 3), but if only a small fraction is, then a major flux of Ca^{2+} from the crust into the mantle may have begun in the Jurassic period. Another aspect is the possible effect on island arc volcanism: if large amounts of $CaCO_3$ are now being subducted, the additional CaO and CO_2 may change the composition or the amount of volcanic rock produced.

Siliceous Organisms. Unlike $CaCO_3$, SiO_2 is precipitated by organisms from solutions far below saturation. Therefore, they should have markedly lowered oceanic H_4SiO_4. Radiolaria first appear during the Cambrian period, and diatoms, which seem to be more efficient silica secreters, are found in Cretaceous and younger sediments. As discussed in Section 1.4.1.4, there is evidence for higher H_4SiO_4 in Precambrian time, but there are no analogous data from the rock record supporting a further decrease in the Cretaceous period, although there is a decrease in the average weight of radiolarian tests at this time (Harper and Knoll, 1975).

C_4 Plants. Some angiosperms, particularly certain grasses, exhibit a distinctive type of photosynthesis, involving intermediates with four-carbon chains instead of three-carbon chains. As discussed in Chapter 4, Section 4.3, they produce organic carbon that is isotopically heavier than normal C_3 plants. Because carbon isotope ratios are unaffected by diagenetic temperatures, there is a good chance that this isotopic signature can be preserved in the rock record. Modern sediments whose carbon is largely of this type are known (Fry et al., 1977), and C_4 residues have been found in Pliocene rocks, but no older occurrences are known (Moore, 1983). This style of photosynthesis seems to be particularly suited to environments in which water is limited, usually deserts, but also salt marshes. For example, *Spartina* and *Thallasia* grasses are of this sort (Smith and Epstein, 1971). Thus, organic-rich fluvial deposits from arid or semiarid paleoenvironments or ancient salt marsh deposits are likely places to find this metabolic fossil. The time of origin of this behavior was presumably in the Cretaceous or later, but not much more is known. Aside from their C-isotopic signature, these plants probably have not left any imprint on the rock record, unless perhaps they have been able to modify the nutrient distribution in salt marshes.

2 GEOCHEMICAL CYCLES OF ATMOSPHERIC GASES

J. C. G. Walker and J. I. Drever

In this chapter we describe the processes that are thought to control the composition of the present and past atmosphere, limiting discussion to the major atmospheric constituents and starting with the examples that are least complicated. The natural systems before human intervention are emphasized, but see Section 2.5.2 and the discussion in Stumm (1977) for a treatment of pollution effects.

2.1 THE MODERN ATMOSPHERE

The approach is to identify the sources, sinks, and reservoirs appropriate to each constituent. The sources are the processes that add a constituent to the atmosphere, whereas the sinks are the processes that remove the constituent from the atmosphere. Associated with sources and sinks are rates, which express the amount of gas added or removed in a given period of time. As a rule, sources and sinks with large rates are more important than those with small rates, but reservoirs and their sizes must also be considered. The atmospheric reservoir represents the amount of the constituent in question that is in the atmosphere. Other reservoirs such as oceans, sediments, or biota are places where the constituent might be if not in the atmosphere. The significance of a source or sink depends to a considerable extent on the size of the reservoirs it links. A sink that transfers atmospheric gas to a reservoir much smaller than the atmospheric reservoir, for example, may not control atmospheric composition even if its rate is large, because there is no room in the receiving reservoir for all the gas in the atmosphere. These ideas can be understood more easily by reference to the specific examples that follow.

2.1.1 Water Vapor

If all the water vapor in the atmosphere were condensed on the earth's surface it would form a layer only 2.5 cm thick (Sellers, 1965). Compared with the oceanic reservoir therefore, the atmospheric reservoir is very small (Fig. 2.1). The source of atmospheric water vapor is evaporation of surface water at a globally averaged rate of 100 cm/yr. Evaporation can fill the atmospheric

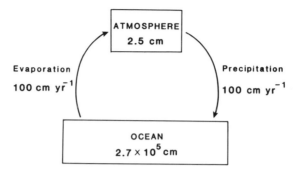

Figure 2.1 The cycle of atmospheric water vapor.

reservoir in a time as short as 9 days. We would say the *residence time* of water vapor in the atmosphere is 9 days. The sink for atmospheric water vapor is condensation followed by precipitation. Because water vapor is not accumulating in the atmosphere, the sink, when averaged over times longer than the residence time, must be equal to the source.

Whether a particular mass of air when exposed to a body of water at the surface will gain water by evaporation or lose water by precipitation depends on its relative humidity. If the air is undersaturated, with relative humidity less than 100%, the rate of evaporation will exceed the rate of condensation and the air mass will gain water vapor. If the air is supersaturated, with relative humidity greater than 100%, the sink will exceed the source and the air mass will lose water vapor. The interaction of source and sink is such then as to drive the atmosphere toward a state of saturation. Because of the variability of weather and because not all air masses are in contact with bodies of water, this state is not achieved for the atmosphere as a whole. The average relative humidity of the air in the troposphere (the lowest 15 km of the atmosphere) is about 50% (Sellers, 1965).

The amount of water required to saturate a given mass of air depends on the temperature of the air. Air at high temperature can hold more water vapor before saturation is achieved than air at low temperature (Fig. 2.2). Because air temperature decreases with altitude in the troposphere, the upper levels contain relatively little water vapor and most atmospheric water is found within a few kilometers of the surface.

Consideration of these factors leads to the conclusion that the average water vapor content of the atmosphere depends on the average temperature of the surface. A rough rule of thumb is that the water vapor pressure at the surface is one-half the saturated vapor pressure at the temperature of the surface. If surface temperatures were higher during the course of earth history, we can confidently predict that the atmosphere contained more water. Conversely, the water vapor content of the atmosphere must have been low during periods of relatively low surface temperatures. There are two interesting implications of

2.1 THE MODERN ATMOSPHERE 57

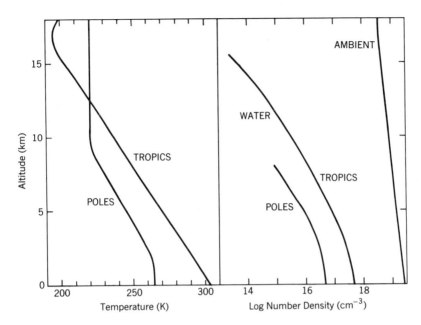

Figure 2.2 The abundance of water vapor in the atmosphere depends on temperature.

this conclusion about the processes controlling the water vapor content of the atmosphere. The first has to do with the influence of atmospheric water vapor on surface temperature by way of the greenhouse effect. Water vapor is transparent to short-wavelength solar radiation, which heats the earth, but it absorbs the long-wavelength infrared radiation that cools the earth, returning some of this radiation to the surface (Goody and Walker, 1972). The result is that increasing amounts of water vapor in the atmosphere yield increasing surface temperatures, all other things being equal. But we have just argued that the water vapor content of the atmosphere depends on its surface temperature. There is therefore a *positive* feedback (Fig. 2.3). If some change, an increase in solar luminosity, for example, causes an increase in surface temperature, atmospheric water vapor will increase and, through the greenhouse effect, will reinforce the effect of the original change. This feedback has received little attention in connection with geologic history, but its magnitude can be deduced from studies of the climatic effects of anthropogenic carbon dioxide. Very roughly, the water vapor–greenhouse feedback serves to double the size of a temperature change caused by some factor other than water vapor (Rodgers and Walshaw, 1966; Henderson-Sellers and Meadows, 1979).

The other implication of the water cycle described above relates to precipitation. More rain can fall from warm saturated air (Fig. 2.4). Other things being equal, global average precipitation should have been high during periods of high average temperature and low during periods of low average temperature. For example, a 10°C rise in temperature should yield an increase in precipitation

58 GEOCHEMICAL CYCLES OF ATMOSPHERIC GASES

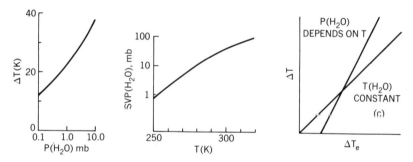

Figure 2.3 Feedback of the water vapor greenhouse increases the sensitivity of surface temperature (T) to temperature changes from other causes (ΔT_e). (*a*) Increment from water vapor greenhouse. (*b*) Saturated water vapor pressure as a function of temperature. (*c*) Surface temperature increment (ΔT) varies more rapidly for constant relative humidity than for constant absolute humidity.

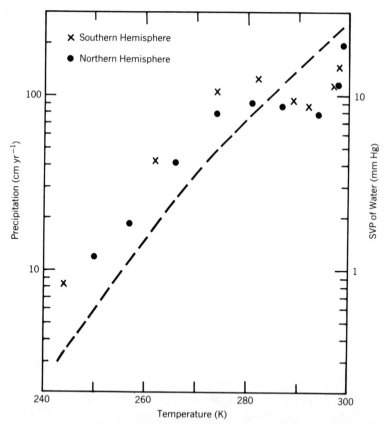

Figure 2.4 Zonally averaged precipitation rate as a function of averaged temperature (Sellers, 1965). The broken line is the saturated vapor pressure of water, with the scale on the right of the figure.

of roughly a factor of 2. Whether the effect of changes in global average temperature on global average precipitation can be detected in the geologic record is not clear. We are not aware of any quantitative exploration of this idea.

2.1.2 Nitrogen

Most of earth's nitrogen is probably in the atmosphere (Walker, 1977); the ocean contains 3000 times less, mostly as nitrate ions or in dissolved organic compounds. Sedimentary rocks contain perhaps one-third as much nitrogen as the atmosphere, largely as a component of organic compounds that were originally synthesized by organisms and became trapped in sediments at the bottom of the sea, but also as ammonium ions on clay exchange sites. The nitrogen content of igneous and metamorphic rocks is uncertain, but concentrations are typically quite low, 1–20 ppm compared to 200–4000 ppm in sedimentary rocks (Wedepohl, 1969, Tables 7-E-2 and 7-K-6).

The geochemical cycles of nitrogen are important to the biota (Burns and Hardy, 1975; Fenchel and Blackburn, 1979) but appear to have little impact on the atmosphere, mainly because of the relatively large size of the atmospheric reservoir (Fig. 2.5). Microorganisms extract nitrogen from the atmosphere in the process of nitrogen fixation in which molecular nitrogen (N_2) is converted into the amino radical (NH_2), an essential constituent of cell material. Nitrogen is released by organisms to the environment as gaseous or dissolved ammonia. In the aerobic world of today, this ammonia is rapidly oxidized by other microorganisms in the process known as nitrification, which produces nitrite or nitrate, NO_2^- or NO_3^-. This process is a source of energy for the nitrifying organisms. The nitrate may be reincorporated into cell material, undergoing reduction back to the level of the amino radical, or the nitrogen may be restored to the atmosphere in the biologic process called denitrification. Denitrification is a process of anaerobic respiration. In environments short of oxygen, certain microbes can obtain energy by oxidizing organic matter with nitrate instead of

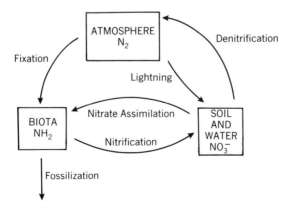

Figure 2.5 The simplified biogeochemical cycles of nitrogen.

with oxygen. Most of this nitrate is converted into molecular nitrogen. The global average rates of this biologic nitrogen cycle are not precisely known, but it appears that the residence time of nitrogen in the atmosphere between trips through the biota is about 20×10^7 years (Walker, 1977).

Biologic nitrogen fixation is supplemented by an abiogenic process: high-temperature chemical reactions that occur in the column of air heated by a lightning bolt. These high-temperature reactions convert atmospheric nitrogen and oxygen into nitric oxide (NO). Subsequent photochemical reactions convert this nitric oxide to nitric acid, which then dissolves in rainwater and enters the hydrosphere. Again the global rates are uncertain, but the lightning sink for atmospheric nitrogen appears to be comparable in magnitude to the biologic sink (Chameides et al., 1977). Humans, of course, have intervened massively in the nitrogen cycle but not enough to affect the nitrogen content of the atmosphere.

The ocean could absorb much more nitrate without becoming saturated, so it is appropriate to ask why the oceanic reservoir of nitrogen is not larger. This subject is controversial, but a possible answer relates to the energy requirements of the organisms that fix nitrogen. Nitrogen fixation is an energetically expensive process that organisms will undertake only if their nutrient supply is deficient in nitrogen. Average seawater today contains as much nitrogen relative to another nutrient element, phosphorus, as does the average cell material of the planktonic microorganisms that fix nitrogen (Broecker, 1974). Thus, as these microbes grow in seawater today, they exhaust nitrogen and phosphorus at the same time (Fig. 2.6). In the absence of phosphorus, no further growth is possible, so there is no further need for nitrogen and therefore no advantage to be gained by fixing nitrogen. At the same time, there is a constant drain on the oceanic reservoir resulting from denitrification and from the burial of nitrogen as part of the organic matter incorporated into sediments. It seems likely that organisms fix just enough nitrogen to make up for deficiencies in supply, which means that the nitrogen content of the sea is related to the phosphorus content by the nutritional requirements of plankton. Since it is not likely that the phosphorus content of the sea was ever much larger in the past than today

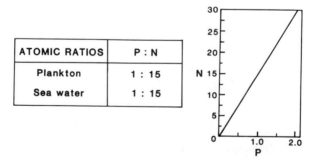

Figure 2.6 Growing organisms exhaust phosphorus and nitrogen at the same time in average seawater. After Redfield et al. (1963).

(phosphate minerals would be more common in the rock record), it is equally unlikely that the oceanic reservoir was ever large enough to absorb a substantial fraction of the nitrogen in the atmosphere.

2.1.3 Carbon Dioxide

The geochemical cycles of carbon dioxide are known in much greater detail than those of any other atmospheric constituent (Walker, 1977, 1984), although there is still debate concerning the processes that control the carbon dioxide partial pressure. The interest in carbon dioxide has derived from chemical oceanography because of the close association of this gas with oceanic bicarbonate, the pH of seawater, and the precipitation of carbonate minerals (Broecker, 1974; Sundquist and Broecker, 1985). The study of the cycles of carbon dioxide has received a further stimulus from concern over the impact of the burning of fossil fuels on atmospheric carbon dioxide and thus on climate (Broecker et al., 1979).

The climatic impact is one reason why carbon dioxide is the most interesting gas from the point of view of atmospheric evolution. Carbon dioxide, like water vapor, contributes to the greenhouse effect that maintains equable temperatures at the surface of the earth. Increased partial pressures of carbon dioxide in the past could have yielded increased surface temperatures. Carbon dioxide is interesting also from the point of view of biologic evolution because it is the carbon source for photosynthetic organisms; fluctuations in atmospheric carbon dioxide may have influenced biologic productivity. Carbon dioxide is important in connection with the photochemistry of ancient atmospheres as well, because its photolysis by solar ultraviolet radiation is a major source of reactive oxygen atoms at middle levels of atmospheres devoid of molecular oxygen (Kasting et al., 1979).

In addition, carbon dioxide is, of all the major atmospheric gases, the one most susceptible to change. Witness the fact that anthropogenic activity is already changing it. This susceptibility to change arises because the atmospheric reservoir of carbon dioxide is much smaller than the other reservoirs to which it is linked. The atmospheric reservoir is closely coupled by the processes of photosynthesis and respiration to a reservoir of organic carbon, about four times larger, consisting of the biota and young organic matter in soils, seawater, and the surface layers of submarine sediments (Fig. 2.7). The size of this reservoir depends on factors associated with the geochemical cycles of atmospheric oxygen, which are discussed below.

Carbon is the central element of life, and the distribution of carbon at the surface of the earth is controlled by biologic activity. Carbon and other vital elements are processed (metabolized) by living systems (organisms) for purposes of (1) providing energy for continued biologic activity and (2) making new biologic cells. Some organisms (heterotrophs) can only metabolize previously synthesized organic matter, whereas others (autotrophs) are capable of metabolizing inorganic compounds such as carbon dioxide if an additional source of energy (sunlight or chemical energy) is available. Green plant photosynthesis is

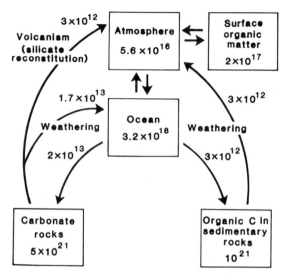

Figure 2.7 Carbon budget for times longer than 10^4 years (reservoirs in moles and rates in moles per year). Double arrows indicate an equilibrium between reservoirs. From Walker (1977).

the most important autotrophic process for converting carbon dioxide into organic matter:

$$CO_2 + H_2O \xrightarrow{light} CH_2O + O_2, \qquad (2.1)$$

where CH_2O is symbolic of reduced carbon and oxygen is an important by-product. Organisms obtain energy by oxidizing this organic matter according to the reaction

$$CH_2O + H_2O \longrightarrow CO_2 + 4e^- + 4H^+. \qquad (2.2)$$

This oxidation reaction (anaerobic decarboxylation) must be coupled with the reduction of a substance (such as oxygen, sulfate, carbon dioxide, or the oxidized portion of an organic molecule) that is capable of accepting electrons, as in the reactions

$$4e^- + 4H^+ + O_2 \longrightarrow 2H_2O, \qquad (2.3)$$

$$4e^- + 4H^+ + \tfrac{1}{2}SO_4^{2-} \longrightarrow \tfrac{1}{2}H_2S + OH^- + H_2O, \qquad (2.4)$$

$$4e^- + 4H^+ + \tfrac{1}{2}CO_2 \longrightarrow \tfrac{1}{2}CH_4 + H_2O. \qquad (2.5)$$

When reaction (2.2) is coupled with a reaction such as (2.3), (2.4), or (2.5), the net result is referred to as respiration, which can be aerobic or anaerobic.

Aerobic respiration is chemically the opposite of photosynthesis:

$$CH_2O + O_2 \longrightarrow CO_2 + H_2O. \qquad (2.2)+(2.3)$$

The most common anaerobic respiration process is sulfate reduction:

$$CH_2O + \tfrac{1}{2}SO_4^{2-} \longrightarrow \tfrac{1}{2}H_2S + HCO_3^-. \qquad (2.2)+(2.4)$$

Methanogenesis (2.5) is less important in modern marine environments; oxygen is the preferred electron acceptor in aerobic environments and sulfate in anaerobic environments. Aerobic respiration generally leads to complete destruction of the organic matter but anaerobic oxidation is usually incomplete. This inefficiency results in burial of organic matter in anaerobic sediments; that is, the carbon is shunted out of a short-term carbon cycle and into a long-term cycle.

Thus, there is a short-term carbon cycle dominated by an exchange of carbon dioxide between the atmosphere and biosphere in the processes of photosynthesis and respiration, and by an exchange between the atmosphere and ocean (solution and dissolution). Radioactive tracer studies demonstrate that a carbon atom will spend about 6 years in the atmosphere, about 20 years in the terrestrial biosphere, about 10 years in the surface waters of the ocean, and about 1600 years in the deep ocean. This short-term carbon cycle is driven almost exclusively by solar energy.

Long-term circulation of carbon in the exogenic cycle is controlled by burial in sedimentary rocks and return to the atmosphere by weathering and volcanism. Once carbon is buried and preserved in sedimentary rocks (either as organic carbon or carbonate carbon), it is kept out of circulation for lengths of time that average about 300×10^6 years. The processes that control the long-term burial and weathering of carbon in rocks are driven largely by heat generated from the decay of radioactive elements in the earth's interior.

At the present time, the carbon cycle appears to be out of balance, as a result of rapid input of carbon dioxide to the atmosphere, primarily from the combustion of fossil fuels but also from deforestation for cultivation and burning of wood. Both of these inputs are related to the rapid expansion of human population and industrial activity since about 1850. This aspect of the short-term carbon cycle presently is undergoing intense study (e.g., Andersen and Malahoff, 1977; J. Williams, 1978; Bach et al., 1979; Bolin et al., 1979; Bach, 1980; Broecker and Peng, 1982). We attempt to evaluate the operation of the carbon cycle in its pre-1850 state in this section and to interpret events in the geologic past that may have involved upsets in its operation.

Over time periods longer than a few thousand years, atmospheric carbon dioxide is closely linked to an oceanic reservoir composed mainly of bicarbonate ions that is 60 times larger than the atmospheric reservoir and 15 times larger than the organic reservoir. The processes of solution and dissociation that link these two reservoirs are rapid and, to a first approximation, the amount of carbon dioxide in the atmosphere is proportional to the amount of bicarbonate in the sea. The proportionality constant depends on such factors as pH, temperature, and biologic productivity, but there has been little or no work that has considered how changes in these factors might have affected the geologic history of the atmosphere. Because of the close coupling of the atmosphere and ocean, however, any change in the size of surface organic reservoirs can easily be accommodated by small changes in the large oceanic reservoir, with little impact on the amount of carbon dioxide in the atmosphere.

Of much greater potential importance is the sedimentary rock reservoir of carbonate minerals and fossil organic carbon. This reservoir is some 2000 times larger than the combination of atmosphere, ocean, and young organic reservoirs. Any imbalance between the rate of formation of carbonate minerals and their rate of destruction by erosion and weathering could exhaust the oceanic reservoir in a time as short as 200,000 years. How is it that such imbalances have not occurred? What processes control the partial pressure of carbon dioxide?

We postpone a quantitative treatment of the carbon cycle until Chapter 4, but it is appropriate here to consider qualitative answers to these questions. At first glance, it might appear that the biologic processes discussed above control the amount of carbon dioxide in the atmosphere (P_{CO_2}). Although fluxes between the biomass and the atmosphere are rapid, the biomass reservoir is so small that it cannot exert an appreciable control. Two other processes appear to be dominant in the long-term behavior of atmospheric CO_2, and both involve coupling of the carbon cycle with cycles of other elements.

The first is burial of organic carbon, which leads to a coupling of the cycles of oxygen and carbon via reactions like (2.1). Furthermore, reactions between sulfur and carbon, like (2.2) plus (2.4), also can affect atmospheric oxygen. Note that increased burial of pyrite would decrease the organic carbon reservoir and return carbon to the atmosphere via oceanic HCO_3^-. Similarly, increased burial of organic carbon would remove CO_2 from the atmosphere and increase O_2. Kump and Garrels (1986) have quantified this coupling and investigated the predicted effect of a 30% increase in the rate of burial of organic carbon. Their model predicts a nearly instantaneous (a few million years) time for adjustment of the carbon cycle, $30-40 \times 10^6$ years for oxygen and over 100 million years for sulfur to attain a new steady state. The difference in the carbon and sulfur responses can be attributed to the much larger proportion of total sulfur present as dissolved sulfate in the ocean compared to the relatively minor amount of oceanic HCO_3^- compared to carbonate in limestone.

The second coupling of CO_2 with other geochemical cycles is with silicate weathering and reconstitution. A number of workers have considered such processes, most recently Berner et al. (1983) and Lasaga et al. (1985). They have developed a computer model for the behavior of a number of substances related to the carbon cycles over the last 100 million years. Their model is based on the idea that reactions of calcium and magnesium-bearing silicates control atmospheric CO_2 via two processes: weathering

$$CaSiO_3 + CO_2 + 2H_2O \longrightarrow CaCO_3 + H_4SiO_4, \qquad (2.6)$$

$$MgSiO_3 + CO_2 + 2H_2O \longrightarrow MgCO_3 + H_4SiO_4, \qquad (2.7)$$

and its reverse, metamorphism

$$CaCO_3 + SiO_2 \longrightarrow CaSiO_3 + CO_2, \qquad (2.8)$$

$$MgCO_3 + SiO_2 \longrightarrow MgSiO_3 + CO_2. \qquad (2.9)$$

The mineral compositions given represent components in more complex phases. The carbonate minerals themselves are also subject to weathering reactions such as

$$CaCO_3 + CO_2 + H_2O \longrightarrow Ca^{2+} + 2HCO_3^-, \qquad (2.10)$$

whose reverse is the precipitation of limestone,

$$Ca^{2+} + 2HCO_3^- \longrightarrow CaCO_3 + CO_2 + H_2O, \qquad (2.11)$$

with analogous reactions for Mg^{2+} and dolomite. In balance, the weathering and precipitation of carbonate minerals do not change the carbon content of the oceans and atmosphere.

A negative feedback response exists to any rise in CO_2. The increased CO_2 will produce an increase in surface temperatures plus an increase in rainfall. Both will accelerate weathering and thus the removal of CO_2 from the atmosphere (Walker et al., 1981).

2.1.4 Oxygen

The presence of abundant free oxygen in the terrestrial atmosphere is an anomaly in a solar system and universe composed predominantly of hydrogen. It is, of course, a direct consequence of the presence of life on earth, particularly the presence of photosynthetic organisms that use water as an electron donor to reduce carbon dioxide to organic carbon, producing molecular oxygen as a waste product (Walker, 1984). The accumulation of this metabolic waste product in the atmosphere is of interest because all aerobic organisms, including all multicellular plants, depend on free oxygen to sustain their vital processes.

The biogeochemical cycles of oxygen, based on the work of a number of authors summarized in Walker (1980), are outlined below (Fig. 2.8). The only significant source of free oxygen is photosynthesis. It is closely linked to a group of processes generally labeled *respiration and decay*, which recombine the oxygen and organic carbon produced by photosynthesis to restore carbon dioxide and water to the biosphere. The cycle of photosynthesis followed by

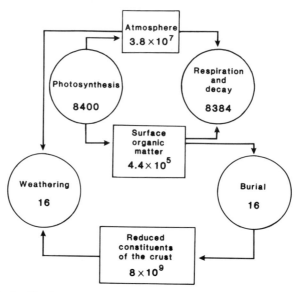

Figure 2.8 A simplified representation of the biogeochemical cycles of oxygen. Circles denote processes, with rates expressed in 10^{12} moles of oxygen per year or equivalent capacity to combine with oxygen. Rectangles denote stores of either oxygen or reduced matter with which oxygen can combine. The units are 10^{12} moles of oxygen or capacity to combine with oxygen (Walker, 1980).

respiration and decay connects a relatively large reservoir of atmospheric oxygen to a relatively small reservoir of organic carbon at the surface of the earth. Stoichiometric considerations suggest that this cycle cannot, by itself, control the amount of oxygen in the atmosphere. Instead, it controls the size of the reservoir of surface organic carbon. If photosynthesis were to stop, surface organic carbon would be consumed by respiration and decay, but atmospheric oxygen would be reduced in the process by only 1% or so.

The abundance of atmospheric oxygen is controlled by a much slower cycle that connects the atmospheric reservoir to a very large reservoir of reduced minerals in the crust of the earth. Oxygen is extracted from the atmosphere when these minerals are exposed at the surface and oxidized during the course of chemical weathering. Corresponding to this relatively slow consumption of oxygen is a net source from the burial of small amounts of organic matter that escape destruction by respiration and decay. This organic matter escapes from the biosphere by settling to the bottom of the ocean and becoming incorporated in sedimentary rocks. The oxygen that was released to the atmosphere when this organic matter was synthesized is left behind, eventually to be consumed by weathering.

The rate at which oxygen is consumed by weathering may depend on the abundance of atmospheric oxygen. The weathering rate may then increase with increasing oxygen partial pressure. On the other hand, the fraction of photosynthetically produced organic matter that escapes oxidation by respiration and decay to become incorporated in sediments should, in principle, decrease with increasing oxygen partial pressure. Therefore, an increase in the amount of oxygen in the atmosphere leads to an increase in the rate of consumption of oxygen by weathering and a decrease in the net rate of production of organic carbon. These interactions are thought to control oxygen abundance, causing an equilibrium to be established at the oxygen level that yields equality between the rates of consumption of oxygen by weathering and net production of oxygen owing to the burial of organic matter in sediments (Holland, 1978, Chapter 6). The geologic history of atmospheric oxygen has received more attention than that of any other atmospheric gas, and our understanding of the controlling mechanisms suggest that fluctuations may have occurred. Attempts have been made to explore these fluctuations from a number of different points of view (Tappan, 1968; McAlester, 1970; Walker, 1977; Holland, 1984; Kump and Garrels, 1986), but as yet no general agreement has emerged. One question that is of great interest and relatively uncomplicated, however, is that of the level of oxygen before the origin of photosynthesis, uncomplicated because it depends largely on well-understood processes of atmospheric physics and chemistry and only weakly on uncertain fluxes involving biota and solid earth. The results of such calculations are summarized in Section 2.2.2.

2.1.5 Minor Atmospheric Constituents

The geochemical cycles of the remaining atmospheric gases, except for the inert gases, are all dominated by photochemical reactions within the atmosphere (Graedel, 1978). Methane, ammonia, nitrous oxide, and hydrogen sulfide are produced by the biota and oxidized within the atmosphere. Carbon monoxide, nitrogen dioxide, and sulfur dioxide are destroyed photochemically and sustained by a combination of photochemical and surface sources and sinks. Methane is potentially the most interesting of these gases at our present level of knowledge of atmospheric evolution. It is produced today in the course of fermentation processes by which organisms extract energy from organic compounds in anaerobic environments (Wolfe, 1971; Ward et al., 1987). This biogenic source of methane was probably larger before the rise of atmospheric oxygen, when aerobic respiration was not available to organisms as a source of metabolic energy (Ycas, 1972). The photochemical destruction of methane occurs today by way of a series of reactions initiated by the attack of an OH radical:

$$CH_4 + OH \longrightarrow CH_3 + H_2O. \tag{2.12}$$

The final products of the methane oxidation chain in the modern atmosphere are water and carbon dioxide, but hydrogen, carbon monoxide, and even ozone are produced along the way or as by-products of oxidation reactions (Walker, 1977). In the anaerobic atmosphere, the oxidation of methane would have yielded hydrogen and carbon dioxide, while hydroxyl radicals would have been produced in abundance by the photolysis of water vapor. Quantitative studies of how much methane might have accumulated in the atmosphere have been published by Kasting et al. (1983) and Zahnle (1986). The question is of interest because methane is a greenhouse gas that could have influenced early climates and because sufficiently high levels of methane will result in its photochemical polymerization to form nonvolatile higher hydrocarbons, the primordial oil slick proposed by Lasaga et al. (1971).

2.2 VARIATIONS WITH TIME

Now let us turn to a consideration of the record of changes with time in the chemistry of the atmosphere. For some substances, we have evidence from the rock record, for others we must infer past behavior from our analysis of the present system (Walker, 1986; Holland et al., 1986).

2.2.1 Biologic Evolution and the History of the Atmosphere

2.2.1.1 Oxygen

The rise of atmospheric oxygen, which occurred between 2.5 and 1.5×10^9 years ago, provides the most striking illustration yet known of the impact of life on the composition of the exogenic system. The prebiologic atmosphere almost certainly contained little free oxygen. A small source of oxygen resulting from the photolysis of water, followed by escape of hydrogen to space, would have been overwhelmed by a much larger source of reduced gases from the interior of the earth released by volcanism and metamorphism (Walker, 1977, 1978). At the same time, geochemical evidence described in Chapter 5 indicates that oceanic composition and presumably oxidation state were strongly influenced by interaction with the sea floor. The result would have been an ocean rich in ferrous ions and other reducing species.

Detailed photochemical calculations by Kasting et al. (1979) and Kasting and Walker (1981) have yielded estimates of the oxygen partial pressure under these circumstances. It would have been about 10^{-14} atm if metamorphic and volcanic sources released reduced gases in amounts exceeding the oxygen provided by photochemistry and escape of hydrogen, as was argued above, but about 10^{-8} atm if there were little or no supply of reduced gases to the atmosphere. More recent work by Canuto et al. (1982) has changed these numbers but not the conclusion that the early atmosphere was anoxic.

The immediate effect of the appearance of oxygen-evolving photosynthesis could have been small. This biochemical process released oxygen largely into the upper, mixed layer of the ocean. Photosynthetic oxygen therefore caused the surface waters to become less highly reduced than before, but these waters remained anoxic as long as the rate of production of oxygen was less than the rate of supply of reactive reduced substances in solution. Either because of an increase in the rate of supply of oxygen (Knoll, 1979) or because of a decrease in the rate of supply of reduced substances (Walker, 1977, 1978), a transition was eventually achieved in which the surface waters became oxidizing and were able to export excess oxygen to the atmosphere and to the deep ocean.

A fully aerobic biosphere could not be achieved, however, until the rate of supply of oxygen by photosynthesis exceeded the rate of supply of reduced species to the entire ocean–atmosphere system, including both reduced gases from volcanic and metamorphic sources and reduced metallic ions in solution, such as Fe^{2+} and Mn^{2+} provided by chemical weathering of the continents and the sea floor. Based on Drever's (1974a) model of large iron-formation deposits, in which these rocks developed at the stage when surface waters were oxidizing but the deep oceans were still reducing, the record of banded iron formations suggests that this second transition was delayed until perhaps 1.7×10^9 years ago, although the first evidence of oxidative weathering on land could be as old as 2.5×10^9 years (see Section 2.2.2; Walker et al., 1981). Once again the respective roles of increasing rate of photosynthesis and decreasing rate of supply of reduced minerals are not known.

Aerobic respiration arose soon after the oceans were finally swept free of reduced species in solution and since that time has buffered the oxygen content of the atmosphere through its control of the rate of burial of reduced organic matter. The impact of oxygen-evolving photosynthesis on atmospheric composition, the redox state of the ocean, and the metabolic potential of life was profound and is clearly revealed in the rock record.

2.2.1.2 Nitrogen

What might have been the situation for nitrogen in the earth's atmosphere before life evolved? Several studies have been conducted of the rate of fixation of nitrogen by lightning in primitive atmospheres (Yung and McElroy, 1979; Chameides and Walker, 1981; Kasting and Walker, 1981). These studies agree that the rate of production of nitric oxide could have approached that of the present day, even in atmospheres devoid of oxygen. The oxygen of NO could have been provided by water or carbon dioxide, although the subsequent photochemical fate of the NO remains a subject of debate. The original study by Yung and McElroy found that most of the NO would have entered the ocean as nitrate at a rate large enough to yield a nitrate-rich ocean in less than a million years. More recent work has identified some reactions that have the effect of converting NO back into molecular nitrogen within the atmosphere. Kasting and Walker argued that the resulting rate of transfer of nitrogen from atmosphere to ocean was negligibly small. The question bears on the selection pressures affecting early metabolic evolution (Olson, 1977). If nitrate were abundant, it is reasonable to suppose that nitrate respiration (denitrification) evolved as an early source of metabolic energy, that it had the effect of lowering the availability of fixed nitrogen, and was therefore followed by the appearance of nitrogen fixation. If, on the other hand, abiotic nitrogen fixation were inefficient, there would have been strong selection pressures in favor of the development of biologic nitrogen fixation. The microbiologic evidence, though debatable, favors the latter point of view. Nitrogen fixation on an anaerobic earth would have produced ammonium ions but not nitrate ions. Ammonium ions would presumably have accumulated in the ocean to a level, dependent on the phosphorus concentration, that was sufficient to meet the nutritional needs of the primitive microbes. A rough guess suggests that the required concentration would have been comparable to the nitrate concentration in modern seawater, an amount too small to deplete the atmospheric reservoir.

According to this hypothesis, nitrate did not appear in abundance until after the biosphere had become aerobic and nitrification had become a profitable metabolic enterprise. Nitrification, the conversion of ammonium ions to nitrate, does not occur in the absence of oxygen. Denitrification is biochemically very similar to aerobic respiration—some microbes can derive energy by either process—and it may well have arisen at about the same time.

These considerations suggest that atmospheric nitrogen has remained at about its present abundance throughout most of geologic history. The abundance would have been different, of course, during the accretion of the earth and the initial growth of the atmosphere, and it is not yet clear when the rate of this growth fell to negligibly small values. It is also possible that the nitrogen partial pressure has declined somewhat as the result of the accumulation of organic nitrogen in sediments, but it seems unlikely that atmospheric nitrogen has varied by as much as a factor of 2 since the effective end of degassing.

2.2.1.3 Carbon Dioxide

In Section 2.1.3, we reviewed the concept of control of atmospheric CO_2 by the balance between weathering, which consumes CO_2, and a combination of carbonate deposition and metamorphism, both of which release CO_2. It is appropriate at this stage to ask: How well balanced is this system? For the Cretaceous period and later we have an excellent record of the earth's tectonic behavior. Thus, it is possible to attempt reconstructions of the response of atmospheric CO_2 to tectonic patterns for this time period, reconstructions like those proposed by Lasaga et al. (1985).

Their model assumes that the rate of sea-floor spreading is the driving function. Weathering and metamorphism respond to a change in spreading rate in several ways. The most direct response is via hydrothermal circulation of seawater through mid-ocean ridges, which releases CO_2 from the atmosphere by exchanging seawater magnesium for calcium in the basalt, with subsequent precipitation of the calcium as calcite (Owen and Rea, 1984):

$$Mg^{2+} + \text{Ca-basalt} \longrightarrow \text{Mg-basalt} + Ca^{2+},$$

$$Ca^{2+} + 2HCO_3^- \longrightarrow CaCO_3 + CO_2 + H_2O.$$

If the spreading rate increases, the rate of circulation of seawater through the ridge should increase, and thus more CO_2 should be released to the atmosphere. However, Kasting and Richardson (1985) have calculated the effect of a sharp rise in oceanic Ca^{2+} on atmospheric CO_2 and found that a fourfold increase in the flux of Ca^{2+} to the oceans would not produce a comparable increase in CO_2 in the atmosphere. Instead, CO_2 would rise by no more than 20% and would readjust to the steady-state value after about 10^6 years.

At the other end of the spreading system, in subduction zones, metamorphism of $CaCO_3$-bearing sediments occurs, which produces a further release of CO_2 [reactions (2.8) and (2.9)]. Kasting and Richardson (1985) argued that it is reasonable to suppose that the release of CO_2 by this process would also be a function of spreading: the faster sea floor is generated at ridge crests, the faster it must be consumed in trenches. They suggested that even if the spreading rate is constant, the total length of mid-ocean ridges could increase, leading to an increase in the rate of metamorphism of carbonates. They calculated that a 15–20% increase in ridge length in the Eocene period would have led to a threefold increase in atmospheric CO_2.

Changes in spreading rate also produce changes in sea level because a faster-spreading and therefore hotter ridge occupies a larger volume and displaces seawater. Higher sea level leads to a reduction in the area of rock

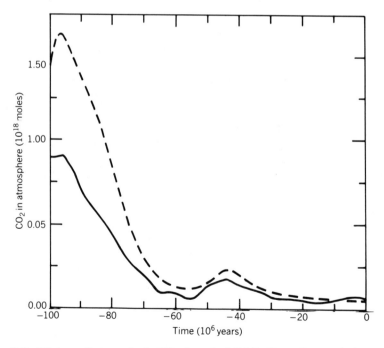

Figure 2.9 History of atmospheric CO_2 since the Middle Cretaceous period according to the model of Lasaga et al. (1985). The dashed line is from a model that considers only carbonate and silicate reactions; the solid line is from a model that adds consideration of reactions of sulfur and of organic carbon.

exposed to weathering and a decrease in CO_2 consumption via reactions like (2.6) and (2.10). Thus, the effect of an increase in spreading rate is to increase atmospheric CO_2 by one or more of three mechanisms.

As we have seen, such a rise in CO_2 is checked by the greenhouse effect. Higher CO_2 levels lead to higher temperatures, which lead to increased rainfall. The combination of higher temperature and higher rainfall greatly accelerates weathering and tends to remove the CO_2 excess.

Lasaga et al. (1985) have written a set of rate expressions for these processes and included a similar set for the sulfur cycle. They then used isotopic data for carbon and sulfur plus sea-level curves for the Cretaceous to Recent periods to generate a computer solution for various parameters including atmospheric CO_2. Their reconstruction (Fig. 2.9) shows CO_2 in the Cretaceous atmosphere several times higher than at present, but falling to values similar to modern ones by the beginning of the Tertiary period. Temperature follows this trend, falling from a Cretaceous high of about 23°C to about 16°C at the Cretaceous–Tertiary boundary and to about 15°C today.

2.2.2 Constraints Imposed by the Rock Record

As we did for the oceans, it is profitable for us to examine what the rock record tells us about different atmospheric conditions in the geologic past. Unfortunately, the evidence is even more meager for atmospheric changes. Evidence for changes in atmospheric oxygen are found in the sedimentary record of iron and uranium, two elements that respond strongly to changes in oxidation state of the environment. The details of the information available and its interpretation can be found in Holland (1984) and in the volumes edited by Armstrong (1981), Holland and Schidlowski (1982), and Schopf (1983). In particular, the paper by Walker et al. (1983) addresses the geologic record of changes in atmospheric oxygen. Such evidence comes in the form of age distributions of three types of sedimentary deposits: banded iron-formations, red beds, and gold–uranium-bearing conglomerates. Banded iron-formations are typically composed of finely interbedded chert and iron minerals. They make up virtually all of the world's iron ore reserves and show a strong time concentration between 2.5 and 1.8×10^9 years ago (for details of the geology of these rocks, see Mel'nik, 1982; Maynard, 1983; Trendall and Morris, 1983). The time constraint and facies characteristics of these rocks are most easily explained by postulating a low-oxygen atmosphere between 2.5 and 1.8×10^9 years ago, one containing enough oxygen to precipitate ferric ion in shallow waters, but too little oxygen to permit oxygenation of the deeper levels of the oceans. This deep water could then serve as a vast reservoir of soluble iron (Drever, 1974a).

The distribution of red beds is more problematic. Red-coloured sedimentary deposits are most commonly developed under strongly oxidizing conditions in fairly dry climates. An abundance of red beds in the rock record should accordingly indicate abundant atmospheric oxygen. Walker et al. (1983, Fig. 11-12) show a transition from nonred to red coloration in such deposits at about 2×10^9 years ago, again suggesting an increase in atmospheric oxygen at that time.

In somewhat older rocks, about 3.0 to 2.5×10^9 years, a unique lithology is found, consisting of quartz-pebble conglomerates with smaller grains of detrital gold, uraninite, and pyrite, apparently deposited on alluvial fans. Because of the ready destruction of uraninite and pyrite on exposure to the atmosphere, the absence of this lithology in younger rocks has led most researchers to view them as evidence for very low atmospheric oxygen before about 2.5×10^9 years ago.

Limits to changes in atmospheric oxygen, albeit only for younger rocks, can be found in the record of terrestrial plant debris. Fossil charcoal is widespread in sedimentary rocks of Carboniferous (340×10^6 years) age and younger, implying that the pressure of oxygen in the atmosphere has not been less than one-third of its present value since that time (Cope and Chaloner, 1980). As an upper limit, the flammability of terrestrial vegetation restricts the atmospheric oxygen pressure to a value no more than about 1.25 times its present level (Watson et al., 1978). Terrestrial vegetation has been present since about 400×10^6 years ago.

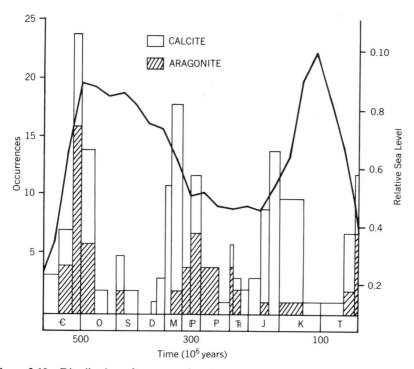

Figure 2.10 Distribution of textures of authigenic carbonates during the Phanerozoic. Times of dominant calcite deposition are thought to have been times of high atmospheric CO_2 (Wilkinson et al., 1985).

Some evidence is also available in the younger rock record for changes in atmospheric CO_2. As noted by Sandberg (1975, 1983), Mackenzie and Pigott (1981), and Wilkinson et al. (1985), the texture of carbonate minerals in limestones shows an oscillating trend through the Phanerozoic (Fig. 2.10). That is, calcareous ooliths deposited during certain episodes of earth history have a well-preserved fine structure, whereas those from other periods are recrystallized. The well-preserved ooliths are thought to have originally been magnesium–calcite, whereas the poorly preserved ones were aragonite. The original mineralogy in turn is thought to be related to fluctuations in P_{CO_2}. Higher atmospheric CO_2, which we have seen is associated with higher temperatures and more equable climates, appears to be correlated with calcitic ooliths, whereas aragonitic ooliths are associated with lower P_{CO_2} and colder temperatures. Thus, calcitic ooliths would imply fast spreading and a high stand of sea level, as discussed in Section 2.2.1.3.

2.2.3 Some Future Perturbations

One disturbance of the atmosphere that has been much discussed lately is the addition of CO_2 to the atmosphere by combustion of fossil fuels. Examination of the eventual fate of this excess CO_2 provides a good example of some of the principles we have been developing (see Walker, 1977).

For simplicity, we assume that the entire fossil fuel reservoir, 600×10^{15} moles, is (geologically speaking) suddenly converted into carbon dioxide. Because there are only 56×10^{15} moles of carbon dioxide in the atmosphere, the immediate effect of such a sudden addition would be to increase the partial pressure of carbon dioxide by a factor of 11 (Fig. 2.11).

In a time less than 2000 years (the mixing time of the ocean), the partial pressure of carbon dioxide and the carbonic acid concentration in seawater achieve equilibrium, the reaction being

$$CO_2 + H_2O \longrightarrow H_2CO_3.$$

Because, under these conditions, the number of moles of carbonic acid in the sea is approximately equal to the number of moles of carbon dioxide in the atmosphere, this equilibrium is achieved when both the carbonic acid concentration and the carbon dioxide partial pressure are approximately five times their present values.

Oceanic pH will decrease as a response to this increase in carbonic acid,

$$H_2CO_3 \longrightarrow H^+ + HCO_3^-,$$

and carbonate ions will be decreased by reaction with the hydrogen ions produced,

$$CO_3^{2-} + H^+ \longrightarrow HCO_3^-.$$

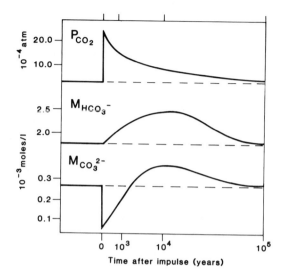

Figure 2.11 Projected changes in the distribution of carbonate species in the atmosphere and oceans as a result of the burning of fossil fuels.

Carbonate ions should decrease in concentration by a factor of about 5, while the pH falls from 8.0 to 7.3. Some increased dissolution of deep-sea carbonate sediments should occur as a result of these changes (a shallowing of the calcite compensation depth), but the time scale is not known.

The reduction in carbonate ion concentration should also reduce the rate of carbonate precipitation in surface waters, while the increase in carbon dioxide partial pressure would, in principle, increase the rate of weathering on land. As a result, cations would accumulate in the sea and still more carbon would be added to the ocean as a result of the weathering of carbonate minerals. The cation content of the sea will continue to rise until most of the original influx of carbon dioxide has been converted into bicarbonate ions. Because the fossil fuel addition amounted to 20% of the initial bicarbonate content of the sea, and because weathering of carbonate minerals adds two bicarbonate ions to the sea for each carbon dioxide molecule extracted from the atmosphere, bicarbonate concentration rises by approximately a factor of 1.4. At present weathering rates, this rise would take about 4×10^4 years.

As oceanic bicarbonate grows at the expense of atmospheric carbon dioxide, however, the carbonate ion concentration in seawater increases, leading to an increase in the rate of precipitation of carbonate minerals in shallow water or a decrease in dissolution in deep water. At the same time, the decrease in the carbon dioxide partial pressure reduces the rate of weathering. In time, therefore, the cation content of the sea ceases to rise, achieving a new equilibrium between the sources of cations (largely the weathering of carbonate minerals) and the sink provided by the precipitation of carbonate minerals. This new equilibrium does not represent the final stage of the system. Not only the calcium content but also the carbonate ion concentration and the carbon dioxide partial pressure are greater than they were before the original perturbation, and the system must ultimately return to its original state.

Silicate weathering provides the mechanism for this return. We presume that the volcanic source of carbon dioxide would not be affected by any of the changes we have described. But the enhanced carbon dioxide level leads to an enhanced rate of silicate weathering. Therefore, the carbon budget is unbalanced. Silicate weathering followed by carbonate precipitation removes carbon from the combined ocean and atmosphere faster than the precipitation of new silicate minerals in the oceans restores it. As a result, the carbon dioxide partial pressure would be reduced and the carbonate ion concentration increased. Carbonate precipitation now exceeds carbonate weathering and the cation content of the sea begins to decrease. As the cation concentration decreases, so too do the bicarbonate and carbonate ion concentrations as well as the carbon dioxide partial pressure. Equilibrium is finally established when all parameters have returned to their preperturbation levels. The time scale for this return to equilibrium is the time for silicate weathering to consume the fossil fuel carbon dioxide. At present-day weathering rates, this time is approximately 2×10^5 years.

There are thus three distinct stages in the response of the atmosphere and oceans to a sudden addition of carbon dioxide. In the first stage, on a time scale of a few thousand years, the carbonic acid concentration of seawater approaches equilibrium with the partial pressure of carbon dioxide, the partial pressure drops by about a factor of 2. The second stage represents the response of the cation content of the sea to imbalance between the rate of weathering and precipitation of carbonate materials. As the cation and bicarbonate concentrations rise, the carbon dioxide partial pressure and the weathering rate fall, while the carbonate ion concentration and the precipitation rate rise. Equilibrium is achieved after a time of a few tens of thousands of years. During the third stage, with a characteristic time approximately ten times as long, the system returns to its initial state, driven by imbalance between the volcanic source of carbon dioxide and the rate of removal of carbon dioxide from the ocean and atmosphere by weathering of silicate minerals followed by precipitation of carbonate minerals.

Evidently, the carbon dioxide content of the atmosphere is determined over time scales exceeding 10^5 years by the requirement that the weathering of silicate minerals consume the CO_2 released to the atmosphere as a result of metamorphic and volcanic processes. The equilibrium carbon dioxide content of the atmosphere should be increased by an increase in volcanic activity, provided other factors remain unchanged. On the other hand, it should be reduced by an increase in the rate of weathering caused, for example, by an increase in the rate at which erosion exposes fresh rock to the atmosphere. There is therefore every reason to suppose that atmospheric carbon dioxide has varied over the course of geologic history, in response to changes in the rates of weathering and precipitation of silicate minerals.

3 INTERACTIONS OF GEOCHEMICAL CYCLES WITH THE MANTLE

T. J. Wolery and N. H. Sleep

In most early studies of geochemical cycling, the addition of material to the crust (including sediments, the oceans, and the atmosphere) was considered to be a one-way process, with crustal and volatile reservoirs forming either continuously or in a short catastrophic event early in the earth's history. With the advent of the theory of plate tectonics, it is now evident that mass fluxes into or out of the mantle cannot be neglected. Recently available chemical and isotopic analyses permit quantitative discussion of exchanges between mantle and exogenic reservoirs. Before discussing the cycling of material into and out of the mantle, we outline movements implied by plate tectonics of which the material fluxes are incidental features. The interactions of uplift, erosion, and subsidence with geochemical cycles are discussed in Chapter 1, Section 1.4.2, Chapter 4, Section 4.6, and Chapter 5, Section 5.5.

3.1. PLATE TECTONICS

The plate tectonic theory unified the earlier theories of continental drift (Wegener, 1926) and sea-floor spreading (Dietz, 1961; Hess, 1962) and subduction (Coats, 1962). The history of the development of this theory is reviewed by Marvin (1973) and Cox (1973). Useful reviews are also given by Le Pichon et al. (1973), Uyeda (1978), Wyllie (1976), and Anderson (1986).

Geometrically, the earth is divided into rigid plates (about 100 km thick) of *lithosphere*, which move with respect to each other. At mid-ocean ridges, the movement is divergent and new oceanic lithosphere is continually created at the ridge axis (Fig. 3.1). Therefore, material becomes progressively older away from the ridge. At trenches or *subduction zones*, the oceanic crust along with the uppermost part of the mantle, down to about 100 km, is returned to the deep mantle (Fig. 3.1). Subduction is a one-sided process in contrast to the spreading at ridges. The third type of boundary is transform or strike–slip faulting, where the motion between two plates is horizontal and parallel to the boundary. By contrast, the motion at ridges is also horizontal but perpendicular to the boundary. Because very little crust is made or destroyed at transform

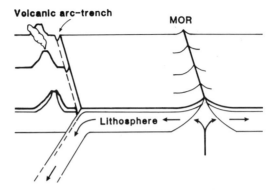

Figure 3.1 Schematic diagram shows the motions implied by plate tectonics. New oceanic lithosphere is formed at the mid-ocean ridge (right) and returns to depth beneath island arcs (left). The initial high temperatures at the ridge promote hydrothermal circulation in the oceanic crust. Almost all the oceanic crust is subducted with the slab, but the extent to which oceanic sediments are subducted is unknown. Island arc volcanoes and emplacement of small amounts of oceanic crust over island arcs provide additional material to the exogenic system. The extent to which subducted sediments and oceanic crust enter island arc volcanics is also unknown.

boundaries, these boundaries are unimportant in cycling material into and out of the mantle.

At the current rate of production of sea floor of about 3 km²/yr, the entire oceanic crust would be replaced in 100×10^6 years. No well-documented oceanic crust older than 200×10^6 years has been found in place. Continental crust is much thicker than oceanic crust (about 35 km versus about 5 km) and less dense than the mantle. Subduction of continental crust has thus been rare and 3×10^9-year-old crust occurs on several continents (see Chapter 5). For bookkeeping purposes, it is convenient to consider the oceanic crust as part of the mantle (to which it almost always returns) and to consider the much less frequently subducted continental crust, sediments, oceans, and atmosphere to be the crust.

Several processes may be involved in the chemical interchanges between shallow crustal or *exogenic* reservoirs and the mantle. At mid-ocean ridges, basaltic rock reacts with seawater. At subduction zones, sediment may become subducted into the mantle, small amounts of oceanic crust and mantle may be added to the continental crust as ophiolites (see Coleman, 1977), or material may be added to the crust by island arc volcanism. Direct determination of the magnitudes of these fluxes even for the present is difficult; these processes are record-destroying or occur at depth or in remote and poorly studied areas. Mass balances from considerations of geochemical cycling are useful in constraining some of the fluxes, although the limits that are obtained may still be fairly broad.

3.1.1 Mid-Ocean Ridge Axis

The reaction of seawater with oceanic crust does not occur by a single well-defined process (see Rona et al., 1983, for several references). Looked at broadly, it proceeds by a low-temperature weathering at shallow depths, by relatively rapid reaction with molten basalt flows, and by reactions at moderately elevated temperatures at depth in hydrothermal systems. Samples of rocks affected by each (and sometimes more than one) of the processes have been obtained and scrutinized (see Andrews, 1977; Scarfe and Smith, 1977; Humphris and Thompson, 1978a,b; Stakes, 1978; Donnelly et al., 1979; Honnorez, 1981; Hart and Staudigel, 1982; Thompson, 1983a,b; Alt et al., 1986a). The major difficulty in trying to use data from these studies to make estimates of geochemical flux between the oceanic crust and the ocean is the lack of definitive data on the volume of crust affected by each process. Furthermore, the samples within any grouping defined as broadly as those above still show considerable heterogeneity. Andrews' (1977) study of low-temperature weathering is particularly instructive in this regard.

Mass transfer in hydrothermal systems in the oceanic crust is not easily constrained because such major controlling parameters as temperature, fluid flow rates, and effective rock/water ratio must all be determined. These are likely to be quite variable in time and space even in a single hydrothermal system. Consequently, material leached in one part of the system may be precipitated elsewhere, resulting in no net flux to the ocean. Similarly, a component such as anhydrite ($CaSO_4$) may precipitate from seawater in the hotter regions of a young hydrothermal system but then redissolve and be transported back to the ocean during subsequent cooler circulation. In view of the global scale of hydrothermal processes at mid-ocean ridges, it also seems probable that differences in the conditions of reaction and hence contributions to the geochemical flux to the ocean may exist among different hydrothermal systems (Stakes, 1978).

Measured conductive heat flow near the ridge axis is less than that predicted by thermal models of the oceanic lithosphere that ignore nonconductive means of cooling. Heat that escapes from hydrothermal vents is not detected by most conventional measurements (Anderson and Hobart, 1976; Wolery and Sleep, 1976; Anderson et al., 1977; Sleep and Wolery, 1978). The anomaly between theoretical and measured heat flow may result from extraction of heat by hydrothermal systems located away from the ridge axis. Alternatively, deeply penetrating hydrothermal systems in the axial region may mine the heat initially present in oceanic crust (and perhaps part of the upper mantle).

For the purposes of this chapter, the heat flow anomaly is important because it affords a means of constraining the flow of heated seawater from the hydrothermal systems of mid-ocean ridges. Globally, the amount of missing heat flow at ridges is about 50×10^{18} cal/yr or 17×10^8 cal/cm^2 averaged over the ocean basins. The total excess heat liberated by sea-floor spreading (i.e., the total heat produced by new oceanic lithosphere less a component equivalent to

TABLE 3.1. Heat Budget of Ridge

Source	Heat per Area of Sea Floor (Average Worldwide) 10^8 cal/cm^2	Global Heat Flux (10^{18} cal/yr)
A. Total heat from sea-floor spreading	39	117
B. Missing heat flow at ridge	17	51
C. Missing heat on crust less than 1×10^6 years old (except latent heat)	4.6	14
D. Latent heat of basalt (5 km, crust)	1.5	5
E. Fast heat loss from extrusive flows (0.5-km section)	0.7	2
F. Heat flow within 1 km of 3-cm/yr ridge axis	1	1
G. Linear cooling to 8-km depth	4.6	14
H. Linear cooling to 30-km depth	17	50

the "background" heat flow observed on very old oceanic crust) is 39×10^8 cal/cm^2 (see Table 3.1). The flow rate is estimated from the heat flow anomaly divided by the temperature at the base of vents (i.e., the point of capture of deep circulation by major passages of upward flow; substitute the exit temperature if venting is isothermal) and the average heat capacity of the fluid from about 0°C to this elevated temperature.

Some observational evidence is available that can be used to estimate these temperatures. Axial hydrothermal vents have been sampled at the Galapagos Ridge and the East Pacific Rise at 21°N. At both locations the (half) spreading rate is 3 cm/yr (Corliss et al., 1979a,b; Edmond et al., 1979a,b,c, 1982; East Pacific Rise Study Group, 1981; von Damm, 1983; von Damm et al., 1983, 1985). The fluid at 21°N exits at 350°C and the fluid at the Galapagos site is now inferred to have been at about 350°C before the onset of shallow subsurface mixing with cold normal seawater within the oceanic crust. Bischoff (1980) suggested that the fluid at some of the 21°N vents may have been somewhat warmer (380°C or so) upon exiting; he further pointed out that if such is the case, the hydrothermal fluid near the top of the magma chamber may have been as high as 420°C. Adiabatic decompression during rapid ascent would cool the fluid to about 380°C. Below about 350°C at depth, however, adiabatic ascent would be nearly isothermal. See Bischoff and Rosenbauer (1984) for the physical properties of hot seawater.

The rate of hydrothermal cooling can also be estimated from the flux of helium isotopes. Jenkins et al. (1978), Edmond et al. (1979a,b), and Corliss et al. (1979a,b) coupled the ^{3}He/enthalpy ratio observed in the Galapagos hydrothermal fluids (the ratio at 21°N on the East Pacific Rise is very similar; Lupton, 1983a) with an estimate of the flux of primordial ^{3}He from the ocean to the atmosphere (Craig et al., 1975) to make an estimate of about 50×10^{18} cal/yr for the rate of hydrothermal cooling. This is the same value obtained from heat flow and thermal modeling discussed earlier. The above authors thus concluded that the bulk of the hydrothermal circulation escaped through axial vents, and they extrapolated the observed vent chemistry to obtain global fluxes. That is, they obtained estimates for the associated geochemical fluxes of components dissolved in seawater using the measured ratios of concentrations to ^{3}He or enthalpy. More recently, the global ^{3}He flux (Craig et al., 1975) has been revised downward by a factor of 3/4 (Welhan and Craig, 1983), which yields a global rate of hydrothermal cooling of 36×10^{18} cal/yr from axial vents (corrected from Turekian, 1983).

The constancy of the ratio of ^{3}He to enthalpy found in the earlier studies is not universal. Mervilat (quoted by Michard et al., 1984) found a higher ^{3}He/heat ratio in samples from the East Pacific Rise at 13°N. Lupton (1983b) found a ^{3}He/enthalpy ratio one-fifth the previous values in the Guaymas Basin in the Gulf of California. Apparently, helium escapes faster than heat, so the global ^{3}He/heat ratio cannot be estimated confidently from only a few samples.

The amount of heat extracted by axial vents can be constrained by considering the heat budget of ridges (Table 3.1). If a large amount of heat were removed by the axial vents, the oceanic crust and mantle would be cooled to a considerable depth at the axis. For simplicity, we assume that this cooling would produce a linear temperature profile starting with 0°C at the surface and continuing to the intrusion temperature at some depth. Axial hydrothermal circulation would have to penetrate a substantial fraction of that depth to remove this heat: cooling to 30-km depth would be necessary if all the missing heat flow at ridges were extracted at the axis (lines B and H, Table 3.1). Axial cooling to 8-km depth would be necessary to explain just the missing heat flow on crust less than 10^6 years old (lines C and G, Table 3.1).

Thus, the upper limit to heat loss by axial hydrothermal systems is the total missing heat of 50×10^{18} cal/yr, although the depth of circulation required makes this seem unlikely. A lower limit is comparable to the heat that would escape in the axial region if only conduction occurred. This amount is approximately the latent heat of crystallization of the basaltic crust (line D, Table 3.1) plus the conductive heat loss from within 1 km of the axis, using a half-space model (line F, Table 3.1), or 2.5×10^8 cal/cm^2. This is about one-half of the missing heat on young crust (less than 10^6 years old) and about one-sixth of the total heat anomaly. The worldwide extrapolation of this estimate would be even lower if hot axial vents are not abundant on slow-spreading ridges and if a significant amount of heat were lost by cooling of surface flows (line E, Table 3.1). Extensive exploration has only recently found hot axial vents on the Mid-Atlantic Ridge (Rona et al., 1986).

There is an additional argument against ascribing most of the missing heat flow to axial hydrothermal systems. The deep cooling of the lithosphere at the ridge axis, which would be necessary if axial vents were to remove large amounts of heat, would decrease the size of the axial magma chamber and cause thermal contraction of the lithosphere. Where the magma chamber has been imaged by seismic reflection (Herron et al., 1978; Hale et al., 1982, Sleep et al., 1983; Morton and Sleep, 1985a; Detrick et al., 1987; Morton et al., 1987), the heat loss can be computed from thermal models. For the East Pacific Rise at 9°N, the back-arc spreading center in the Lau Basin, and the Juan de Fuca Ridge, fluxes between 1.7×10^8 and 3.3×10^8 cal/cm^2 were obtained by Morton and Sleep (1985b). For other regions, the cooling rate can be estimated from bathymetry, although some of the topography is probably associated with subcrustal processes (Sleep et al., 1983; Madsen et al., 1984). A contraction of 300 m is implied by the missing heat within 10^6 years of the axis and a contraction of 1 km by the total heat flow anomaly. An axial elevation anomaly of this magnitude has not been found in studies of fast-spreading ridges (Sleep and Rosendahl, 1979). However, Lewis (1982) has contended that the density decrease from cooling and freezing is more than offset by the additional creation of porosity by cracking and that the entire magma chamber freezes rapidly at the ridge axis.

Our estimate for axial flux (2.5×10^8 cal/cm^2) is used for our preferred estimates in this chapter because it is supported by the seismic reflection studies. We also note that preliminary data on Ge/Si ratios (Froelich et al., 1985), modeling of boron and lithium cycles in the ocean (Seyfried et al., 1984; but compare Spivack and Edmond, 1987), and modeling of oxygen isotopes (Bowers and Taylor, 1985) support this lower estimate. Chemical fluxes based on all the heat loss coming from axial vents are given for comparison.

If (as this reasoning indicates) all hydrothermal cooling cannot be ascribed to vents near the axis, then off-axis vents must be postulated. There is some direct evidence that off-axis vents exist, but not enough to estimate directly chemical fluxes. Off-axis circulation has been detected at depth in a drill hole (Anderson and Zoback, 1982; Williams et al., 1986). In addition, sediment mounds on 0.7×10^6-year-old crust near the Galapagos spreading center (Corliss et al., 1978) and metal-rich sediments on the Mid-Atlantic Ridge have been attributed to off-axis activity (R. B. Scott et al., 1974). Rocks altered at depth by 20–80°C hydrothermal flow have been recovered from three drill sites (Hart and Staudigel, 1978).

A significant amount of the off-axis circulation enters and leaves the basaltic crust by porous flow through sediments (Abbott et al., 1981, 1983, 1984). Where the sediment cover is thin, the flow is rapid enough that the pore water in the sediment is basically seawater with downward circulation and water reacted with basalt on the upward circulation. The normal diagenetic sequence in thicker sediments is precluded and there are significant chemical fluxes between the seawater, the basalt, and the sediments. Although this process is probably significant to global mass balances, the chemical fluxes cannot now be estimated reliably.

If we accept that much of the heat escapes through off-axis vents, how can the helium data be explained? Craig and Lupton (1981) and Lupton (1983a) present comprehensive reviews of helium in the oceans. They point out that measurements of helium in fresh basaltic glasses, which should best reflect the concentration of helium in the oceanic crust at the time of emplacement, show considerable scatter. Using an average value of 400×10^{-8} cm^3 STP (total He)/g (basalt), the global spreading rate of 3 km^2/yr, and a crustal thickness of 5 km, they calculated a maximum outgassing equal to about one-tenth the primordial helium flux observed oceanographically (although the calculation is done for total primordial helium, the conclusion is also true for ^{3}He).

One possible explanation not offered by Craig and Lupton for the excess helium in the ocean is that ocean helium is recording a recent spurt, about tenfold, in the rate of sea-floor spreading. The associated geochemical fluxes would presumably be proportionately increased, thus perhaps accounting for the seemingly high estimates based on helium presented by Edmond et al. (1979b,c). However, the present understanding of the helium cycle outlined by Craig and Lupton (1981) would require that this excursion have gone on for a time roughly equal to the residence time of helium in the atmosphere, $0.5-1.0 \times 10^6$ years.

An alternate explanation discussed by Craig and Lupton (1981) seems much more likely. They point out that significant amounts of helium may have escaped the glass samples during the formation of vesicles, thus resulting in the observed scatter. If so, the highest measured values probably best represent the emplacement value. Fisher (1979) and Kurz and Jenkins (1979) report values as high as 2400×10^{-8} cm^3 STP/g. Within the remaining uncertainties, it appears that there is enough helium in the oceanic crust at the time of emplacement to account for the oceanographic flux at the normal spreading rate.

Most of the primordial helium in the water column in the Pacific Ocean appears traceable to release from the axes of spreading centers (Craig and Lupton, 1981; Lupton, 1983a), with no readily obvious contributions from off-axis localities. We do not believe that this should be taken as evidence that there is no hydrothermal activity at the latter. Rather, it appears from the calculation above that helium is largely stripped out of the oceanic crust very close to the ridge axis, so that little is left to escape from off-axis hydrothermal systems.

As discussed earlier, extrapolation of the ^{3}He/enthalpy ratios observed at the two Pacific vent sites yields a hydrothermal cooling rate of about 50×10^{18} cal/yr, which matches the estimate from heat flow studies. However, the helium method appears capable of detecting only heat vented from on-axis hydrothermal systems because the helium is removed faster than the heat (Lupton, 1983b). The heat flow method detects both axial and off-axis hydrothermal heat loss and requires that a substantial fraction occur off-axis. We therefore consider it likely that the helium method estimate, applied to the axial component, is too great.

In summary, sampling biases complicate determination of the effect of basalt–seawater reaction on exogenic cycles. On a small scale, the flow of hydrothermal seawater, and hence the nature of reaction, is likely to be very heterogeneous. Major passages are likely to carry the bulk of the flow. The wall rocks of these passages would react extensively, but the composition of the water might be altered only slightly. In other regions, the water may be nearly stagnant, and the composition would be modified strongly. A sample of rock or of fluid taken at random from the crust would probably reflect the latter situation. Weathering, axial hydrothermal flow, reaction with stagnant water, and off-axis hydrothermal flow are likely to affect the same rock at various times. Any basalts altered at high temperature and collected from the sea floor have had a chance to weather. Major off-axis hydrothermal flow may use the conduits of previous axial systems. The wall rocks in these conduits would already be extensively altered and thus react less than fresh basalt.

3.1.2 Subduction Zones

The other point of interaction between exogenic and endogenic cycles is in subduction zones. Subducted oceanic crust carries elements added by hydrothermal circulation at the ridges back to the interior of the earth. The sediments that accumulated on this crust may be carried down as well. The depth to which the material is carried is very important to geochemical cycling considerations: material that is underplated onto the accretionary wedge is only transferred from one reservoir of sedimentary rocks to another. This material is not considered to be subducted because these sediments, though metamorphosed, would be recognized as such on their return to the surface. Such rocks are not unusual in island arcs.

Material that penetrates below crustal depths can generally appear at the surface again only through igneous processes. It has been proposed that subducted sediments, oceanic crust, and hydrous exhalations of the oceanic crust are incorporated into island arc volcanics (e.g., Kay, 1980). The relative importance of these ingredients and whether they are essential parts, minor contaminants, or generally absent are still controversial points. So is the fraction of oceanic sediment that is subducted and the fraction of the subducted material that is remobilized into island arcs rather than incorporated into the deep mantle reservoir (Karig and Sharman, 1975; Scholl et al., 1977). The recycling of elements within the mantle must largely be inferred from geochemical considerations since the process is hidden from view.

The composition of material added to the crust at island arcs either by volcanism or by obduction of sea floor can be analyzed. The global rate at which such material is added to the crust, however, is poorly constrained because the emplacement is episodic, and the simple geometric constraints that exist at spreading axes do not apply. (See Kay, 1980, for estimates of the volume of arc igneous activity.) The igneous rocks now exposed at the surface have been studied extensively, and the general abundance and composition of rock types are thus known (see Chapter 5; Ronov and Yaroshevskiiy, 1969, 1976). We return to these considerations when we deal with the question of calcium cycling.

3.2 EXPERIMENTAL AND THEORETICAL INVESTIGATIONS

Studies of the few samples of venting hydrothermal fluid from two localities in the Pacific and of the many samples of altered rocks representing most of the major ocean basins comprise only part of the data available for studies of geochemical fluxes. Seawater–basalt interactions have been the focus of numerous laboratory investigations conducted over a wide range of conditions. These studies are significant because they provide a great deal of direct information on the nature and roles of the major controlling parameters. Also, the changes in seawater composition observed in these experiments can be used in conjunction with estimates of the global flow rate (Wolery and Sleep, 1976; Sleep and Wolery, 1978) to make estimates of geochemical fluxes associated with mid-ocean ridge hydrothermal circulation.

The first experimental investigations (Bischoff and Dickson, 1975; Hajash, 1975; Mottl and Holland, 1978; Mottl et al., 1979) focused on closed systems with rock/water ratios of 0.1 g/g or greater and temperatures of 200°C and well above. Many of these experiments, especially some by Mottl and co-workers, showed a close correspondence to the seawater–basalt interactions observed at the subaerial geothermal brine field at Reykjanes, Iceland (described by Bjornsson et al., 1972).

Later studies examined conditions at lower rock/water ratios (in the range of 0.016–0.020 g/g; Seyfried, 1976; Seyfried and Bischoff, 1977; Seyfried and Mottl, 1977; Hajash and Archer, 1980; Hajash and Chandler, 1981). Under these conditions, the seawater was more acidic and the solubilization of such metals as iron, manganese, and zinc at temperatures below 300°C was greatly increased. Bischoff and Seyfried (1978) investigated the hydrothermal chemistry of seawater alone and discovered an important new mineral phase, a magnesium hydroxysulfate, that precipitates between 250 and 300°C. This phase has not been reported from any seawater–basalt experiments, but it would be significant in systems with a sufficiently low rock/water ratio. Seyfried and Bischoff (1977) also studied seawater–basalt interactions at 70 and 150°C, where reaction is more sluggish on a laboratory time scale.

Several caveats must be placed on the experimental results. First, it is doubtful that any of the experiments proceeded to a state of thermodynamic equilibrium. For example, some of Hajash's (1975) experiments yielded one phase assemblage at the bottom of his capsules and another at the top. In the studies that used the Dickson hydrothermal apparatus (those reports cited above in which either Dickson or Seyfried are co-authors), fluid was sampled during the course of the experiments and agitation was applied so that diffusion of dissolved components was not a control on the overall interaction. Even in these, however, a steady-state fluid composition was never attained, which precludes the existence of equilibrium. The results reported by Hajash (1975), Mottl and Holland (1978), and Mottl et al. (1979) are somewhat clouded by the fact that fluid sampling could only be done after quenching the experiment. Comparison with results from the studies that used the Dickson apparatus indicates that back-reaction during quenching occurred in the former experiments.

The experimental investigations, under the best of conditions, reflect metastabilities pertinent to laboratory time scales. For example, sulfate persists at temperatures below about 300°C when the thermodynamic driving force for its reduction to sulfide is great, and phases such as quartz and epidote often fail to form when thermodynamically favored. These experiments have produced phase assemblages with varying degrees of correspondence with samples of actual hydrothermally altered rock. However, such differences may be due in part to the fact that the experiments were done on essentially closed systems, whereas the real systems may occasionally approach such a state but are in fact open systems.

Wolery (1978) investigated seawater–basalt interactions by means of computerized simulations based on the reaction-path concept established by Helgeson (1968). The database that supports these simulations is described by Wolery (1978) and is limited to a maximum temperature of about 300°C. Similar calculations using a more realistic flow-through model have been made by Bowers and Taylor (1985). Some important minerals that in reality exhibit extensive solid solution (especially smectite and chlorite) were represented by a range of fixed compositions. In addition, the thermodynamic functions of a number of the sheet silicates were estimated by various methods. Despite these caveats, and those above concerning experimental studies, the simulations showed a qualitatively good agreement with the experimental study at 150°C reported by Seyfried and Bischoff (1977), that at 200°C by Seyfried (1976), and that at 300°C by Seyfried and Mottl (1977). Agreement was only close to being quantitative (for some dissolved components under some conditions), but the major trends in the mass transfer between rock and aqueous solution were in good accord. Two significant metastabilities in the experiments stood out in this comparison: the failure of quartz to precipitate when the fluid was supersaturated and the persistence of sulfate in the face of driving forces for its reduction.

A complete discussion of the results of all experimental and theoretical studies of seawater–basalt reaction is beyond the scope of this section. Some of the results, however, are used in the following sections in discussing the geochemical fluxes of particular components.

3.3 ELEMENTAL FLUXES

It should be evident from the preceding sections that fluxes of material between the mantle and the exogenic system are quite complex and involve most areas of geology. A review of all related literature is clearly not feasible, but the limited amount of hard data precludes definitive calculation of global values of chemical fluxes. For example, the volume, nature, and degree of alteration in the oceanic crust must be extrapolated from a few drill holes and emplaced bodies of oceanic

3.3 ELEMENTAL FLUXES 87

TABLE 3.2. Exogenic–Endogenic Flux Estimates by Various Authors[a]

Component	Low-Temperature Alteration	Off-Axis Hydrothermal	Axial Hydrothermal	Total Hydrothermal	Net Flux	Reference
Na	−0.35 to −0.04	0.48	?	?	0.13–0.44	This book
	0.009	—	—	−1.96	—	Maynard (1976)
	1.39	—	—	−10.39	—	R. Hart (1973)
	—	—	—	—	—	Mottl (1976)
	—	—	−8.6 to 1.9	0 to −3.83	—	von Damm (1983)
	−0.47	0.17	—	—	—	Thompson (1983a,b)
K	−0.51 to −0.10	0.33	0.21	0.54	0.03–0.44	This book
	−0.31	—	—	0.61	—	Maynard (1976)
	−1.54	—	—	0	—	R. Hart (1973)
	—	—	—	<0.26	—	Mottl (1976)
	0 to −0.17	—	—	—	—	Seyfried (1976)
	—	—	1.9–2.3	1.28	—	Edmond et al. (1979b)
	—	—	—	—	—	von Damm (1983)
	−0.26	−0.56	0.13–1.26	−0.43 to 0.69	−0.69 to 0.44	Thompson (1983a,b)
Ca	0.25–1.25	1.20	0.60	1.80	2.05–3.05	This book
	0.06	—	—	0.38	—	Maynard (1976)
	4.68	—	—	15.58	—	R. Hart (1973)
	—	—	—	—	—	Mottl (1976)
	0.15–0.70	—	2.05–4.30	2.58–7.05	—	Seyfried (1976)
	—	—	—	3.80	—	Edmond et al. (1979b)
	0.20–1.25	—	—	1.63–3.38	—	Wolery and Sleep (1976)
	—	—	0.24–1.5	—	—	von Damm (1983)
	0.32	1.18	0.32–3.25	1.50–4.43	1.83–4.75	Thompson (1983a,b)

Mg	0.29–1.89	−1.23	1.27	−1.50	−2.21 to −0.61	This book
	0.05	—	—	−1.73	—	Maynard (1976)
	4.98	—	—	−8.39	—	R. Hart (1973)
	0.41–0.58	—	—	−4.52	—	Seyfried (1976)
	0.04–0.45	—	—	−2.71	—	Wolery and Sleep (1976)
	—	—	−7.5	—	—	von Damm (1983)
	1.21	0.46	−7.79 to −4.17	−7.33 to −3.71	−6.13 to −2.5	Thompson (1983a,b)
SiO$_2$	0.08–0.47	0.37	0.52	0.89	0.97–1.36	This book
	0.87	—	—	3.27	—	Maynard (1976)
	4.83	—	—	6.67	—	R. Hart (1973)
	0–0.63	—	—	—	—	Seyfried (1976)
	—	—	3.10	—	—	Edmond et al. (1979b)
	0.50–1.33	—	—	—	0.50–1.50	Wolery and Sleep (1976)
	—	—	2.2–2.8	—	—	von Damm (1983)
	1.88	0.71	0.31–3.10	1.02–3.81	2.96–5.71	Thompson (1983a,b)

[a]Flux in units of 10^{12} moles/yr. Fluxes from basalt into seawater are positive.

TABLE 3.3. Comparison of Basalt–Seawater Exchange with Other Fluxes in the Exogenic System, 10^{12} moles/yr

	Basalt–Seawater Flux	River Dissolved Flux	Flux to Pelagic Sediment Reservoir	Total Exogenic Reservoir (4×10^9 years)
Na	0.28 ± 0.16	5.91	0.35	0.52
K	0.24 ± 0.20	1.17	0.44	0.41
Ca	2.55 ± 0.50	12.36	3.03	1.58
Mg	-1.41 ± 0.80	4.85	0.78	0.66
SiO$_2$	1.17 ± 0.19	6.47	6.87	7.17

crust on land. Only a few axial and off-axis vents of active mid-ocean ridge hydrothermal systems have been sampled. The potential fluxes of several elements obtained by apparently reasonable extrapolations of data from axial hydrothermal vents and alteration of sea floor are excessive in regard to the present understanding of the exogenic system discussed elsewhere in this book. The philosophy of this chapter is to use mass balances of the exogenic system to obtain upper limits on these fluxes rather than to accept the flux estimates at face value. We do this because it is felt that the exogenic system has been better sampled than the oceanic crust at depth. An estimate of the extent of alteration of the oceanic crust accrues from this exercise. The fluxes of various elements along with previous estimates are summarized in Table 3.2. These estimates are compared with exogenic fluxes and the sizes of exogenic reservoirs in Table 3.3. The methods by which the estimates were obtained are discussed in the remainder of this section.

As the categories for fluxes are different among authors, the estimates within a column in the table are not strictly comparable. Not all sources of flux are estimated in each paper. The papers published before the discovery of high-temperature vents did not recognize separate axial fluxes. The categories of Thompson (1983b) on the basis of rock samples are most similar to those used below. Thompson (1983b) distinguished true weathering in the upper 10 m from the deeper low-temperature alteration in his flux estimates. As the fluxes from this shallow layer are fairly small, we have regrouped these estimates in Table 3.2.

3.4 LOW-TEMPERATURE ALTERATION OF SEA-FLOOR BASALTS

Low-temperature (0 to perhaps 60°C) alteration in the oceanic crust appears to extend to a depth of several hundred meters. The process is sometimes called weathering although the flow is driven by temperature gradients. R. Hart (1976) summarized seismic velocity data, which he interpreted as indicating geographically widespread low-temperature alteration in the oceanic crust. Subsequent deep drilling into the oceanic crust on DSDP Legs 37, 45, and 51 to 53 showed that the oceanic crust was 5–15% weathered to depths in excess of 600 m into the basalt (Muehlenbachs, 1977a,b, 1980). However, this deep drilling was done on a slow-spreading ridge with rough topography, so that the weathering depth there may be higher than the global average.

More recently, a 1-km-deep borehole was drilled into crust formed at the intermediate-spreading Costa Rica Rift (Newmark et al., 1984; Alt et al., 1986a). The upper 100 m is highly permeable and has large open cracks and drained pillows. The next 550 m consists of small pillows and is extensively fractured and altered. The underlying dike complex is less altered and less fractured. The clay content is between 5 and 15% as in the Atlantic holes.

Basalts altered at low temperatures can be treated as a mixture of two components: fresh material and alteration products with much higher $\delta^{18}O$ (Muehlenbachs, 1977b). Thus, the degree of weathering can be characterized and evaluated from oxygen isotopic composition of bulk samples. Although the drilling data tell us how much basalt is weathered, the associated major element flux cannot easily be evaluated directly from oxygen isotopic data, because few weathered basalts have been analyzed for both $\delta^{18}O$ and major element chemistry. It is also difficult to determine how much of the material leached from the basalt is deposited elsewhere in the oceanic crust as veins. However, once a weathering flux is obtained, $\delta^{18}O$ balance can be used to complete the amount of high-temperature alteration (Muehlenbachs and Clayton, 1976; Bowers and Taylor, 1985).

An alternative approach is to use the H_2O content as an index of weathering in the basalt, because water is commonly included in bulk chemical analyses. When one component of weathered basalt is plotted against another for a large suite of analyses, an essentially linear relation is found, in agreement with Muehlenbach's model. This relation holds for $\delta^{18}O$–H_2O (Muehlenbachs, 1977a, 1980), other pairs consisting of a major element oxide and H_2O (R. Hart, 1973), and also boron or $\delta^{11}B$ and H_2O (Spivack and Edmond, 1987). To estimate weathering fluxes, we use Hart's alteration trends and assume an average degree of weathering equivalent to 2–2.5% H_2O in bulk weathered rock. Deep drilling on moderately young oceanic crust shows this degree of weathering (Andrews, 1977; Muehlenbachs, 1977a,b; Alt et al., 1986a,b). Assuming the average depth of weathering is 200–800 m, we calculate the mass of weathered basalt to be $1.8–7.0 \times 10^{15}$ g/yr.

The resulting fluxes are given in Table 3.2. We may have counted some mass transfer twice by inferring the flux owing to weathering from rock samples and that owing to hydrothermal flow from fluid composition and volume. That is, descending hydrothermal fluids may "weather" the crust on the way down. This consideration is likely to be most important on the flanks of the ridge where the flow is slower. Also, at least some of the weathered basalt may have reacted at high temperatures earlier in its history.

3.4.1 Potassium and Axial Flow

A useful limit on the amount of basalt reacted by axial hydrothermal flow can be obtained by considering potassium, following the approach of Edmond et al.

(1979b) and Mottl (1983). The potassium in primitive basalts is about 0.06% (Langmuir et al., 1977) and the potassium enrichment (over local seawater) in the 350°C end member Galapagos fluid is about 0.035% (Corliss et al., 1979a). A minimum rock/water ratio of 0.58 is thus implied if the axial fluid totally extracts potassium from primitive basalt or an average oceanic crust. Using this value, the water flux of 1.4×10^{17} g/yr for axial vents used by Edmond et al. (1979b) would require reaction of at least 9 km of basaltic crust, an excessive amount. The flow of one-sixth this amount (which was obtained from the sum of latent heat and heat flow within a kilometer of the axis in Table 3.1) used to obtain the axial entries in Table 3.2 would require 1.5 km of crust—a large but possible amount.

We believe this amount to be a reasonable upper limit. If the basalts have higher original potassium contents, the calculated flow would be correspondingly less. For example, if the uppermost basalts have undergone fractional crystallization and have become enriched in potassium to 0.24%, the thickness of basalt implied by the total extraction of potassium by the hydrothermal fluid would be a minimum of 0.4 km with a minimum rock/water ratio of 0.15. This minimum rock/water ratio of 0.15 g/g can be compared with the minimum of 0.05 obtained below by assuming the nearly complete depletion of magnesium in hydrothermal fluids. A rock/water ratio that is somewhat larger than the value obtained from using potassium concentrations was deduced by consideration of rubidium and lithium, which are enriched in the Galapagos fluids relative to seawater (Edmond et al., 1979b; Seyfried et al., 1984), although it is possible that rocks enriched in these elements and boron at low temperatures are then leached, giving excessive rock/water ratio values (Spivack and Edmond, 1987).

Pervasive alteration of the oceanic crust is indicated if the axial hydrothermal fluids obtained in the Galapagos are representative of even one-sixth the flow through ridges. At fast ridges where high-temperature axial alteration can occur at shallow depths, the zone altered by axial hydrothermal activity would be later altered by lower temperature off-axis hydrothermal activity and low-temperature weathering. This complicates calculation of fluxes for the later processes.

The potassium enrichment in the axial vent waters that were sampled may be higher than in average axial flow because the vent fields are quite young (Edmond et al., 1979b). Radioactive dating indicates an age of 20 years or so for the Galapagos vents (Turekian, quoted in Corliss et al., 1979b). The heat flux at 21°N is also too high to be in steady state (Macdonald et al., 1980; Converse et al., 1984). Therefore, caution is warranted in extrapolating the features of the Galapagos and 21°N vents to all axial hydrothermal flow. Also, dissolution of previously formed hydrothermal deposits and boiling (Welhan and Craig, 1979) could change the composition of the venting fluids without producing any net flux to the ocean. The importance of these processes is uncertain.

At temperatures of 350°C, as observed at axis vents, potassium may be quantitatively leached from basalt (see also Mottl and Holland, 1978). Below

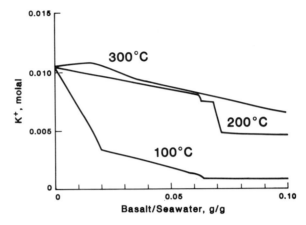

Figure 3.2 Concentration of K^+ in seawater during reaction with fresh basalt to 0.1 g (rock)/g (seawater) at 500 bars of pressure and at 100, 200, and 300°C. At low temperatures there is significant removal of K^+ from seawater, in contrast to the release seen at temperatures above 300°C. After Wolery (1978).

300°C, however, it may be strongly abstracted from seawater into phases including smectite, celadonite, and philipsite. Figure 3.2 shows results from theoretical simulations, which suggest that moderately warm circulation (about 200°C) should lead to removal of substantial quantities of potassium from seawater. The flux of K^+ out of basalt at higher temperatures is probably small compared to the opposite movement at warm temperature. Boron behaves similarly (Spivack and Edmond, 1987). Overall, we believe that the oceanic crust is a net sink for potassium, not a net source (see Mottl and Holland, 1978), with respect to hydrothermal processes.

3.4.2 Magnesium and Hydrothermal Flux

Magnesium concentration in the Galapagos hydrothermal fluids is strongly depleted compared to local seawater (Corliss et al., 1979a,b); experimental hydrothermal fluids show the same effect. Computed values of the amount of magnesium remaining in seawater at various temperatures and rock/water ratios are shown in Fig. 3.3. After reaction of 0.06 g (rock)/g (water) the magnesium is essentially removed from seawater, indicating that the mid-ocean ridges might be an important sink for magnesium.

Consideration of the geochemical balance of magnesium reinforces the hypothesis that hydrothermal alteration of basalt is a significant magnesium sink. Drever (1974b) carefully studied the sources and sinks of magnesium in the ocean and concluded that the current river input could not be accounted for if the oceans were in steady state. Almost immediately, workers investigating mid-ocean ridge hydrothermal phenomena claimed that these systems were the

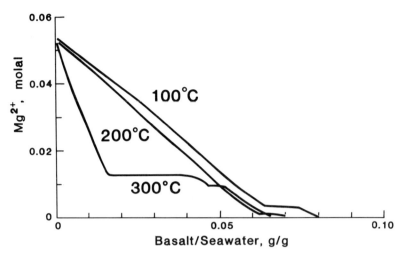

Figure 3.3 Concentration of Mg^{2+} in seawater during reaction with fresh basalt to 0.1 g (rock)/g (seawater) at 500 bars of pressure and at 100, 200, and 300°C. Note that Mg^{2+} is more strongly depleted than K^+. After Wolery (1978).

missing sinks (e.g., Maynard, 1976; Mottl, 1976; Wolery and Sleep, 1976; R. Hart, 1973, had proposed this earlier). If we use our estimate that one-sixth of the hydrothermal heat loss occurs from axial vents, 1.27×10^{12} moles of magnesium are removed (Table 3.2). If we assume that off-axis greenstone formation removes the remaining magnesium anomaly, the magnesium loss computed in Fig. 3.2 can be used to place limits on the effective rock/water ratio in the off-axis systems. The exact value of the magnesium anomaly in the exogenic system is uncertain. For the sake of illustration, we use 2.5×10^{12} moles/yr. The result is a ratio of around 0.0050 g (rock)/g (water); see Table 3.4. The fluxes of several elements at 0.005 g (rock)/g (water) are given for a temperature range of 80–250°C in Table 3.5. The ratio is equivalent to reaction

TABLE 3.4. Global Discharge Rates of Off-Axis Hydrothermal Fluids to Balance Magnesium in Ocean[a]

Temperature (°C)	Discharge Rate[b] (10^{14} kg/yr)	Magnesium Loss (molal)	Calculated Rock/Water Ratio (g/g)
80	5.37	0.0025	0.0030
120	3.57	0.0037	0.0043
160	2.66	0.0050	0.0051
200	2.10	0.0064	0.0063
240	1.74	0.0077	0.0026

[a] From Wolery (1978).
[b] Necessary discharge for all off-axis flow is at given temperature.

TABLE 3.5. Calculated Global Flux of Various Elements in 10^{12} moles/yr by Off-Axis Hydrothermal Activity at Various Temperatures and a Ratio of 0.005 g (basalt)/g (water)[a]

Temperature (°C)	Na	K	Ca	Mg	SiO$_2$	Mn
80[b]	0.62[c]	−0.81	1.75	−2.22	0.75	0.015
120	0.41	−0.58	1.18	−0.95	1.07	0.010
160	0.31	−0.24	0.88	−0.78	0.57	0.008
200	0.25	−0.01	0.45[d]	−0.62	0.45	0.006
240	0.73	0.04	1.70[d]	−1.56	1.14	0.022
Average[e]	0.48	0.33	1.20	−1.23	0.79	0.012
Flow-through[f]	0.78	0.79	1.30	−1.69	1.29	N.D.[g]

[a] Adapted from Wolery (1978).
[b] All flow is assumed to be at the given temperature.
[c] Fluxes into the exogenic system are positive.
[d] All anhydrite, CaSO$_4$, is assumed to redissolve.
[e] Equal heat is assumed vented at each temperature.
[f] Based on two theoretical simulations for seawater–basalt reaction in a flow-through (open) system for which temperature increases with the amount of basalt reacted. The temperature goes from 0 to 160°C and the final rock/water ratio is 0.008 g/g (Wolery, 1980).
[g] N.D. = not determined.

of an average thickness of 170 m of basaltic crust by off-axis hydrothermal activity. The actual alteration would be laterally variable and spread out over a large depth range.

It is difficult to estimate with any precision the actual hydrothermal flux of magnesium or any other component. However, the following points seem relevant. First, a significant amount of magnesium appears to be leached from oceanic crust by low-temperature palagonitic alteration (Table 3.2). This represents a source of endogenic magnesium that is not part of the exogenic cycle (i.e., river input). This flux must be added to those considered in Chapter 1, Section 1.2.1. It seems likely that the hydrothermal fixation of magnesium is at least of the same magnitude as the low-temperature alteration flux. If these fluxes were equal but opposite in sign, then basalt–seawater interaction would produce no net exogenic–endogenic magnesium flux. However, if hydrothermal circulation accounted for both the low-temperature alteration flux and one-half the river input, this would, extrapolating over 10^9 years, remove 650×10^{20} g of magnesium from the sedimentary cycling system, which is about the total reservoir (Table 3.6). Models for the long-term behavior of the exogenic system show no significant magnesium imbalance (see Section 3.4.3). This reasoning indicates that basalt–seawater interaction has not been a major net sink of magnesium.

3.4.3 Calcium and Effects Over Geological Time

In Section 1.1, it was shown that there is a notable excess of calcium in sedimentary rocks. The exogenic system currently contains about 0.24×10^{24} g of

3.4 LOW-TEMPERATURE ALTERATION OF SEA-FLOOR BASALTS

TABLE 3.6. Inventory of Exogenic System and Concentration of Elements into Crust

Element	Exogenic Mass (10^{20} g)[a]	Calculated Igneous Rock (Brotzen, 1966) (10^{20} g)	Percentage of Element in Crust[b]
Mass	—	—	0.5[c]
Si	8033	8115	0.7[d]
Al	2099	2099 (fixed)	1.6[d]
Na	473	670	2.9[d]
K	642	749	2.9[e]
Ca	2535	960	2.1[d]
Mg	655	476	0.05[d]
Fe	1318	1258	0.4[d]
Mn	28	—	—
C	907	—	69[f]
S	106	—	3.6[g]
Cl	339	—	61[f]
H	1627	—	64[f]
Sm	—	—	8.3[h]
^{40}Ar	0.64	—	66[e]

[a] Adapted from Ronov and Yaroshevskiy (1969, 1976) by Wolery (1978). The exogenic system here includes low-grade metasediments and volcanogenic sediments. The total exogenic mass is uncertain. Excess digits are retained so that numbers will give correct ratios.
[b] The percentage of the element in the crust (including oceans and air) over the total terrestrial amount (excluding the core).
[c] For 35-km thick continental crust.
[d] Mantle amount from Ringwood (1975); average igneous rock used for elements not strongly concentrated in crust.
[e] Sleep (1979, Table 1).
[f] Using potassium, exogenic masses, and mid-ocean ridge basalt ratios from Anderson et al. (1979).
[g] Mantle estimate is 1/5 times 0.1 FeS in mid-ocean ridge basalts.
[h] O'Nions et al. (1979).

calcium in contrast to the 0.096×10^{24} g expected on the basis of average igneous parent rocks (Table 3.6). An improved choice of parent rocks is not likely to remove this anomaly since the other major elements, except possibly sodium (see Chapter 1, Section 1.1) show no large anomalies. The apparent two- to threefold excess of calcium in the exogenic system has been attributed to subduction of large volumes of ancient calcium-poor abyssal sediments (Sibley and Vogel, 1976; Sibley and Wilband, 1977) or to selective leaching of calcium by weathering and hydrothermal circulation at ridges (Fig. 3.4; see also Wolery and Sleep, 1976; Andrews, 1977; Humphris and Thompson, 1978a).

In the former hypothesis, the excess calcium results because sediments that have been subducted are not included in a present-day mass balance. Pelagic sediments were probably very low in calcium until the evolution of calcareous plankton (specifically the calcareous foraminifera) in the mid-Mesozoic (see Section 1.4.2.3). At that time, the locus of carbonate deposition is thought to

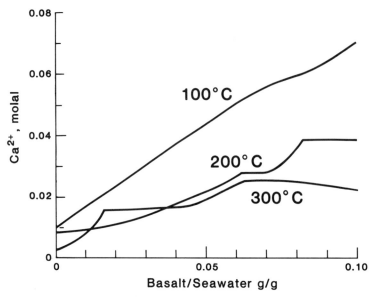

Figure 3.4 Concentration of Ca^{2+} in seawater during reaction with fresh basalt to 0.1 g (rock)/g (seawater) at 500 bars of pressure and at 100, 200, and 300°C. The initial concentrations at 200 and 300°C are depleted by anhydrite precipitation from heated seawater. After Wolery (1978).

have shifted from continental shelves and shallow seas to the deep ocean (Hay and Southam, 1975). Accordingly, most of the subducted sediment mass has been anomalously low in calcium.

If this hypothesis is correct, a mass equal to about two-thirds that of the current exogenic sediments has been subducted (Sibley and Wilband, 1977). Using data from Wolery (1978), we calculate that this amount would be 2.2×10^{24} g. In Section 3.1, we calculated a time of 100×10^6 years to subduct an area of oceanic crust equal to the present sea floor. The pelagic sediments on this crust have a mass of about 0.12×10^{24} g, half of which is $CaCO_3$ (see Chapter 1, Section 1.1). Thus, the rate of subduction of material in an ocean free of pelagic carbonate would be 0.6×10^{15} g/yr, or 2.5×10^{24} g in the 4×10^9 or so years since the origin of the ocean. If Hay and Southam (1975) are correct about pelagic carbonates, Sibley's mechanism could account for the apparent calcium excess in existing sediments, but only if almost all pelagic sediment is subducted.

Removal of calcium from basalt by seawater can also potentially account for the excess calcium (Table 3.2), although the evidence from the exogenic balance suggests not. The main difficulty with this hypothesis is that a potassium or magnesium anomaly would be produced by the hydrothermal and weathering fluxes. The average igneous rock model works well for these elements and aluminum and silicon, which are not as mobile during alteration. However, the weathering, moderately elevated, and high-temperature fluxes of potassium, sodium, and magnesium may largely cancel.

Another problem is that the off-axis hydrothermal flux and to some extent the weathering flux were computed assuming that only fresh basalt reacted with the circulating water. At fast ridge axes, temperatures over 300°C may exist a few hundred meters below the surface. If the global flow rates and potassium enrichments used in Table 3.2 are not in gross error, at least a few hundred meters of the upper crust are strongly altered by the axial circulation. Subsequent off-axis hydrothermal activity would tend to react less with the already altered basalt than with fresh basalts. In particular, the ratio of potassium, calcium, and magnesium fluxes might be significantly different from those in Table 3.5. Thus, a clear resolution of the origin of the calcium anomaly in sediments is not obtained from the hydrothermal flux estimates.

3.4.4 Oxygen Reservoirs

The earth's atmosphere and oceans are strongly oxidizing because of the buildup of oxidants produced by creation of organic matter that is now buried in sediments (Section 2.1.4). Basalt (and the mantle) is much more reducing than the ocean. Reaction of basalt with seawater is thus a drain on the exogenic oxygen reservoirs.

The principal oxidant in seawater, dissolved sulfate, is quantitatively removed in the Galapagos hydrothermal fluid (Edmond et al., 1979b). Dissolved oxygen is also removed but is only 0.5% of the total oxidant capacity originally present. Methane and H_2 are present (Welhan and Craig, 1979) and together represent about three times as much oxidant flux (in the negative sense) as the loss of dissolved O_2. A sulfide concentration of 10–20% of the original sulfate is found in the fluid (Edmond et al., 1979a). In addition, sulfide minerals are forming in the mixing area beneath and around the vent (Edmond et al., 1979c; Arnold and Shepard, 1981; Styrt et al., 1981). The source of the sulfide in the hydrothermal fluid is unclear. The percentage of the hydrothermal sulfide that originates from seawater sulfate is probably low, about 14%, but could be conceivably as high as 69% (Kerridge et al., 1983). Enrichments of arsenic, selenium, zinc, copper, silver, gold and platinum in the hydrothermal fluid also indicate a substantial basaltic component (Bischoff et al., 1983; von Damm, 1983). It is unclear whether the remaining 80–98% of the sulfate entering hydrothermal systems is reduced or precipitated as sulfates, which may later redissolve. See McDuff and Edmond (1982) for a view favoring reduction and Seyfried et al. (1984) for a view favoring precipitation. The latter view is supported by relics and pseudomorphs of anhydrite in deep core samples (Alt et al., 1986a). A net oxidant sink occurs whether the sulfide in the fluid comes from sulfate reduction or sulfide in the rock because in either case the sulfide is oxidized once it is mixed with the ocean. Oxidation of sulfides within the oceanic crust is also a sink for exogenic oxidants.

If all the sulfate in the circulating seawater is reduced to sulfide, a loss of 15% of the total exogenic oxidants would occur in the 10×10^6 years needed to circulate all the ocean's water through the hydrothermal systems of mid-ocean ridges. This would change 5% FeO to Fe_2O_3 in a 5-km section of crust. If sulfate in the ocean has not been rapidly depleted with time, additional sulfate would have to be supplied from $CaSO_4$ deposits in evaporites and by reaction of sulfide with atmospheric oxygen. A large compensating production of organic matter would then be necessary to balance the sinks for oxidants if such an extent of sulfate removal in hydrothermal systems persisted even 100×10^6 years. New sulfide would have to be supplied by volcanism and the additional organic matter would build up through time unless it were subducted.

The variation of sulfur and carbon isotopic ratios throughout the Phanerozoic can be associated with closed processes in the exogenic system (see Chapter 4). It is unlikely that this correlation would occur if large amounts of sulfate were lost at ridges and replaced by other sulfate from the mantle, since sulfur and carbon behave differently in magmatic and hydrothermal processes. A maximum loss of 2% of exogenic oxidants at ridges during the 10×10^6-year time to cycle seawater through ridges, in contrast to the potential loss of 15%, was obtained considering upper limits for the buildup of organic matter and subduction of sediments (Wolery and Sleep, 1976). A loss of only 0.2% is needed to explain the excess of exogenic reductants over oxidants. The upper limit is equivalent to a change of 1% FeO to Fe_2O_3 in a 1-km section of oceanic crust or a recycling of exogenic sulfur in 800×10^6 years. Because reaction of only a small fraction of the currently circulating sulfate would cause an excessive flux, the above arguments are not strongly dependent on the composition of the ocean in the geologic past.

If oxidation is a necessary part of the alteration of oceanic crust, the amount of altered crust can be limited by the above arguments. The kinetics of sulfate reduction are slow below about 300°C (see Section 3.2). It is thus possible that slow kinetics limits oxidation of the crust except by oxygen dissolved in seawater and the production of minor reductants such as dissolved hydrogen.

3.4.5 Metal-Rich Sediments

Hydrothermal processes extract manganese and iron from basalt, which precipitates to form basal manganese- and iron-rich sediments near ridge axes. From observation of actual vents, Edmond et al. (1979c) confirmed that these elements are initially enriched in axial hydrothermal fluid and that subsurface mixing with normal seawater preferentially removes iron because oxidation of iron is faster than that of manganese. Cooling without mixing removes iron and base metal sulfides but not manganese.

Although these hydrothermal deposits are useful for finding vents, they do not provide a useful constraint on the amount of interaction between basalt and seawater. The global flux of 0.9×10^{12} g/yr of manganese (Lyle, 1976) could be obtained from only a 60-m section of basalt. Thus, the Galapagos hydrothermal fluids are saturated with respect to manganese compounds (Edmond et al., 1979c). The manganese flux given above would equal the exogenic reservoir of manganese in about 3×10^9 years. It is likely, however, that the long-term effect of this process on the exogenic system is much less because the iron and manganese precipitate within or at the surface of the oceanic basalt and thus are geometrically most likely to be subducted.

At least some of the iron and manganese in metal-rich sediments is supplied by quick leaching from hot pillow basalts on the sea floor (Corliss, 1971). Wolery and Sleep (1976) and Elderfield (1976) recognized this as a hydrothermal process separate from large-scale cooling in their compilations. Iron appears to be more effectively mobilized than manganese during quick leaching.

A significant drain on exogenic oxidants results from Fe_2O_3 formation in the metal-rich sediments. Taking the iron flux into these sediments to be 30×10^{12} g/yr, we calculate that 0.13×10^{12} moles/yr of O_2 are removed by these sediments. This is one-tenth the maximum permissible oxidant flux quoted in Section 3.4.4.

3.4.6 Water and Other Volatiles

Water is added to oceanic crust by both low- and high-temperature alteration. The initial water in primitive basalt and thus in average crust is about 0.12% (Bryan and Moore, 1977). Hydration to a few percent is not uncommon. For example, hydration of 0.1 g (water)/cm^3 (basalt) (R. Hart, 1973) to an average thickness of 0.5 km would recycle about one-third of the exogenic water through geologic time. The water added to the crust probably remains in the subducted slab at least to the region beneath the island arc volcanoes (Wyllie, 1976; Delaney and Helgeson, 1978). The dehydration of partial melting of the oceanic crust may return this water to the surface through island arc volcanoes. Alternatively, the water in island arc magmas may come from normal mantle above the slab and the water in the slab is carried to great depths.

In the latter case, hydration of the oceanic crust would cause a net loss of water from the exogenic system and a fall of sea level through geologic time. The demonstrably low rate of change in sea level in the Phanerozoic (less than 300 m; see Wise, 1972) could then be used to constrain the amount of hydration in the oceanic crust. The volcanic flux of water is poorly constrained. Large offsetting volcanic and hydration fluxes of water are possible but unlikely unless the volcanic flux is linked directly to dehydration at depth. Using this reasoning, Wolery and Sleep (1976) calculated that on the average about 1 km (or less) of basalt becomes hydrated at ridges.

This reasoning could eventually be improved by considering the ratio of water to chlorine, CO_2, or other volatiles. Water is preferentially added to the oceanic crust relative to chlorine during alteration (Ito et al., 1983). Chlorine is added to basalt around 410°C, but the phase redissolves (Seyfried et al., 1986). The Cl/H_2O ratio in arc volcanics is not strongly different from the ratio in ridge volcanics or the exogenic system, and the mass flux from arc volcanoes may be insufficient to balance the flux of water into hydrated oceanic crust (Anderson, 1974, 1975; Ito et al., 1983). It thus appears that the Cl/H_2O ratio in the exogenic system should increase with time and the mass of exogenic water should decrease.

Unfortunately, CO_2 is not very soluble in magma at crustal pressure and thus it is difficult to determine its ratio to other volatiles. The CO_2 isotopic ratios in uncontaminated mid-ocean ridge basalts fall in a small range (Des Marais and Moore, 1984). This indicates that extensive amounts of recycled material are not sampled by these basalts or that there is so much CO_2 in the mantle compared with the exogenic system that the recycled material is not easily detected.

If the values in Table 3.6 are not in gross error, H_2O, CO_2, and chlorine largely reside in the exogenic system and not in the mantle. Potassium is the only major rock-forming element that is strongly concentrated in the crust. This concentration implies that large volumes of ordinary mantle must be processed to balance volatiles or potassium lost to subduction. An ^{40}Ar from decay of ^{40}K would be degassed along with the volatiles. If this ^{40}Ar remains in the exogenic system (mainly in air), an upper limit on the amount of potassium and volatile subduction can be obtained (Sleep, 1979). From Table 3.6, 66% of the ^{40}Ar calculated from the potassium abundance is in the crust. If the exogenic reservoir of potassium or water had been cycled several times through the mantle by subduction followed by replacement of potassium or water from the general mantle reservoir, the ^{40}Ar degassing should be close to 100%. Sleep (1979) estimated that close to the total present amount of exogenic potassium could be subducted as sediments (Sibley and Wilband, 1977) or a hydrated crust without causing an ^{40}Ar or heat balance problem. Note that this amount is similar to the amount of recycling required if the calcium anomaly is due to subduction of calcium-poor pelagic sediments. Thus, the balance of volatiles is at least consistent with the model of Sibley and Wilband (1977).

The behavior of argon and also neon, krypton, and xenon is sufficiently controversial that the assumptions in the above argument are open to question. See Craig and Lupton (1981) and Fisher (1986) for reviews of rare gases in the ocean and oceanic basalts. It has been contended that (1) ^{40}Ar in air is primordial, not radiogenic, and that $^{40}Ar/^{36}Ar$ is the same in the mantle as the air (Manuel, 1978); (2) the rare gases in mid-ocean ridge basalts come from near-surface contamination (Saito, 1978); or (3) ^{36}Ar and other nonradiogenic rare gases are strongly depleted from the mantle and the $^{40}Ar/^{36}Ar$ in the mantle is many times that in the atmosphere (Dymond and Hogan, 1978; Schwartzman, 1978). The first two hypotheses imply that the potassium is very low in the part of the mantle sampled by ridges and thus the H_2O and potassium abundances inferred for the mantle from ridge basalts are not necessarily representative of the mantle. The third hypothesis (assumed above) implies early degassing and is compatible with the lack of an obvious buildup of water through the rest of geologic time. However, argon and xenon isotopes indicate that both degassed and undegassed reservoirs exist in the upper mantle (Allegre et al., 1987).

The assumption that argon does not get subducted is open to question because significant amounts of atmospheric argon are found in the few analyses of weathered basalt. The apparent potassium–argon ages (without atmospheric correction) are around 1.8×10^9 years (Dymond and Hogan, 1973, 1978). If this value is representative, 17% of the argon generated by potassium in the oceanic crust through geologic time is returned. This amount would modify but not invalidate the argument that the mantle has not been reprocessed several times. Very rapid recycling of argon into and from the bulk of the mantle would cause $^{40}Ar/^{36}Ar$ to become homogenized to the atmospheric ratio. As noted above, the mantle ratio is subject to controversy.

Hydrogen/deuterium ratios are another potential tracer for material recycled into the mantle at island arcs (Magaritz et al., 1978). However, the uniformity of D/H ratios from various ridge and off-axis lavas indicate that the water in the mantle is not greatly affected by recycling (Craig and Lupton, 1981). More study is needed.

3.4.7 Radiogenic Isotopes

Helium can be used to make additional constraints on mantle degassing and the origin of the volatiles in island arc volcanoes (Craig and Lupton, 1981). ^4He is produced through geologic time by uranium and thorium decay; ^3He is mainly primordial. Recycling is not a problem because helium quickly escapes from the earth. The existence of ^3He in ridge basalts precludes extreme recycling of the mantle through volcanic source regions—else the mantle would have been totally depleted—but is difficult to quantify because the primordial ^3He abundance is unknown.

The $^3He/^4He$ ratio in island arc volcanics is similar to but slightly lower than that in ridge basalts (Craig et al., 1978). Hotspot magmas have somewhat higher $^3He/^4He$ ratios than ridge basalts (Craig et al., 1978; Tolstikhin, 1978). A significant component of normal mantle rather than down-going slab (which is degassed of ^3He) is thus implied in island arc volcanics. The argument is difficult to quantify because degassed components from the slab would also contain little ^4He and thus not change the ratio greatly. Thus, some involvement of the slab in island arc volcanism is neither required nor precluded by the helium data.

^{87}Sr is produced by ^{87}Rb decay. The $^{87}Sr/^{86}Sr$ ratio in seawater and other crustal reservoirs is higher than in the mantle because the Rb/Sr ratio is higher in the crust. Reaction between oceanic crust and seawater thus increases the $^{87}Sr/^{86}Sr$ ratio in the basalt (e.g., Spooner et al., 1977; O'Nions et al., 1979). Weathering and off-axis hydrothermal activity, but apparently not axial hydrothermal activity (Edmond et al., 1979b), account for the exchange. By analogy to calcium, a net flux of strontium from the oceanic crust is expected. In analogy to potassium, rubidium is removed from seawater by low-temperature processes and added by axial vents (Edmond et al., 1979b). A meaningful average age of continental crust formation may not be obtained from Rb/Sr systematics if these fluxes are significant.

The exogenic reservoir of strontium, which is mainly in carbonates, is cycled through the ocean and back into sediments on the order of every 10^8 years (Brass, 1976). A flux (or isotopic exchange) from ridges of only 1/40 the strontium flux in rivers would dominate the isotopic composition of exogenic strontium over geologic time but would not influence shorter-term variations in the strontium isotopic composition of the ocean. See Chapter 5 for more on the balance of strontium in rivers and the oceans.

The $^{87}Sr/^{86}Sr$ ratio forms a potential tracer for subducted sediments or altered oceanic crust in island arc volcanics. For normal mantle sources such as ridges and oceanic islands, the ratio of $^{145}Nd/^{144}Nd$ correlates linearly with the $^{87}Sr/^{86}Sr$ ratio. The $^{145}Nd/^{144}Nd$ ratio is not affected by hydrothermal alteration of basalt so that altered basalt has excess $^{87}Sr/^{86}Sr$ value over the normal correlation. Island arc lavas either have $^{87}Sr/^{86}Sr$ and $^{145}Nd/^{144}Nd$ ratios on the mantle correlation line or show excess $^{87}Sr/^{86}Sr$ values. This has been attributed to a component of subducted oceanic crust in island arc basalts (DePaolo and Wasserburg, 1977; DePaolo and Johnson, 1979; Hawkesworth et al., 1979).

Contamination of the lava at a high level in the island arc by crustal material is also a possible explanation for the enriched $^{87}Sr/^{86}Sr$ in arc volcanics and intrusions (DePaolo and Johnson, 1979; DePaolo, 1980a). Nearly assimilated limestone, for example, occurs in the gabbroic part of a magma chamber at a paleodepth of over 20 km (Burns, 1985). If the contamination is by wall rocks of the magma chamber, increased contamination would occur during fractional crystallization in the chamber. The $^{87}Sr/^{86}Sr$ ratio would therefore correlate positively with indices of fractional crystallization such as SiO_2 or strontium. This correlation is observed in samples from the Marianas (DePaolo and Wasserburg, 1977) and New Britain (DePaolo and Johnson, 1979). Contamination by a small fixed amount of strontium-rich material, such as limestone, would affect those lavas having little strontium more than it would affect the fractionated strontium-rich lavas. This correlation is observed on Grenada (Hawkesworth et al., 1979). More study is needed to establish any real contribution to island arc volcanics from the slab.

Lead isotopes can be used in a similar way to strontium and neodymium isotopes. A significant contribution from subducted sediments is not indicated in most cases (Oversby and Ewart, 1972). Sediment components have been suggested for Aleutian (Kay et al., 1978) and Caribbean (Armstrong and Cooper, 1971) lavas. Again the problem of discriminating between contamination at high levels and slab components exists. The high-level contamination problem is bad for island arc volcanics because the rocks erupt in zones of compression in small volumes over long periods of time. Ocean ridge and ocean island volcanism are much more short-lived at any point on the crust and do not erupt through thick piles of accreted sediment and deformed oceanic crust.

The cosmogenic isotope ^{10}Be is a potentially useful tracer for subducted sediments in island arc volcanics. The half-life of 2.7×10^6 years is sufficiently short that this isotope is present in young sediments but absent in older crustal rocks. The isotope ^{10}Be has been detected in several samples of arc volcanics but is absent in others (Brown et al., 1983). The abundance is low enough that the subducted sediment is not a major ingredient in arc volcanics.

3.5 SUMMARY

It can be seen from the above discussion that there is potentially a large exchange of material between the mantle and the exogenic system, either at mid-ocean ridges or in subduction zones. Quantifying the amounts, however, cannot yet be done with confidence. At the ridges, interaction of basalt with seawater can take up about 30% of the river flux of magnesium, depending on how much of the magnesium taken up in high-temperature alteration is returned by low-temperature reactions.

The effect of subduction is even more poorly constrained. It is clear that a considerable amount of water and magnesium must be carried down toward the mantle in subducting oceanic crust. On the other hand, much of this material may be returned to the surface by island arc volcanism. Another crucial question is how much pelagic sediment is carried down with the subducting slab and how much is scraped off against the inner wall of the trench. It is possible that subduction of sediments explains the excess of calcium in sedimentary compared with igneous rocks, but only if the process is very efficient.

4 GEOCHEMICAL CYCLES OF CARBON AND SULFUR

W. T. Holser, M. Schidlowski, F. T. Mackenzie, and
J. B. Maynard

The biogeochemical cycles of carbon and sulfur play an active role in the biosphere, hydrosphere, atmosphere, and sedimentary lithosphere. The two cycles have been studied intensively and a great deal is known about their reservoirs and fluxes. These studies have also revealed a number of important questions in geochemistry. Aspects of these cycles have been summarized in a number of earlier publications (e.g., Garrels and Perry, 1974; Ronov et al., 1974; Garrels et al., 1975, 1976; Ronov and Yaroshevskiy, 1976; Ronov, 1976, 1980; Holland, 1978, 1984; Claypool et al., 1980; Holland, 1984; Sundquist and Broecker, 1984; Berner, 1987); these publications were a starting point for this chapter.

The cycles of carbon and sulfur are dominated by their biologically mediated reduction. In the carbon cycle, oxidized carbon (CO_2) is reduced to organic carbon by photosynthesis in plants and sooner or later reoxidized by processes that are, for the most part, mediated by aerobic bacteria. In the sulfur cycle, sulfate ions are reduced to sulfide by bacteria, which oxidize organic carbon in the process. The sulfide is eventually reoxidized by the oxygen of air and water, usually with the assistance of sulfur-oxidizing bacteria. Small fractions of the organic carbon and sulfide sulfur produced are stored in sedimentary rocks as bitumen plus kerogen and as pyrite.

The carbon and sulfur cycles are both involved with the atmospheric subcycles of oxygen and carbon dioxide. In the carbon cycle, the reduction of carbon dioxide produces oxygen and organic carbon. If the organic carbon is buried, there is an addition to the store of oxygen in the atmosphere and dissolved in the ocean. In the sulfur cycle, the sulfur-reducing bacteria produce carbon dioxide as they consume organic carbon. Both the carbon and sulfur cycles consume oxygen while organic carbon and sulfide sulfur are being oxidized.

The biochemical reduction of both carbon and sulfur produces large, negative isotopic fractionations. Therefore, the remaining oxidized species are enriched in the heavy isotopes ^{13}C and ^{34}S. Consequently, the measurement of stable isotope ratios in the reservoirs of carbon and sulfur provides a check of our estimates of the balance of the present cycles and is a clue to changes in the fluxes and reservoirs of carbon and sulfur during geologic time.

106   GEOCHEMICAL CYCLES OF CARBON AND SULFUR

Our study of the cycles of carbon and sulfur begins with estimates of the sizes of the present reservoirs and the fluxes between them, insofar as these can be estimated by observing the cycles as they operate now. A survey of the isotopic effects generated by the fluxes is followed by a description of observed variations in the isotopic composition of the reservoirs through geologic time and an interpretation of these in terms of varying masses and fluxes. The chapter concludes with some speculations on the geologic origins of the changes in fluxes and reservoirs with time.

This chapter should illustrate the myriad of facts, processes, and concepts that must be evaluated to define the geochemical cycle of an element and to apply it in elucidating the history of the earth. Some of what follows is still controversial. However, we have tried to be cautious in our statements and to point out areas of difficulty.

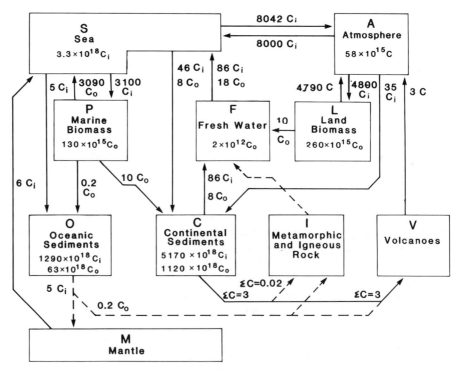

Figure 4.1 The main reservoirs and fluxes of the exogenic carbon cycle. Details of the reservoirs are given in Table 4.1a and of the fluxes in Table 4.3, where the fluxes are designated by subscripts of source and sink corresponding to the symbols designating each reservoir. Reservoirs are aggregated as continental if they are generally on continental crust—arbitrarily including Ronov's platform, synclinal, and subcontinental (shelf and continental slope) sediments—and as oceanic if they are generally on oceanic crust (see Table 4.1a; also, e.g., Ronov, 1980, Table 10). Reservoir units are moles C, flux units are 10^{12} moles C/yr; C_i = inorganic carbon (carbonate or carbon dioxide) $C_o = C_{org}$ = organic carbon.

4.1 THE CARBON CYCLE

4.1.1 Reservoirs of Carbon

Calculating the sizes of geochemical reservoirs is nearly always done by estimating separately the mass of a physical unit (rock, ocean) and the concentration of the element in that unit. The mass of the reservoir is the product of these two numbers. In the present instance (Fig. 4.1a, Tables 4.1a and 4.1b), we have elected to subdivide the rock record into both tectonic units (platform, geosynclinal, marginal, and pelagic sediments) and rock types (sandstones, shales, carbonates, evaporites, and effusives). The masses of sedimentary rocks presently in the four tectonic units have been examined critically by Southam and Hay (1981) using measurements and estimates of previous authors, and their evaluation is accepted for the present compilation. They emphasized the large contributions, only recently appreciated, of submarine sedimentary rocks to the total volume, both in the pelagic region and on the continental slope, rise, and shelf. The proportions of rock types within the tectonic units and the concentrations of reduced and oxidized carbon in each are taken from Ronov (1980). An exception is the composition of pelagic sediments, for which the value reported by Southam and Hay (1981) is used.

The most important result of the compilation of reservoirs is the estimate of the total masses of reduced and oxidized carbon and particularly their ratio. At least 15% of the carbon in the exogenic cycle is present in the reduced state, as organic carbon. Previous determinations of these quantities were reviewed by Garrels and Lerman (1981). Comparison with our estimates (Table 4.2) reveals variations that perhaps indicate our collective ignorance of the true values. Many of these variations are caused by divergent estimates of the masses of the tectonic units, which Southam and Hay (1981) have illustrated, and of the proportions of rock types within each tectonic unit. The most important unknowns by far are the mass and composition of the marginal sedimentary wedge. As illustrated by Southam and Hay (1981), estimates of its mass vary by a factor of 3. For the composition of these sediments, we have had to follow Ronov (1980) in the unsupported assumption that it is equivalent to that of the average sediment on the continents. It is more difficult to estimate possible errors in the concentration of reduced and oxidized carbon in each rock type. The present compilation has been formulated using the determinations of Ronov (1980), because this work is the only extensive program that couples chemical analysis with mass determinations. No measure of precision is stated by Ronov, however, and inasmuch as he relied heavily on the analysis of aggregate samples, it would be very difficult to arrive at such a measure, even given his original data.

Furthermore, it is certain that the size of the reservoirs has not been constant, even during Phanerozoic time, to say nothing of the early Precambrian: wide variations are evident in Ronov's compilation of masses for stratigraphic stages of the Phanerozoic (Ronov, 1976). Arthur (1979) and others have found major

TABLE 4.1a. The Reservoirs of Carbon in the Exogenic Cycle

	Reservoir	Reservoir Mass[a] (10^{21} g)	Carbon Reservoirs			
			Mean C (wt %)[b]		Mass C (10^{18} moles)	
			Reduced	Oxidized	Reduced	Oxidized
I	Atmosphere	5.1	—	0.011	—	0.06
II	Hydrosphere					
	A. Fresh waters	4.	—	0.0012	—	0.004
	B. Oceans	1420.	—	0.0028	—	3.31
III	Biosphere					
	A. Living	0.0012[c]	45.	—	0.05	—
	B. Dead	0.0077[c]	50.	—	0.32	—
IV	Lithosphere					
	A. Pelagic ocean sediments	250.	0.3[d]	6.2[d]	63.	1290.
	B. Marginal (rise, slope, and shelf) sediments	850.	0.52[e]	2.41[e]	370.	1710.
	C. Geosynclinal sedimentary rocks					
	1. Sandstones	274.	0.38	1.06	87.	242.
	2. Shales	552.	0.96	0.76	442.	350.
	3. Carbonates	259.	0.18	8.6	39.	1860.
	4. Evaporites	3.	—	0.08	—	0.2
	5. Effusives	262.	—	0.04	—	9.
	Total	1350	0.5	2.18	5.68	2460.
	D. Platform sedimentary rocks					
	1. Sandstone	83.	0.37	0.7	26.	48.
	2. Shales	171.	0.92	1.36	131.	194.
	3. Carbonates	90.	0.33	10.1	25.	758
	4. Evaporites	9.	—	0.36	—	3.
	5. Effusives	16.	—	0.04	—	1.
	Total	370.	0.59	3.24	182.	1004.
	E. Fossil fuels[f] (includes oil shale)	0.036	30.	—	0.95	—
	Total continental and marginal sediments	2570.	0.52	2.41	1120.	5170.
V	Total exogenic sediments and seawater	4250.	—	—	1180.	6460.

[a] Masses of reservoirs from Southam and Hay (1981) and percentages of rock types from Ronov (1980), unless otherwise indicated.
[b] Mean contents from Ronov (1980), unless otherwise indicated.
[c] Ajtay et al. (1979); Mopper and Degens (1979).
[d] Pelagic sediment types of El Wakeel and Riley (1961) weighted by the volume proportions of Poldervaart (1955) as suggested by Southam and Hay (1981); carbon contents of rock types from El Wakeel and Riley (1961); sulfur content from a survey of literature data.
[e] Assumed mean composition of continental sediments, as in Ronov (1980). According to Southam and Hay this may underestimate the amount of evaporites (and hence sulfate), but, on the contrary, Ronov (1980) may already have overestimated sulfate in the nonevaporite sediments of the continents, from which this mean was derived. For alternative interpretations see Table 4.2.
[f] Averitt (1975); Meyer (1977); Ziman (1978).

TABLE 4.1b. The Reservoirs of Sulfur in the Exogenic Cycle

		Reservoir Mass[a] (10^{21} g)	Sulfur Reservoirs			
			Mean S (wt %)[b]		Mass S (10^{18} moles)	
Reservoir			Reduced	Oxidized	Reduced	Oxidized
I	Atmosphere	5.1	—	—	—	—
II	Hydrosphere					
	A. Fresh waters	4.	—	0.001	—	0.04
	B. Oceans	1420.	—	0.09	—	40.
III	Biosphere					
	A. Living	0.0012[c]	0.5[d]	—	—	—
	B. Dead	0.0077[c]	0.5[d]	—	—	—
IV	Lithosphere					
	A. Pelagic ocean sediments	250.	0.1[e]	—	8.	—
	B. Marginal (rise, slope, and shelf) sediments	850.	0.18[f]	0.19[f]	48.	50.
	C. Geosynclinal sedimentary rocks					
	1. Sandstones	274.	0.161	0.037	13.8	3.2
	2. Shales	552.	0.227	0.086	39.1	14.8
	3. Carbonates	259.	0.088	0.01	7.1	0.8
	4. Evaporites	3.	—	10.	—	9.4
	5. Effusives	262.	0.037	0.004	3.	0.3
	Total	1350	0.15	0.068	63.	28.5
	D. Platform sedimentary rocks					
	1. Sandstone	83.	0.3	0.136	7.7	3.5
	2. Shales	171.	0.41	0.232	21.9	12.4
	3. Carbonates	90.	0.17	1.08	4.8	30.3
	4. Evaporites	9.	—	9.74	—	27.3
	5. Effusives	16.	—	0.024	—	0.1
	Total	370.	0.3	0.64	34.4	73.6
	E. Fossil fuels[g] (includes oil shale)	0.036	2.3	—	0.03	—
	Total continental and marginal sediments	2570.	0.18	0.19	146.	152.
V	Total exogenic sediments and seawater	4250.	—	—	153	192

[a] Masses of reservoirs from Southam and Hay (1981) and percentages of rock types from Ronov (1980), unless otherwise indicated.
[b] Mean contents from Ronov (1980), unless otherwise indicated.
[c] Ajtay et al. (1979); Mopper and Degens (1979).
[d] Orr (1974).
[e] Pelagic sediment types of El Wakeel and Riley (1961) weighted by the volume proportions of Poldervaart (1955) as suggested by Southam and Hay (1981); carbon contents of rock types from El Wakeel and Riley (1961); sulfur content from a survey of literature data.
[f] Assumed mean composition of continental sediments, as in Ronov (1980). According to Southam and Hay this may underestimate the amount of evaporites (and hence sulfate), but, on the contrary, Ronov (1980) may already have overestimated sulfate in the nonevaporite sediments of the continents, from which this mean was derived. For alternative interpretations see Table 4.2.
[g] Averitt (1975); Meyer (1977); Ziman (1978).

TABLE 4.2. Comparison of Estimates for the Reservoirs of Reduced and Oxidized Carbon and Sulfur in the Exogenic Cycle[a]

Reservoir Mass (10^{18} moles)

Author	Carbon			Sulfur		
	Reduced	Oxidized	Reduced/Total	Reduced	Oxidized	Reduced/Total
Holser and Kaplan (1966)				84	202	0.29
Li (1972)	930	3712	0.20	231	189	0.55
Garrels and Perry (1974)	1042	5083	0.17	294	240	0.55
Schidlowski et al. (1977)	1078	3505	0.23	184	238	0.44
Nielsen (1978)[b] (A)				178	205	0.46
(B)				178	81	0.69
Garrels and Lerman (1981)	1300	5200	0.20	178	150	0.54
This chapter[c] (A)	1180	6460	0.15	153	192	0.44
(B)	1180	6460	0.15	180	166	0.52

[a] Including seawater.
[b] (A) assumes the evaporite mass deduced by Holser and Kaplan (1966), whereas (B) assumes an evaporite mass suggested by Otto Braitsch (K. H. Wedepohl, personal communication, 1976).
[c] (A) takes Ronov's (1980) analyses for sulfate and sulfide at face value (Table 4.1b), whereas (B) assumes that all sulfur in geosynclinal clastic sediments is sulfide.

accumulations of organic carbon during anoxic events of the Cenozoic era, and Hay and Southam (1975) have emphasized the dramatic increase in pelagic carbonate related to the emergence of calcareous nannoplankton in Cretaceous time (Selection 1.4.2.3). Substantial variations with time in the ratio of reduced to oxidized carbon can also be deduced from the variations in carbon isotope ratios of sediments with time (e.g., Veizer et al., 1980), as discussed below.

4.1.2 Carbon Fluxes and Their Balance

Fluxes in the exogenic carbon cycle are dominated by the short-term action of the biosphere (Figs. 4.1 and 4.2, Table 4.3). Central to this scheme is the carbon dioxide reservoir of the atmosphere, which is available directly for large-scale photosynthesis in forests on the continents and indirectly through sea–air exchange for marine photosynthesis on nearly as large a scale. Almost all the organic carbon so formed returns by bacterial respiration in a short time—days to years—to the atmospheric/oceanic pool of carbon dioxide. The input of carbon dioxide into the atmosphere by human activities has focused attention on the balance of inputs to and outputs from the atmosphere. Although it is clear that the CO_2 content of the atmospheric reservoir is presently increasing and that a large fraction of this increase is attributable to the burning of fossil fuels, reductions of continental phytomass and soil carbon may also contribute (e.g., Kohlmaier et al., 1980). Even today it is difficult to determine how effectively an increase of atmospheric carbon dioxide would drive an increase in photosynthesis, which would in turn tend to damp the rate of increase of atmospheric CO_2 (Van Keulen et al., 1980).

The atmospheric CO_2 reservoir is also tied to the reservoir of dissolved bicarbonate in the sea by an exchange at the air–sea interface so vigorous that

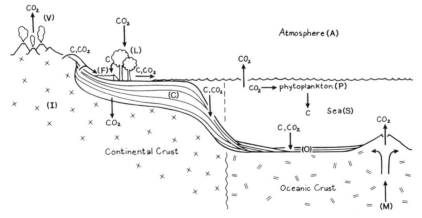

Figure 4.2 A qualitative overview of the reservoirs and fluxes of the exogenic carbon cycle, illustrating the box model of Fig. 4.1. Letters in parentheses refer to the boxes in Fig. 4.1; C corresponds to organic carbon or C_o, while CO_2 corresponds to inorganic carbon or C_i.

TABLE 4.3. Fluxes in the Carbon Cycle

Symbol		Present (Units: 10^{12} moles C/yr)		Preanthropogenic (Units: 10^{12} moles C/yr)	
		Range of Estimate	Best estimate	Range	Best estimate
	Short-Term Biosphere Subcycle				
$F_{SA} \simeq F_{AS}$	Air–sea exchange		8,300[a]	7,000–9,000[b]	8,000[a]
$F_{AL} \simeq F_{LA}$	Terrestrial photosynthesis and respiration	4,000–6,500[a]	5,000	3,700–6,400	4,800[c]
$F_{SP} \simeq F_{PS}$	Marine photosynthesis and respiration	1,000–10,000[d]	3,000[d]	1,000–10,000	3,100[e]
	Medium-Term Marine Subcycle				
F_{CF}	Weathering into C_i		53[f]	40–60[g]	86[h]
	C_o		5[f]	5–7[g]	8[g]
F_{LF}	Soluble C_o into rivers		10[f]	7–13	10[c]
$F_{FS}(= F_{CF} + F_{LF})$	Rivers to sea				
F_{SO}	Carbonate to deep sea	5–15[i]	13[i]	5–25[i]	5[j]
F_{SC}	Shelf deposition: C_i		?		46[k]
	C_o		?		8[l]
F_{PO}	C_{org} to deep sea	0–10[m]	0.6[m]	0–10[m]	0.2[n]
F_{PC}	C_{org} to shelves	5–50[o]	10[o]	5–50[o]	10[o]
	Long-Term Crustal Subcycle				
F_{MS}	MOR to sea		6[p]	2–20[p]	6[p]
F_{OM}	Subduction of oceanic sediments:				
	C_i				5[q]
	C_o				0.2[q]

112

F_{CI}	Metamorphism of continental rocks		0.02^r
F_{CV}	Volcanic regeneration of continental carbon		3
F_{VA}	Volcanic and metamorphic emissions	$0.1-0.6^s$	0.3^s

		0.02^r
	$1-6^s$	3^s

[a] Bolin et al. (1979a); Ajtay et al. (1979); Kohlmaier et al. (1980).
[b] Allowing for nominal variations of P_{CO_2}.
[c] Mean estimate assumes present stand of forests, which account for 75% of present terrestrial productivity, on a land area decreased (to allow for present sea level lower than usual) by the $4 \pm 10\%$ measured by Barron et al. (1980) for the past 180×10^6 years. Substitution of tundra or grasslands for forests would decrease productivity for those land areas by half; substitution of desert would decrease it by a factor of 10 or more (Ajtay et al., 1979). These variables were certainly important on a long-term basis, because latitudinal distribution and other climatic factors shifted with plate movements (Thompson and Barron, 1981), and land plants took over in the early Paleozoic (Gray and Boucot, 1978) and went through major evolutionary changes (C. C. Delwiche, personal communication, 1981). Anthropogenic impact is debatable (Hampicke, 1979).
[d] Range of estimates by DeVooys (1979), Mopper and Degens (1979), and Duce and Duursma (1977).
[e] Mean estimate of marine productivity increased over present to take into account the shift of land area to shelf area (see footnote c), which is three times as productive as the open ocean (Ajtay et al., 1979).
[f] Present river burden of C_i, reevaluated by Kempe (1979a) from new data on river flow and content. Thirty percent is particulate calcite, and 70% is bicarbonate in solution from weathering, the latter requiring the participation of a like amount of CO_2 input from the atmosphere (F_{AC}). Following an analysis by J. C. G. Walker (personal communication, 1980), we estimate that substantially all the particulate C_o in rivers may be inert material that is recycling. Hence, in the present table and in Fig. 4.1, the particulate C_o in rivers is all arbitrarily counted as old carbon reeroded from continental rocks, whereas dissolved C_o in the rivers is counted as new carbon formed in the present terrestrial biomass. By a charming coincidence, the ratio of C_o/C_i in weathering equals the ratio of C_o/C_i in the reservoir (Tables 4.1a and 4.1b). However, the origins of C_o delivered to the ocean are still in considerable dispute (e.g., Handa, 1977; Tissot and Pelet, 1981).
[g] Range and mean of recycling rates of surficial sediments (which contribute most of the erosional load of rivers) as calculated by Veizer (Chapter 5). Estimates of erosion rates based on the absence of human activity, glaciers, and the present unusually high continental relief would probably be near the lower end of the stated range. Recycling rates of buried sediments are about one-quarter to one-fifth those of surficial sediments.
[h] Compromise between (1) the input to the sea of precipitable dissolved bicarbonate from the rivers, 14×10^{12} moles C/yr, derived from their dissolved calcium load, recalculated by Kempe (1979b), as compared with (2) output precipitated to both the oceanic and continental shelf reservoirs. As emphasized by both Holland (1978) and Berger (1976), we have neither a direct estimate of precipitation on the shelves nor the proportion of the total carbonate precipitation that it represents. Holser estimates the latter by the following precarious chain of assumptions: (1) shelves have three times the primary productivity of the deep seas (Ajtay et al., 1979), and the 20% of the deep sea that is shallow enough to preserve carbonate is about half the area of the shelves; (2) primary productivity is correlated with standing stock of phytoplankton, which in turn is correlated with shell generation, and shell carbonate suffers insignificant dissolution during

fall to the areas of deposition (Berger, 1976). Consequently, the fraction of precipitated carbonate found in the deep sea is about 0.15. Then the total carbonate precipitation should be $F_{SO} + F_{SC} = F_{SO}/0.15 = (13 \times 10^{12}$ moles C/yr)/0.15 = 87×10^{12} moles C/yr. In order to attain the necessary short-term balance of carbonate in the sea, the geometric mean of the disparate estimates of input and output is taken as the best estimate of total input equals output: $(14 \times 87)^{1/2} = 35 \times 10^{12}$ moles C/yr. To complete the new value for total C_i in rivers, the bicarbonate is doubled to allow for release on precipitation, and the particulate carbonate river load is added, for a total of $F_{FS} = (2 \times 35) + 16 = 86 \times 10^{12}$ moles C/yr. Holland has already remarked that "the current calcium balance of the oceans is difficult to evaluate quantitatively. The uncertainty in the calcium input by rivers may be as large as 50%, and the uncertainty in the rate of $CaCO_3$ removal from the oceans is at least as large" (Holland, 1978, p. 165).

[i] Reservoir of oceanic carbon (Table 4.1a) divided by a mean age of oceanic sediments of 10^8 years. Range of sedimentation rates of carbonate in DSDP cores compiled by Southam and Hay (1981). Their curve suggests that present-day rates may actually be near the top of the range indicated.

[j] From footnote h: $F_{SO} = 0.15 \times 35 = 5 \times 10^{12}$ moles C/yr.

[k] Sum of fraction calculated (footnote i): $(0.85 \times 35) \times 10^{12} = 30 \times 10^{12}$ moles C/yr precipitate plus 16×10^{12} moles C/yr particulate = 46×10^{12} moles C/yr total.

[l] Particulate organic carbon carried by rivers is probably redeposited on the shelf (Handa, 1977) although some reaches the slope (Tissot and Pelet, 1981) or the deep sea (Summerhayes, 1981). It is also claimed that most of the dissolved organic matter from rivers is precipitated in estuaries or shelves (Handa, 1977), but this process is lumped here with the marine biomass and its shelf deposition. No one seems to have determined uncontroversially how much of the river C_o is oxidized to CO_2 rather than being redeposited.

[m] Calculated as in footnote j. Other estimates are higher: 5–10×10^{12} moles C/yr (Mopper and Degens, 1979, p. 306) and for the total river input of organic matter ($= F_{PO} + F_{PC}$) 25–27×10^{12} moles C/yr (Handa, 1977). The average analysis of C_{org} in deep-sea sediments, on which the preferred value of flux is based, may be an underestimate, see footnote d to Table 4.1a.

[n] Reduced in proportion to F_{SO} carbonate carbon (footnote i).

[o] C_{org} deposition on the shelves is as poorly known as the corresponding deposition of carbonate (footnote h), but the two are not proportional because organic carbon (nearly all particulate) is oxidized logarithmically with depth everywhere below the mixed layer (Suess, 1980), whereas carbonate (mostly microfossil shells) is virtually untouched until it passes the lysocline. Maps of both primary production (Berger, 1976) and C_{org} concentration in sediments (Kempe, 1979b) suggest that not only production but deposition is much higher on the shelves, especially when the high C_{org} concentrations are multiplied by the greatly increased sedimentation rates (Ibach, 1982). One estimate of primary production in the continental shelf area is 800×10^{12} moles C/yr (Ajtay et al., 1979). The fraction that reaches a 200-m depth is estimated at from 10% (Suess, 1980) to less than 1% (Mopper and Degens, 1979; DeVooys, 1979). A poorly measured but large fraction of this is oxidized by bacteria after burial (Mopper and Degens, 1979). An average might be 10×10^{12} moles C/yr; Berner (1982) has estimated 11.4×10^{12} moles C/yr.

[p] Delaney et al. (1978) make a rough estimate of 1100 ppm C in microvesicles trapped in microphenocrysts of basalt glass erupted in the deep sea. Most MOR basalts are erupted at water pressures such that the saturation CO_2 content corresponds to 40 ppm C (Harris, 1981). Assuming that all such vesicle CO_2 is finally delivered to seawater, we find that over 1000 ppm C is lost from basalts to seawater at the sea floor and correspondingly smaller amounts certainly escape from magmas that come to rest at high pressures under the sea floor. An average over all depths might be half that or 500 ppm. Following the suggestion of Delaney et al. (1978), we find that the estimated amount of inclusion CO_2 can be used to estimate a depth of entrapment of about 4 km. This is a minimum thickness of magma effervescing CO_2 into the sea. Multiplied by a present rate of sea-floor generation of 3 km²/yr (Wolery and Sleep, Chapter 3), this gives a production

of 6×10^{12} moles C/yr, within the range of the estimate by Des Marais (1985). Our estimate is subject to the three fold variation of ridge generation (Pitman, 1978) and the assumptions of this calculation. Earlier work on this subject was reviewed by Anderson (1975) who estimated $F_{MO} = 0.2 \times 10^{12}$ moles C/yr.

^{q}The process of subduction may return most oceanic carbon to the mantle, as the flux of carbon involved would be undersaturated (data reviewed in Delaney et al., 1978) in the accompanying recycled basalt. Some may be baked out of obducting or subducting sediments (dashed line for $F_{OM} = F_{SO} + F_{PO}$).

^{r}Veizer (Chapter 5) calculates a rate of *cratonization* of sediments of 4×10^{-10} year^{-1}, which applied to the total reservoir of continental carbon of 6600×10^{18} moles (Table 4.1) gives a flux of 3×10^{18} moles total C/yr. Comparing the carbon concentrations in this reservoir of 2.6% (Table 4.1) with that in igneous and metamorphic rocks of a few hundred ppm C (Hoefs, 1969), it seems that less than 1% of this carbon flux is retained in igneous and metamorphic rocks and must appear in volcanoes, hot springs, or as cooler ground emanations (Irwin and Barnes, 1980).

^{s}Cadle (1980) suggests after a review of data on the emissions in volcanic eruptions that the present flux of CO_2 is about 0.3×10^{12} moles C/yr. However, in view of the saturation of magma that fails to retain most of its CO_2 even under deep-sea pressures (footnote p), it seems likely that in continental volcanoes an even larger proportion of magmatic CO_2 would be lost through other avenues before eruption. Furthermore, when carbonate (or orgaric-carbon) rocks are buried, they are likely to begin evolving CO_2 at even low grades of metamorphism, leaving little carbon for incorporation in any magma formed in a later and higher-temperature milieu in the continental (including island-arc) crust. Irwin and Barnes (1980) describe a worldwide association of springs rich in bicarbonate within tectonically active zones and imply that an unstated large fraction of these cool emissions represent release of CO_2 from metamorphism of carbonate rocks. In terms of the gross chemical reservoirs and fluxes intended in Fig. 4.1, these cooler fluxes of gaseous carbon are arbitrarily included in volcanicity. It is difficult to estimate the flux of CO_2 from these additional sources (both metamorphic and early magmatic) to the surface, but with no other numbers to rely on it seems reasonable that they might raise the carbon output of continental rocks by an order of magnitude. Holser likes this result because it just happens to equal the F_{CV} calculated in footnote p.

it virtually can be considered an equilibrium. Seawater bicarbonate is in turn controlled to a great extent by the saturation of carbonate minerals, so that an increase in carbon dioxide should be buffered by dissolution of carbonate sediment exposed on the sea floor below the carbonate compensation depth (CCD). This is about as good a feedback mechanism as may be found in the whole system, but even so it is actually controlled by kinetics of precipitation and dissolution rather than by equilibrium (Berger, 1976). The lesson in this for us is that substantial changes of the reservoirs can be produced by even minor shifts in the fluxes. So in Fig. 4.1, inputs and outputs for the atmosphere, sea, and the two biomass reservoirs have each been balanced to indicate a steady-state system that is possible within the poorly known numbers, both present and past, but not to imply the reality of such perfection.

In contrast to the active biologic cycle, the main marine subcycle is a longer-term operation. Its principal elements are the delivery of carbonate and organic carbon through rivers to the oceans and to their marine sediments. The fluxes of carbonate and organic carbon need to be divided into particulate and dissolved components, although for simplification the division is not made explicit in Table 4.3 and Fig. 4.1. The particulate and dissolved fluxes of organic carbon represent, in one respect, a small "leak" from the large fast flux of the biosphere subcyle.

The flux of carbon in the word's rivers requires further attention (Kempe, 1979a). For instance, what fraction of the organic carbon in rivers (particulate or dissolved) is new carbon from the present land biomass, and what fraction results from weathering of old organic carbon in sedimentary rocks? Whereas for modern sediments, isotopic carbon and other evidence indicate that most land-derived particulate organic matter is deposited within a few kilometers of the shoreline (Gardner and Menzel, 1974; Hedges and Parker, 1976; Duce and Duursma, 1977; Handa, 1977), for older sediments, significant transport into deeper water seems to have occurred often (Summerhayes, 1981; Tissot and Pelet, 1981). A new survey of carbon in world rivers (Degens, 1982) may provide answers to such questions.

In order to complete the calculation of the carbon cycle in Fig. 4.1, it has therefore been necessary to make some arbitrary assumptions. (1) All *dissolved* organic carbon in rivers comes from the land biomass and none from erosion of sedimentary rocks. (2) All *particulate* organic carbon in rivers comes from pre-existing sediments and none from the land biomass. (3) All *particulate* organic and inorganic carbon is delivered to estuaries or near-shore areas in the continental reservoir and none to deep-sea sediments. (4) All *dissolved* inorganic and organic carbon from the rivers is mixed into the main reservoir of the sea and hence distributed to both continental and deep-sea sediments. (5) Some transfers between organic and inorganic carbon (such as oxidation of organic to inorganic carbon in rivers or the sea) are neglected (cf. Degens and Ittekkot, 1987).

A prominent feature of the marine subcycle (Fig. 4.1, Table 4.3) is the dominance of continental shelf and slope sediments, from which the carbon recirculates to the sea only after a shift of relative sea level or tectonic uplift exposes these sediments to erosion. Inasmuch as erosion is mainly from newer, softer sediments that are closest to the surface, most of the fluxes into and out of the continental reservoir involve relatively short residence times on a geologic scale. Smaller fractions of the fluxes, not separated in this schematic view, involve marine rocks that reside in the continental mass for hundreds of millions of years.

Deposition of marine sediments also occurs onto oceanic crust and is a key aspect of both the marine subcycle and the long-term subcycle involving the crust and mantle. Hay and Southam (1975) and Sibley and Vogel (1976) have emphasized that the overwhelming increase of pelagic carbonate sedimentation since 200×10^6 years ago may represent a prominent and even permanent leak of calcium carbonate from the exogenic cycle (see Chapter 3, Section 3.4.3). A large proportion of the flux of carbonate deep-sea sediments may be subducted into the mantle or boiled off into the volcanoes above subduction zones. The fate of this carbon is crucial to the overall cycle of carbon, but the new calculation given here suggests that the "loss" of this carbon is not certain. First, a rough calculation of the division of carbonate sedimentation between oceanic and continental regimes (Table 4.3, footnote *h*) indicates that although a small fraction of the calcium input is diverted to deep-sea sediments most of it is deposited on the continental shelf. A compromise between this output and direct estimates of deep-sea carbonate sedimentation halves the latter. A new assessment of the flux of CO_2 to the sea during mid-ocean ridge formation (Table 4.3, footnote *o*) greatly increases that value to a level that seems to match the sedimentation rate of carbonate carbon. Of course this balance, as given in Table 4.3, is somewhat misleading even if true: the carbon dioxide that might be delivered to the mantle by subduction at active continental margins is not likely to be returned for a very long time to resupply the mid-ocean ridge. The carbon "dilemma" (to paraphrase Sibley and Vogel, 1976) has thus been swept into the mantle, where it may still be with us.

The other possible fate of deep-sea carbon is release in volcanic or related cooler effusions on the continents. However, a calculation of the amount of carbon released by the slow metamorphism of continental rocks (Table 4.3, footnote *r*) is an order of magnitude more than sufficient to supply the present estimated rate of carbon release in volcanoes (Cadle, 1980). Even allowing for a much greater amount of cooler, tectonically released emission of carbon dioxide (Irwin and Barnes, 1980), for which there are no published estimates, it looks as if subducting carbon may not be needed on the continents. This conclusion is of course as tentative as the pyramid of assumptions on which it is based.

In summary, the available data and calculations *allow* the possibility of a balanced cycle for carbon, as shown in Table 4.3 and Fig. 4.1, but they are so approximate that they *do not require* a balance in either the short or long term. Evidently, we must look to the geologic and geochemical record, as it has varied through time, to assess what changes may actually have occurred in this key geochemical cycle of the surface regime.

4.2 THE SULFUR CYCLE

4.2.1 Overview

Just as for carbon, many fluxes, reservoirs, and isotope relations of the sulfur cycle are dominated by the effects of biologic reduction. In the atmosphere, the cycle of sulfur is characterized by large fluxes but very short residence times (Kellogg et al., 1972; Friend, 1973; Granat et al., 1976). The atmosphere does not have a major impact on the long-term cycle of sulfur, but its large fluxes tend to obscure the long-term activity and so it will be discussed here before proceeding further. This then allows consideration of the main long-term subcycle, which includes large components of sedimentation onto continental crust, both of sulfide-bearing sediments and of evaporites, and the eventual erosion of both. We then evaluate the effects of deep-sea sedimentation and igneous activity on sulfur in the ocean. Only then are we in a position to consider the more obscure subject of volcanic sulfur and its relation to subduction and metamorphism.

The biologic cycle of sulfur has been reviewed effectively by Goldhaber and Kaplan (1974, 1975), Orr (1974), and Ivanov (1981). Its main features are outlined in Fig. 4.3. The system is dominated by the activity of anaerobic sulfur-reducing bacteria (*Desulfovibrio desulfuricans*, see Section 4.3.2). Although sulfate is thermodynamically unstable with respect to various reduced sulfur species in the presence of organic carbon compounds (C_{org}), reduction is inoperative at surface temperatures unless mediated by enzymatic action of sulfur bacteria. The sulfur-reducing bacteria also require organic carbon as an energy source, although they are selective in choosing degradable fractions of the total pool of organic carbon (Lyons and Gaudette, 1979). Intermediate products such as sulfite and other ions are important as steps in the process, and the final product may be distributed among H_2S, HS^-, S^{2-}, and S^0, depending on conditions in the medium. The overall reaction can be represented schematically as

$$(SO_4)^{2-} + 2CH_2O = H_2S + 2HCO_3^-, \qquad (4.1)$$

where the CH_2O is representative of any (degradable) organic carbon compound, and the H_2S is representative of any (completely reduced) sulfide.

Sulfur-reducing bacteria vigorously pursue sulfide manufacture in a wide variety of anaerobic situations (Eh $< +100$ mV; Orr, 1974). The rate of reduction is virtually independent of sulfate concentration (Goldhaber and Kaplan, 1975). The bacteria tolerate high concentrations of the sulfide they produce and a wide range of pH (but they are most active near neutral). Different strains are productive under increasing salt concentration, to which they gradually adapt, and are still active at concentrations up to at least 12 wt % NaCl (Dostalek and Kvet, 1963). Isotopic evidence indicates that sulfur-reducing bacteria have been quantitatively important in the sulfur cycle for the last 2.7×10^9 years (see Section 4.6.2).

4.2 THE SULFUR CYCLE   119

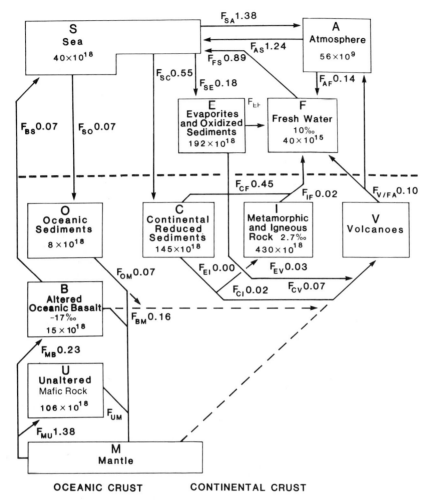

Figure 4.3 The preanthropogenic exogenic cycle of sulfur. The atmospheric subcycle is on the upper right, and the oceanic subcycle is on the left. The dashed fluxes in the lower right indicate possible intermittent transfers of sulfur from oceanic sediment, basalt, or mantle to volcanic and metamorphic reservoirs. Sulfur above the heavy dashed line is mostly oxidized to sulfate; sulfur below the line is mostly reduced to sulfide (pyrite). Data are assembled for the reservoirs in Table 4.1b, column 5, in moles, and for the fluxes in Table 4.4 in units of 10^{12} moles/yr, using the balanced estimate.

The present sea has only a few anaerobic locales, such as the Black Sea and the Cariaco Trench, in which bacterial sulfate reduction occurs in the water column. Immediately below the seawater–sediment interface, however, anaerobic conditions are widespread in both shelf and abyssal regimes. A high rate of supply of degradable organic material to the sediment (owing to a combination of high pelagic productivity and minimal oxidation in the water column;

Tissot, 1979), a high rate of sedimentation (Berner, 1984), and a low rate of stirring by burrowing organisms favor the anaerobic milieu and the abundant metabolizable organic carbon necessary for high activity of the sulfur-reducing bacteria. Differing intensities and patterns of deep-sea circulation at certain times and places in past geologic periods increased the extent of the anaerobic regime in oceanic sediments (e.g., Arthur, 1979). It is not always clear to what extent these reducing conditions also extended above the sediment surface into the overlying water column.

Two sources of sulfate are of importance for sulfur-reducing bacteria: interstitial seawater in marine sediments and sulfate-rich evaporite rocks. In marine sediments, the amount of sulfate reduced far exceeds the amount initially deposited in pore water with the sediment. Sulfate concentration gradients, isotope data, tracer experiments, and other evidence show that most of the sulfate reduced by the bacteria comes by diffusion of sulfate from the overlying seawater, through the sediment–water interface and solution-filled pores. Although diffusion and reduction may extend to several meters, the bulk of reduction takes place within a few centimeters of the interface. On the continents, sulfate dissolved from buried gypsum or anhydrite rocks is occasionally reduced rather than discharged directly to rivers (Davis and Kirkland, 1978).

In order for reduced sulfide to be significant in the long-term geochemical cycle of sulfur, it has to be sequestered as a metallic sulfide, usually pyrite, or as native sulfur. Pyrite typically forms within a few meters of the surface in modern marine sediments (Goldhaber and Kaplan, 1974). Deposits in which native sulfur is the dominant form of reduced sulfur are locally prominent and an important economic resource but they are quantitatively minor. The source of the iron required for sulfide fixation is usually the goethite or hematite coatings on clastic grains of clay and silt. Consequently, this ferric iron must also be reduced by the H_2S or C_{org} and dissolved during the depositional process. The overall reaction may be written

$$8(SO_4)^{2-} + 2Fe_2O_3 + H^+ + 15CH_2O \longrightarrow 4FeS_2 + 15HCO_3^- + 8H_2O.$$

(4.2)

Of the substances required for sulfate reduction, supplies of both sulfate and iron are usually in excess in marine clastic sediments. Under nonmarine conditions, the supply of sulfate may at least temporarily limit pyritization; in some marine carbonate sediments the fixation of sulfide may be limited by a low concentration of iron (Raiswell and Berner, 1985). We may speculate that lack of iron may also be the reason for the predominance of native sulfur over pyrite produced in the bacterial reduction of sulfate in salt domes and other evaporite rocks.

In normal sedimentation of marine clastics, the parameter that seems to limit sulfate reduction is the supply of organic carbon in a form usable by the reducing bacteria (Goldhaber and Kaplan, 1974, 1975). Actually, the range of organic molecules that can be used in laboratory cultures is limited to a few short-chain carboxylic acids (reviewed in Goldhaber and Kaplan, 1974, 1975) or simple hydrocarbons (Davis and Yarbrough, 1966). Goldhaber and Kaplan (1974, 1975) speculated that edible carbon may also be supplied by a symbiotic community of fermentative bacteria breaking down some of the more complex organic molecules. In any case, there is always a residue of presumably unusable organic carbon that passes through the zone of sulfate reduction to be incorporated eventually in sedimentary rocks.

Surprisingly, pyritic sulfur and residual carbon are in a relatively constant proportion of about 0.12 (in moles) in Recent marine sediments (Goldhaber and Kaplan, 1974). Ancient marine shales show some variation in C/S ratio as a function of age (Dimroth and Kimberley, 1976; N. Williams, 1978, p. 1041; Leventhal, 1979, 1987), but shales of a given age have a characteristic C/S ratio (Raiswell and Berner, 1986). A constant ratio results only under the circumstance that both the fraction of carbon that is usable is constant *and* that the fraction of produced H_2S that is fixed as pyrite is also constant. This relation was derived by Sweeney and Kaplan (Goldhaber and Kaplan, 1974, p. 600) but is more clearly developed by N. Williams (1978, p. 1039).

Berner (1978, 1984) has demonstrated that, for modern marine sediments in contact with oxic bottom water, the rate of sulfate reduction by bacteria increases as the rate of sedimentation increases because the bacteria have less competition for metabolizable organic matter from the oxygen-using organisms near the sediment surface. As the sedimentation rate increases, the bacteria work faster, thereby fixing a larger fraction of the available sulfur, while more refractory organic matter, which sulfate-reducing bacteria cannot use, is accumulating because of the decreased time available for the oxic organisms to attack it. In this way, carbon and sulfur should both increase together with increasing sedimentation rate (Berner and Raiswell, 1983).

Raiswell and Berner (1986) reported substantially lower C/S ratios for Cambrian and Ordovician marine shales than for younger rocks. They attributed lower C/S ratios to an absence of higher plants on land. Vascular land plants, which produce lignin, are a major source of refractory organic matter, substantial amounts of which can be carried by rivers into the oceans (Hedges and Parker, 1976). The proportion of H_2S that is fixed as pyrite could vary widely. Goldhaber and Kaplan (1974, p. 602) assumed that the fraction of H_2S lost in basinal sediments off California was "relatively small," whereas field experiments by Jørgensen et al. (1978) on tidal flats show a 90% loss.

4.2.2 Reservoirs of Sulfur

The reservoirs of sulfur are summarized in Table 4.1b and Fig. 4.3, based on the data collected by Ronov's group (e.g., Ronov, 1980), with the volumes of tectonic units modified according to the critical evaluation by Southam and Hay (1981). An important difference from earlier representations is an attempt to separate explicitly those sediments that are on oceanic crust and therefore at risk of an early demise by subduction.

The key numbers in the sulfur reservoirs of Table 4.1b and Fig. 4.3 are the totals for reduced and oxidized sulfur, indicating that perhaps half the total sulfur in the exogenic cycle is in the reduced state. Considered in another way, the 1180×10^{18} moles of C_{org} formed 1180×10^{18} moles of O_2 by photosynthetic utilization of primordial CO_2, and in turn $2 \times 192 \times 10^{18} = 384 \times 10^{18}$ moles (33%) of this O_2 were used to oxidize 192×10^{18} moles of primordial sulfide.

Table 4.2 indicates that other estimates of the fraction of sulfur in the reduced state vary from 0.29 to 0.69. As was the case with carbon, this variation is considered to be more a measure of our ignorance than it is of any gradual approach to the truth of the matter. Because of the importance of these numbers for the overall cycles of both sulfur and oxygen, some points are worth discussing further. A fundamental difference between the present estimates and those previously published relates to the particular method by which we treated the data of Ronov. The usual approach in estimating geochemical reservoirs has been to take some list of rock types and determine their relative importance by making a geochemical balance of their major constituents against an average igneous rock. Then independently published analyses of rocks given the same name are averaged to get a mean concentration. A different approach was necessary in using Ronov's data, because they were not from individual rock types but from stratigraphic formations, each of which was designated by its dominant rock type (Ronov and Ermishkina, 1953). Consequently, Ronov's category of carbonate rock is an average of formations that were generally limestones, but which probably contained shaley beds and gypsum, rather than being a selection of individual samples that a geologist would recognize as "pure" limestones. A second consideration that applies particularly to sulfur is that, when sampling geosynclinal rocks in the Caucasus, Ronov's group necessarily had to depend on outcrop samples rather than the cores that they used so extensively for the sediments of the Russian Platform. Weathering at the outcrop leads to local oxidation and early dissolution of pyrite so that total sulfur in such samples would tend to be low and the ratio of sulfate to sulfide high in comparison with the fresh rock, a problem that Ronov recognized (Ronov et al., 1974). This conclusion is verified by sulfur isotope analyses, which, for example, show the Caucasus Alpine geosynclinal shales to be 0.25% S_{pyr} and 0.21% $S_{sulfate}$, with nearly identical $\delta^{34}S \approx -11.3‰$ (Ronov et al., 1974). A third problem is that in composing his mean composition of evaporites, Ronov departed from his usual method of formation aggregation and arbitrarily *assumed* that evaporites were 50% sulfates and 50% salt (Ronov, 1980, Table 10), whereas detailed measurements by Holser et al. (1980) and Zharkov (1981) indicate about 25% sulfates and 75% salts. Finally, Kinsman (1974) and Southam and Hay (1981) independently pointed out that the passive continental margins today contain massive deposits of evaporite rocks—about 65% of the known calcium sulfate in evaporites (Holser et al., 1980)—whereas Ronov was forced by lack of data to assign a composition to marginal sediments equal to that of all geosynclinal plus platform sediments. Although these errors of estimation cut both ways, our conclusion is that in Table 4.1b (and Table 4.2, Fig. 4.3) both the total amount of sulfur and the fraction that is sulfide have been underestimated. An alternative estimate of the distribution of sulfide and sulfate may be made as follows. Assume in line with the above discussion that *all* sulfur in geosynclinal clastic sediments is actually sulfide (although in part analyzed as sulfate); then the sizes of the sulfur reservoirs are as calculated in case B of Table 4.2. Unfortunately, the discrepancies in the estimates are not only undetermined but probably indeterminant in the foreseeable future. Because of definitional questions, it is impossible to use Ronov's proportions of rock types—the only extensive measurements—without also using his mean analyses.

4.2.3 Sulfur Fluxes and Their Balance

Estimates of fluxes between reservoirs are listed in Table 4.4 and Fig. 4.3. The second column lists present fluxes, but only insofar as they can be more or less directly ascertained. The "measured" preanthropogenic values (column 3) are given as a range that can be deduced from the present values as modified by the absence of human activity and by the probable range of geologic factors operating during the last billion years, among which the "direct estimate" (column 4) is the best mean choice from this range.

In spite of intensive investigation, the atmospheric cycle of sulfur is still incompletely understood (Meyer, 1977). From the geologic standpoint, a basic difficulty is that the atmosphere is far from filling any useful definition of a reservoir, and it does not help much to split it into upper and lower, northern and southern, or oceanic and continental. A calculated approximate mean residence time of 1 month tells us only that whatever amount of sulfur goes up must come down, without storage. Sulfur may occur in the atmosphere in a variety of forms: droplets of sea-salt spray or of acidic sulfite/sulfate solutions, crystalline sea salt (gypsum?), $(NH_4)_2SO_4$, gaseous SO_2, SO_3, H_2S, or dimethyl sulfide ($CH_3SO_2CH_3$). Transfers and exchanges among these phases continually shift atmospheric burdens of sulfur. Volcanic SO_2 that is blasted to the stratosphere and oxidized to sulfate may not settle out for years, but an overwhelming proportion of the flux of SO_2 emissions from fuel burning and smelting returns to the surface within a few days and within a couple of hundred kilometers of the source. Removal of sulfur is not only through precipitation but also via the less easily and less commonly measured dry particulate fallout and finally as direct adsorption of SO_2 or other gases. The balance of these inputs and outputs has been elusive and has generally shown a large deficit of input to the atmosphere, which has been balanced by postulating emissions of H_2S from coastal areas (Hitchcock, 1976; Jørgensen et al., 1978) or dimethyl sulfide from the ocean surface. The amount of these additional inputs suggested for balance has decreased over the years with increasing recognition of the importance of anthropogenic and volcanic sources. The atmospheric section of Table 4.4 is

TABLE 4.4. Fluxes in the Sulfur Cycle[a]

Flux	Present	Preanthropogenic, Long Term		
		Measured	Direct Estimate	Balanced Estimate for Phanerozoic
Atmospheric Subcyle				
F_{SA}	1.38[b]	1.38	1.38[c]	1.38
F'_{CA}	2.03[b]	—	—	—
F_{AS}	0.91[d]	—	1.24[c]	1.24
F_{AF}	1.67[b]	—	0.14[c]	0.14
Main Exogenic Subcycle				
F_{SC}	0.19[e]	0.2–0.8[f]	0.55[f]	0.55
F_{SE}	0	0–0.3[g]	>0.07[g]	0.18
F_{CF}	1.88[h]	0.8–1.2[i]	0.98[i]	0.45
$F'_{IC/F}$	1.03[j]	—	—	—
F_{EF}	0.63[h]	0.25–0.38[i]	0.31[i]	0.17
F'_{EF}	0.88[k]	—	—	—
F_{FS}	5.41[l]	0.7–16[m]	1.43[n]	0.89
Oceanic Subcycle				
F_{SO}	0.045[f]	0.02–0.5[f]	0.07[f]	0.07
F_{BS}	0.065[o]	0.02–0.2[o]	0.065[o]	0.07
F_{MB}	0.23[o]	0.1–0.7[o]	0.23[o]	0.23
F_{MU}	1.38[o]	0.5–4.[o]	1.38[o]	1.38
F_{OM}	—	0.4–3.5[p]	0.07[p]	0.07
F_{OI}	—	—	—	0
F_{OV}	—	—	—	0
F_{BM}	—	0.06–0.5[p]	0.16[p]	0.16
F_{UM}	—	0.5–4.[p]	1.38[p]	1.38
Metamorphism/Volcanism Subcycle				
F_{CI}	—	—	—	0.02
F_{CV}	—	—	—	0.07
F_{EI}	—	—	—	0
F_{EV}	—	—	—	0.03
F_{IF}	—	—	—	0.02
$F_{V/FA}$	0.14[q]	0.1–1.0[r]	0.2	0.1

[a] Units are 10^{12} moles S/yr. See Fig. 4.3 for definitions; anthropogenic F values are primed.
[b] Granat et al. (1976). Meyer (1977) succinctly compared the full atmospheric cycles of Granat et al. (1976) with five earlier estimates of the cycle. The values given in Table 4.4 are principally cyclic sea-salt spray and exclude natural cryptic emission of H_2S (F_{CA}) and absorption of SO_2 (F_{AC}), which has traditionally been calculated by difference so as to balance the atmospheric sulfur cycle (Granat et al., 1976, and references therein). This H_2S is thought to be released to the atmosphere by bacterial reduction in anoxic coastal sediments, and new measurements (Jørgensen et al., 1978; Jaeschke et al., 1980) substantiate the local efficacy of such a source. However, it is difficult to estimate the importance of coastal H_2S for the worldwide atmospheric sulfur cycle. New extensive and detailed measurements of SO_4/Cl ratios in rainfall in isolated oceanic areas indicate that most of this sulfur comes from seawater. Kroopnick (1977) found a mean weight ratio of $SO_4/Cl = 0.20 \pm 0.11$ compared to 0.14 in seawater. Duce and Hoffman (1977) point out that such comparisons are more appropriately made in terms of ratios to sodium, because of gaseous reactions

involving chlorine that reduce Cl/Na by about 8%. Consequently, Kroopnick's measured levels of sulfate are even more like those of seawater than he thought. Therefore, if anoxic coastal sediments do generate substantial H_2S at the present time, the resulting SO_2 or SO_4^{2-} must be returned to freshwater or seawater reservoirs near to the source rather than entering a worldwide cycle. Coastal H_2S may have been important in some earlier geologic times when shallow seas were transgressive on the continental platforms. At present, atmospheric fluxes are out of balance in the opposite direction. See also Petrenchuk (1980).

^{c}Sea spray salts, of which 10% falls on the continents; this estimate by Eriksson (1960) should be re-examined, especially to take account of the data of Kroopnick (1977) and Petrenchuk (1980).

^{d}Recent extensive data on the sulfate content of oceanic rainfall measured by Kroopnick (1977) weighted by amount of rainfall represented by each sample ($= 1.80$ ppm S), normalized to mean oceanic rainfall (Holland, 1978, p. 60). An earlier estimate by Junge (1963) that has been adopted by all subsequent compilers, including Granat et al. (1976), is more than twice as large, but was admittedly based on sparse data. Following Friend (1973) the precipitation is increased by 20 percent to allow for dry fallout.

^{e}Berner, 1972.

^{f}From an unpublished compilation by Holser of DSDP and other recent data.

^{g}Range of rates of deposition of evaporites taken by geological periods, and mean rate for Phanerozoic (Holser, et al., 1980). Of course, the short-time variations of evaporite deposition are much larger because the process is markedly episodic. The value is a minimum, as it is based on evaporites now present, not allowing for those deposited and since eroded.

^{h}Calculated by Holland's (1968, p. 96) method, using the same (Holeman, 1968) present rate of erosion but with sulfur contents modified according to Table 4.3. Previous estimations of the partition of weathering between sulfide and sulfate were: a) by a roundabout geochemical calculation that gave the same answer of 23% sulfate (Berner, 1971); and b) by an arbitrary assumption of 50% sulfate (Holland, 1978, p. 111).

^{i}Range and mean of two similar recycling rates, fitted by Veizer and Jansen (1985) to distribution with time of areas of sedimentary rock, multiplied by present masses of sulfur (Table 4.1).

^{j}From production of sulfuric acid and sulfur (Shelton, 1978), manufactured from pyrite, and as metallurgical and petroleum by-products, assuming all end products (mostly fertilizer) are eventually exposed and dissolved.

^{k}From 1976 production of gypsum (Pressler, 1978) and native sulfur (Shelton, 1978), assuming all end products (mostly fertilizer plant tailings, plaster, etc.) except gypsum in Portland cement are eventually exposed and dissolved.

^{l}Holland (1978, p. 96); previous lower estimates (e.g., Friend, 1973) were based on same river analyses but lower runoff—both from Livingstone (1963). The analyses are old and should be updated.

^{m}Assuming same sulfur contents (Table 4.1b) and range of rates of denudation among the continents at present time (Holland, 1978, p. 87).

^{n}A sum of $F_{AF} + F_{CF} + F_{ER}$. A similar result (1.81×10^{12} moles) may be derived by the ratio of Gregor's (1970) average denudation rate of the continents to the present rate of denudation (g).

^{q}Cadle (1980) has critically discussed most recent estimates, and the value adopted is representative of those estimates based on either direct measurement of volcanic gases or on production of volcanic material. Other measurements by Taylor and Stoiber (1973) and by Johnston (1980) are similar to those considered by Cadle (1980). However, recent measurements suggest to Stoiber (Cadle et al., 1979) that the above level may be low by an order of magnitude, and this argument is supported by data from Mt. St. Helens (Casadevall et al., 1981).

^{r}Range to take account of the range of estimates of present volcanic output (footnote q) and range of volcanicity through geologic time (Ronov, 1976).

^{o}Difference of the sulfur contents of unaltered and altered deep-sea basalt, multiplied by a present rate of sea-floor generation of 3 km^2/yr (Wolery and Sleep, Chapter 3). For unaltered basalt the content of 0.085% sulfur is a grand mean of eight means of glassy rims of oceanic pillow basalts: Moore and Fabbi (1971), Keays and Scott (1976), Mathez (1976), Delaney et al. (1978), Garcia et al. (1979), and Muenow et al. (1980). For altered basalt the concentration of 0.065% sulfur is a grand mean of six means of altered whole rock oceanic basalts: Moore and Fabbi (1971), Keays and Scott (1976), Moore and Lewis (1976), Bachinski (1977), Aumento (in Naldrett et al., 1977), and Andrews (1979). Additional data (Naldrett et al., 1978; Puchelt and Hubberten, 1980) suggests that this grand mean is even somewhat high.

[p] Sulfur contents of oceanic sediment or basalt (Table 4.3) multiplied by a subduction rate of oceanic crust assumed equal to generation rate.

simplified by assuming that the H_2S flux to the continent is negligible, leaving mainly pollution and sea spray (and volcanism—see below) as inputs and rainfall (including dry fallout) as output.

The main subcycle is the depositional–erosional system, with the departure that sediments on the oceanic crust (as approximately defined by the abyssal plain and continental rise) are separated. The *present* rate of sulfide deposition in marginal sediments, F_{SC}, is poorly known. Berner (1972) summarized available data and obtained 0.27×10^{12} moles S/yr. A time-averaged rate of sulfide accumulation for the Phanerozoic of 0.55×10^{12} moles S/yr was estimated by comparing Ronov's (Ronov et al., 1974) sulfur contents and sedimentation rates (for platform and geosynclinal sediments).

Essentially no evaporites are forming today, as has seemingly happened in past periods such as the Ordovician. Berner (1972) proposed that seawater sulfate is presently grossly out of balance, with its concentration increasing because seawater receives much more input from river water (F_{FS}) than it deposits (F_{SC}, F_{SO}). Rates of sulfate deposition in Table 4.4 are based on a new census of known evaporite rocks of all ages (Holser et al., 1980; Zharkov, 1981). Evaporitic sulfate is mainly of marine origin (although a few examples of non-marine deposits exist; Holser, 1979). It occurs in a variety of paleographic situations: prograding sabkha shorelines, lagoons, and in deep basins of cratonic, Mediterranean, or rift-valley origin, where the brine may have varied from hundreds of meters to centimeters deep. All these environments are very restrictive and are created by specific and short-lived tectonic events. Once precipitation begins, it proceeds rapidly, typically at a rate of about 70 mg S/cm^2 · yr (Holser, 1979, p. 266), three orders of magnitude faster than sulfide precipitation in muds. Available basins are quickly filled, especially if halite precipitation is involved. But the special conditions necessary for evaporite concentration generally do not last long—hundreds of thousands of years—before tectonic conditions again shift to change shorelines, barriers, and basinal circulation to either fully marine or nonmarine conditions. While evaporite deposition *is* going on, a large fraction of the sulfate in the affected seawater (90% by the beginning of the halite facies; Holser, 1979, p. 245) is precipitated. Rees (1970) modeled the sulfur cycle with the ad hoc assumption that sulfate deposition is proportional to sulfate concentration in the ocean. Whereas this is a likely consequence for the particular seawater that enters the evaporite regime, the actual occurrence of such regimes, as controlled by tectonics, is a secular variable.

Sulfide has been deposited much more widely in time and place than sulfate, but the range of rates is hard to estimate. Ronov et al. (1974) indicate a variation of one order of magnitude in the time-rate of accumulation of C_{org} averaged over intervals of $7-40 \times 10^6$ years through the entire Phanerozoic, which would translate to a similar variation in the sulfide based on an assumed relative constancy of S/C_{org}. On a shorter time scale, the deposition of sulfide and C_{org} in black muds of both the deep ocean and the continental shelves may also have been increased sharply in short *anoxic events* with changes of oceanic circulation and oxygen content (e.g., Arthur, 1979). Analyses of C_{org} contents of sediments deposited during the past 100×10^6 years vary with time by a factor of 1000 when averaged over an ocean and by a factor of 100 when averaged over the world (Southam and Hay, 1981, Fig. 4-6).

Tectonic uplift, necessary to expose both sulfide and sulfate sediments to recycling (Holland, 1978, p. 146), has varied substantially through geologic time (Ronov et al., 1974; Ronov, 1976, Vail et al., 1977; Southam and Hay, 1981). The assertion that evaporites should be eroded faster than other sediments (Garrels and Mackenzie, 1969) would suggest a bias in the amounts of remaining evaporite sulfate toward more recent periods, which would lead to underestimating the amount of sulfate deposited in the Phanerozoic. But although evaporite deposition is sharply episodic both within and between geologic periods, no long-term trend is apparent in the remaining distribution among the periods of the Phanerozoic (Holser et al., 1980). Consequently, the range of rates of sulfate deposition (F_{SE}, column 3, Table 4.4) is probably of the correct order of magnitude.

Interactions with mid-ocean ridges are a potential source or sink for oceanic sulfur. Circulation of seawater in sea-floor hydrothermal systems on the one hand may deliver sulfur to the oceanic crust by precipitation of anhydrite because of its retrograde solubility or by precipitation of iron sulfide by reduction with the ferrous iron of silicates; on the other hand, it may leach sulfide from the intruded basalts. Perhaps each of these processes is active at some time or place, but for the sulfur cycle the question is whether the total flux of sulfur in either direction is significant. Specifically, Edmond et al. (1979b) measured a sharply decreasing level of sulfate in approaching the Galapagos deep-sea hydrothermal vents, which indicated complete removal of sulfate from the circulating seawater at $380°C$. They extrapolated this result to infer a flux of 3.8×10^{12} moles/yr of sulfur into mid-ocean ridges worldwide. Andrews (1979), Holser et al. (1979, p. 10), and Wolery and Sleep (Chapter 3, Section 3.4.4) have argued against massive transfer of sulfur from seawater to ridge basalts. The points they make are that major seawater circulation for even a few million years would have reduced the sulfate concentration of seawater to a value corresponding to the very low solubility of anhydrite at high temperature, and if the sulfate underwent reduction it would have resulted in oxidation of all the FeO in several kilometers of crust. It remains to be seen whether sulfide and sulfate found at greater depth in oceanic basalts (Alt et al., 1986) indicate a significant flux. The analytical data appear to indicate transfer of sulfur from oceanic crust to seawater rather than the reverse (Table 4.4, footnote o). This amount of sulfur added to the sea by alteration of basalt is an appreciable but not dominant component of the seawater balance (Fig. 4.3).

The amount of sulfur contributed to oceanic sediments by bacterial reduction is even more problematic than the corresponding flux of organic carbon, because of a lack of analyses of sulfide in these sediments. As for carbon, the ultimate fate of sulfur in the mantle or in volcanic emanations is also an open question. It is clear that magmas are at least saturated with sulfur shortly before eruption (Mathez, 1976; Delaney et al., 1978) and that in subaerial volcanoes most of this sulfur is lost to the atmosphere (Cadle, 1980) or adsorbed on ash (Taylor and Stoiber, 1973). Whether most of this sulfur comes from metamorphism of continental sediments (as assumed in Fig. 4.3) or from subducted oceanic sediments is completely unknown. Thus, although a steady-state balance for the deeper cycle of sulfur is *allowed* by the data (Table 4.4 and Fig. 4.3), just as it is for carbon (Table 4.3 and Fig. 4.1), the facts are so inadequately known that either system may actually be significantly out of balance.

4.3 ISOTOPIC FRACTIONATIONS IN THE GEOCHEMICAL CYCLES OF CARBON AND SULFUR

Many elements entering geochemical cycles undergo significant isotopic fractionation because of the general dependence of their thermodynamic and kinetic properties on their masses and quantum characteristics. An understanding of these fractionations is important for the interpretation of isotopic data and their relevance to earth history. Thus, at this point, we consider carbon and sulfur isotopic fractionation in some detail. Fractionations are either *thermodynamically controlled* (for processes at equilibrium) or *kinetic* (with different isotope species reacting at different rates). In the case of carbon and sulfur, such redistributions principally affect the ratios $^{12}C/^{13}C$ and $^{32}S/^{34}S$ whose crustal averages are close to 89.4 and 22.2, respectively. Isotope fractionations between two substances A and B may be conveniently expressed by a fractionation factor K:

$$K = \frac{(^{12}C/^{13}C)_B}{(^{12}C/^{13}C)_A}.$$

In a thermodynamically controlled equilibrium process, K is the temperature-dependent *equilibrium constant*, whereas in a unidirectional reaction, it reflects the *kinetic isotope effect*, which equals the ratio k_1/k_2 of the rate constants for the reactions of the individual isotope species:

$$^{12}C_A \xrightarrow{k_1} {}^{12}C_B; \quad {}^{13}C_A \xrightarrow{k_2} {}^{13}C_B$$

If $k_1 \neq k_2$, the isotopic composition of product B is necessarily different from that of source material A.

Differences in the isotopic compositions of various substances are usually reported in terms of the δ-notation, expressing the content of the less abundant isotope in a sample (sa) as per mill (‰) difference relative to a standard (st); for instance,

$$\delta^{13}C_{sa} = \left(\frac{(^{13}C/^{12}C)_{sa}}{(^{13}C/^{12}C)_{st}} - 1\right) \times 10^3 \text{ [‰]}.$$

In the case of carbon, the usual standard is Peedee belemnite (PDB) with $^{12}C/^{13}C = 88.99$. For sulfur, it is Canyon Diablo troilite (CDT) with $^{32}S/^{34}S = 22.22$. The $\delta^{13}C$ and $\delta^{34}S$ values of these standards are 0.0‰ by definition.

The geochemical cycles of carbon and sulfur are outstanding examples of element cycles involved significantly with the earth's biosphere. It is well known that the principal isotopic fractionations in these cycles are biologically mediated. Because biochemical reactions are mostly enzyme controlled, it is generally believed that the observed fractionations are the result of kinetic rather than equilibrium effects. There have been claims, however, that the inter- and intramolecular carbon isotope distributions in biologic systems often display regularities approaching those of thermodynamically ordered states (Galimov, 1980). Because the biologic effects are preserved in sediments with but minor alterations, the isotopic fractionations are faithfully recorded. Biologic processing of carbon and sulfur through the ages has, accordingly, left an isotopic signature on the crustal inventories of both elements that can be traced back to almost the beginning of the rock record (Section 4.6.1).

4.3.1 Isotopic Fractionations in the Carbon Cycle

The earth's primordial carbon is generally believed to have a $\delta^{13}C$ value close to -5‰ as obtained from analysis of the principal forms of carbon of presumed mantle derivation, that is, diamonds, carbonatites, and carbonate constituents of kimberlites (Deines and Gold, 1973). This average comes close to the values for the bulk of carbonaceous meteorites (Smith and Kaplan, 1980; Libby, 1971) and to the average obtained for the inventory of exogenic reservoirs (Table 4.5). Isotopic fractionation during subsequent geochemical cycling of this primordial carbon has resulted in the formation of crustal reservoirs of varying isotopic composition.

4.3.1.1 Biologic Fractionations

The most important fractionations in the earth's carbon cycle are linked to processes of photosynthetic carbon fixation, notably C3 or *Calvin cycle* photosynthesis employed by the majority of higher plants, algae, and autotrophic bacteria. The biochemical background of carbon isotope fractionation in different photosynthetic and chemosynthetic pathways has been discussed by numerous investigators (e.g., Wong et al., 1975, 1979; Benedict, 1978; O'Leary, 1981). Most of the isotopic fractionayon produced by the C3 pathway occurs in the enzymatic carboxylation of CO_2 by ribulose-1.5-bisphosphate (RuBP)

TABLE 4.5. Balance of Carbon Isotopes in the Exogenic Cycle

Author	Mass (10^{18} moles)		$\delta^{13}C$ (‰) (PDB)		
	Reduced	Oxidized	Reduced	Oxidized	Total
Li (1972)	930	3712	−25	0	−5.0
Garrels and Perry (1974)	1042	5083	−26	0	−4.4
Galimov et al. (1975)	1000	5170	−27	−0.4	−4.7
Garrels and Lerman (1981)	1300	5200	−26	0	−5.2
This chapter	1180	6460	−27[a]	−0.4[a]	−4.5

[a] The values of Galimov et al. (1975) were assumed, inasmuch as these values were determined on aggregate samples from Ronov's material, which is also the main basis for the reservoirs of Tables 4.1a and 4.1b.

carboxylase. Fractionations attributable to this step typically amount to −25 to −35‰. Other plants follow an alternate reaction sequence, referred to as the C4 pathway. C4 plants are all angiosperms and thus this pathway developed relatively recently in the earth's history. They exhibit the Krantz-type leaf anatomy (Reibach and Benedict, 1977) and produce much less of a fractionation than do C3 plants (Fig. 4.4). Another group of plants employing the C4 pathway, those with Crassulacean acid metabolism (CAM), can also operate in the C3 mode and so produce a wide range of fractionations intermediate between C3 and C4 plants (Fig. 4.4).

In contrast to the relative wealth of data available for the principal photosynthetic pathways, few figures have been reported for isotope fractionations in the less common (and less understood) assimilatory sequences. Preliminary work on green sulfur bacteria, presumably operating the reductive carboxylic acid cycle unique to photosynthetic prokaryotes, has yielded fractionations of about −12‰ between source carbon dioxide and resulting cell material (Sirevåg et al., 1977). Data pertaining to isotope discriminations by various methanogens (cf. Fig. 4.4) have been reported recently by Fuchs et al. (1979) and Belyaev et al. (1983).

Figure 4.4 shows the isotopic compositions of some major groups of extant higher plants, algae, and autotrophic bacteria as compared to the terrestrial "feeder" pools of atmospheric CO_2 and marine HCO_3^-. Because the C4 syndrome has arisen in angiosperms no earlier than late Mesozoic times, and with C4 plants constituting only a fraction of the total angiosperm population confined mainly to tropical and subtropical regions, the small fractionations brought about by this group have not exercised a major impact on the carbon cycle as a whole. Accordingly, the fossil carbon record basically displays the isotopic signature of C3 (or Calvin cycle) photosynthesis (Section 4.6).

4.3.1.2 Inorganic Fractionations

Compared to biologic fractionations, isotope discriminations imposed on the carbon cycle by equilibrium processes are of minor importance. Thermodynamically

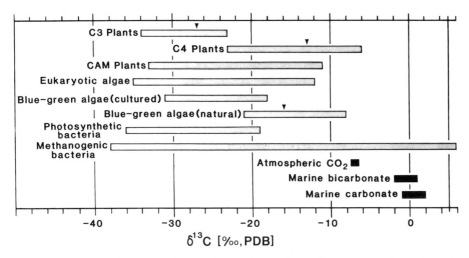

Figure 4.4 Isotopic composition of higher plants, algae, photosynthetic bacteria and chemosynthetic bacteria (methanogens) compared to the inorganic feeder reservoirs of atmospheric CO_2 and marine HCO_3^- (adapted from Schidlowski et al., 1983). Approximate means for some groups are indicated by black triangles. The average biomass is depleted by about $25 \pm 5‰$ in heavy carbon (^{13}C) as compared to marine bicarbonate, the main reservoir of inorganic carbon in the environment. $\delta^{13}C$ ranges for individual groups according to Behrens and Frishman (1971), Benedict (1978), Calder and Parker (1973), Belyaev et al. (1983), Fuchs et al. (1979), O'Leary (1981), Pardue et al. (1976), Seckbach and Kaplan (1973), Sirevåg et al. (1977), Smith and Epstein (1971), Wong and Sackett (1978), and Wong et al. (1975). The higher fractionations for cultured blue-green algae (*cyanobacteria*) as compared to natural communities were obtained at partial pressures of $CO_2 > 0.5\%$. The positive part of the spectrum accrued for methanogens was obtained at very low P_{CO_2} where preferential conversion of CO_2 to CH_4 with a strong bias in favor of ^{12}C is assumed to give rise to unusually heavy residual CO_2 subsequently incorporated into cell material (Fuchs et al., 1979). All fractionations observed in culture experiments have been recalculated for atmospheric carbon dioxide ($\delta^{13}C = -7‰$) as the carbon source to facilitate comparison with ranges obtained from natural communities.

controlled exchange equilibria primarily govern the isotopic compositions of the various oxidized carbon species in the exogenic compartment of the cycle, notably in the system

$$CO_{2(gas)} \leftrightarrow CO_{2(aq)} \leftrightarrow HCO_3^- \leftrightarrow CO_3^{2-},$$

which constitutes the link between atmospheric carbon dioxide and sedimentary carbonate. Following the pioneering studies by Craig (1953), isotope exchange equilibria in these reactions have been investigated by several workers (e.g., Vogel, 1961; Deuser and Degens, 1967; Wendt, 1968; Mook et al., 1974). Over the temperature range of 0–30°C, CO_2 dissolved in water ($CO_{2(aq)}$) is enriched in heavy carbon (^{13}C) between 1.0 and 1.2‰ compared to the gaseous phase, whereas fractionations between $CO_{2(aq)}$ and HCO_3^- result in a 7–11‰ increase of ^{13}C in bicarbonate ion (Mook et al., 1974, Table 4). Fractionation between bicarbonate and carbonate is very small, causing an increase in heavy carbon of only 1‰ in the carbonate phase (CO_3^{2-}).

Comparison of the experimentally obtained values with isotopic compositions of oxidized carbon phases in nature shows fair agreement: the δ^{13}C value of atmospheric carbon dioxide is about -7‰, the average for oceanic bicarbonate lies between -1 and 0‰, and the mean for carbonate is shifted by perhaps one more per mill in the positive direction (Fig. 4.4). Accordingly, carbonate rocks have essentially the isotopic composition of the parent marine pool of bicarbonate, thus monitoring the state of the oxidized carbonate reservoir through time (Keith and Weber, 1964; Schidlowski et al., 1975).

Equilibrium fractionations also control isotopic exchange among the carbon phases in magmatic and related high-temperature systems (Bottinga, 1969; Ohmoto and Rye, 1979). However, because such systems represent side stages of the global carbon cycle, the geochemical relevance of these fractionations is rather limited. A quantitatively more important high-temperature process is the isotopic reequilibration between sedimentary carbonate and organic carbon in metamorphism. The pronounced temperature dependence of the isotope exchange between coexisting calcite and graphite has made this system a convenient and rather sensitive geologic thermometer (Valley and O'Neil, 1981). The magnitude of fractionation between the two phases decreases with increasing metamorphic temperature, with complete equilibrium usually attained at $t \gtrsim 650°C$ in the granulite facies (Fig. 4.5). Knowledge of these equilibria is crucial for a conclusive intepretation of those parts of the oldest carbon isotope record that bear a metamorphic overprint (Section 4.6).

4.3.2 Isotopic Fractionations in the Sulfur Cycle

The earth's primordial sulfur is believed to have had an isotopic composition like that in meteoritic sulfur, or $\delta^{34}S = 0$ relative to the Canyon Diablo troilite (CDT) standard (Nielsen, 1978). Mantle-derived rocks such as oceanic basalts usually have sulfur close to this value.

4.3.2.1 Biologic Fractionations

As in the case of carbon, the geochemically important fractionations in the sulfur cycle are exclusively biologic, taking place primarily during *assimilatory* and *dissimilatory* reduction of inorganic sulfate. In assimilatory reduction, the resulting sulfur is incorporated into the cell, whereas in dissimilatory reduction, sulfate is used as an oxidizing agent and is not permanently incorporated. Whereas the assimilatory pathway entails isotope discriminations on the order of only a few per mil, fractionations in dissimilatory sulfur metabolism are large. Accordingly, the decisive control of the isotopic geochemistry of crustal

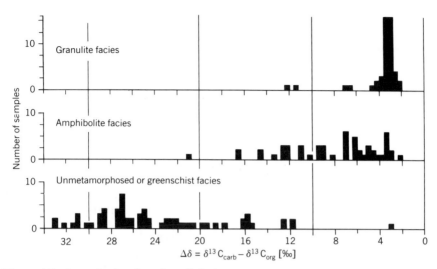

Figure 4.5 Isotopic fractionation ($\Delta\delta$) between coexisting sedimentary carbonate (calcite) and organic carbon (graphite) as a function of metamorphic grade. Note that the magnitude of fractionation decreases with increasing metamorphic temperature (amphibolite facies: 450–650°C; granulite facies: ~650°C). The narrow spread of values in the granulite facies indicates that thermodynamic equilibrium is closely approached at this grade. Adapted from Valley and O'Neil (1981).

sulfur is by the process of dissimilatory sulfate reduction, which is performed by a small group of sulfate-reducing bacteria (Baas-Becking, 1925; Postgate, 1979; Trüper, 1982). Bacterial sulfate reduction is the principal agent responsible for low-temperature conversion of sulfate (SO_4^{2-}) to sulfide (S^{2-}) in the exogenic system (Section 4.2.1). The reduction of sulfate constitutes a crucial link in the sulfur cycle but has never been conclusively demonstrated to proceed at temperatures less than 150°C unless biologically mediated (Chambers and Trudinger, 1979). The process releases large quantities of hydrogen sulfide (H_2S) to near-surface environments, some of which ends up as sedimentary sulfide (mostly pyrite). Hence, dissimilatory sulfate reduction is ultimately responsible for the observed sulfate–sulfide dichotomy of sulfur in the crust.

The pathway of dissimilatory sulfate reduction may conveniently be summarized in four principal steps (Trudinger, 1969; Roy and Trudinger, 1970; Rees, 1973):

$$SO_4^{2-}(\text{ext}) \longrightarrow SO_4^{2-}(\text{int}) \xrightarrow{\text{ATP}} \text{APS} \xrightarrow{2e^-} SO_3^{2-} \xrightarrow{6e^-} S^{2-}. \quad (4.3)$$
$$\quad\quad (1) \quad\quad\quad\quad (2) \quad\quad\quad (3) \quad\quad\quad (4)$$

After the uptake of external sulfate in step (1), it is activated by expenditure of ATP (2). The activated species (APS, adenosine-5-phosphosulfate or adenylyl

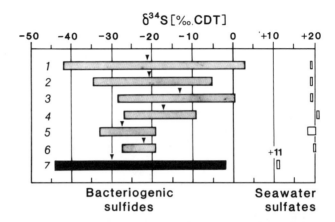

Figure 4.6 Isotope fractionation between bacteriogenic sulfide and seawater sulfate in present-day anoxic marine sediments (stippled) and in a fossil euxinic facies (Permian Kupferschiefer, black). Means for groups are indicated by black triangles. $\delta^{34}S$ values of modern seawater sulfate are close to 20‰ (Pacific, California: 19.7; Spencer Gulf, Australia: 21.0; Black Sea: 18.2–20.2; Baltic Sea: 20.0). Numbered occurrences: (1) Pacific basins off California (Kaplan et al., 1963; Sweeney and Kaplan, 1980); (2) Pescadero Basin, Gulf of California (Goldhaber and Kaplan, 1980); (3) Carmen Basin, Gulf of California (Goldhaber and Kaplan, 1980); (4) Spencer Gulf, Australia (Chambers, 1980); (5) Black Sea (Vinogradov et al., 1962); (6) Kiel Bay, Baltic Sea (Hartmann and Nielsen, 1969); (7) Kupferschiefer, Central Europe (Marowsky, 1969; note that $\delta^{34}S$ for Permian seawater sulfate is 11‰).

sulfate, respectively) is then enzymatically reduced to sulfite by APS reductase (3). Subsequent reduction of SO_3^{2-} by sulfite reductase in step (4) yields sulfide as the end product of the reaction series. Over the whole process, the oxidation state of sulfur changes from $6+$ to $2-$, necessitating a transfer of eight electrons from an external (organic) donor.

Inherent in these reactions are kinetic isotope effects of variable magnitude that have been shown to sum up to a negative change in $\delta^{34}S$ of $-46‰$ in culture experiments and as much as $-60‰$ in the natural environment; that is, the $\delta^{34}S$ values of bacteriogenic sulfide are markedly displaced to the negative side of the scale as compared to those of the parent sulfate pool (Fig. 4.6). Because the bias in favor of light sulfur (^{32}S) in this reaction series is the sum of the discriminations inherent in the individual reactions, the magnitude of the isotope shift may vary depending on which of steps (1)–(4) becomes dominant or rate controlling in the specific instance (see Kaplan and Rittenberg, 1964; Kemp and Thode, 1968; Rees, 1973; Chambers and Trudinger, 1979). Although kinetic isotope effects assigned to the single reactions are still subject to debate, there is widespread agreement that discriminations inherent in steps (1) and (2) are on the order of a few per mill only. Values suggested for step (3) are $-25‰$ (Rees, 1973) and -10 to $-15‰$ (Chambers and Trudinger, 1979), whereas the effect associated with step (4) was found to range anywhere between 0 and −33‰.

Experiments performed with bacterial cells using organic electron donors and unlimited sulfate have shown that the isotope fractionations in the reaction series of Eq. (4.3) are a function of the rate of reduction, with the largest fractionations occurring at the slowest reduction rates. At sulfate concentrations less than 1 mM, a sharp decrease in the magnitude of fractionation has been noted, indicating that the small discrimination inherent in step (1) becomes rate controlling for the overall process. The direction of the effect may even be inverted to the positive side (toward ^{34}S enrichment in the sulfide) if concentrations drop below 0.01 mM (Harrison and Thode, 1958; McCready and Krouse, 1980).

In the natural environment, additional factors are imposed on the biochemical fractionations inherent in the pathway of Eq. (4.3). Being obligate anaerobes, desulfobacteria are confined to anoxic habitats, either euxinic basins where anoxia occurs in the water column or sediment pore waters where anoxia occurs a few centimeters below the sediment–water interface. In such environments, sulfate reduction may be coupled with sulfate depletion. Given a constant kinetic fractionation with preferential removal of light sulfur, the δ^{34}S values of both residual sulfate and late-formed sulfide are apt to increase with increasing extent of the reaction. Treatment of these relations in terms of a fractional distillation (Rayleigh) model permits the calculation of a theoretical ("apparent") fractionation factor $K = (^{32}S/^{34}S)_{\text{sulfide}}/(^{32}S/^{34}S)_{\text{sulfate}}$ for the specific sedimentary environment (Goldhaber and Kaplan, 1974). Using this and other approaches (see Chambers and Trudinger, 1979), fractionation factors between 1.015 and 1.045 have been obtained for sulfate reduction in modern sediments, implying that bacteriogenic sulfide is depleted in ^{34}S by about 15–45‰ as compared to the parent seawater sulfate. An *open system* model that makes allowance for diffusion of SO_4^{2-} into pore waters has yielded factors up to 1.060 (Goldhaber and Kaplan, 1980), which is about the maximum fractionation observed for sulfide in contemporaneous low-oxygen environments (Pacific basins off California, Black Sea; cf. Fig. 4.6).

Because some of this bacteriogenic H_2S is preserved as sulfide minerals (mostly pyrite), the δ^{34}S patterns of sedimentary sulfides broadly reflect the magnitude of sulfur isotope fractionation in bacterial sulfate reduction (Fig. 4.6). In general, bacteriogenic sulfide patterns show (1) a marked shift toward negative δ^{34}S values relative to the parent sulfate and (2) an extended spread of these δ^{34}S values. These criteria allow the identification of bacteriogenic sulfides in ancient sedimentary basins. Furthermore, the extent of the negative shift can be related to features of the depositional environment, particularly the rate of sedimentation (Berner, 1978; Maynard, 1981).

As pointed out above, biologic fractionations of sulfur isotopes other than in bacterial sulfate reduction are of minor geochemical importance. Assimilatory reduction of sulfate for primary synthesis of sulfur-containing cell constituents entails fractionations between 0 and −4.5‰ only (Kaplan et al., 1963; Kaplan and Rittenberg, 1964; Mekhtieva, 1971). Besides, the turnover rates of sulfur in assimilatory (sulfur-supplying) metabolism are smaller by several orders of magnitude than rates in dissimilatory (electron-supplying) reactions. Photosynthetic oxidation of reduced sulfur compounds (notably H_2S) by green and purple sulfur bacteria (Chlorobiaceae, Chromatiaceae) only yields fractionations of a few per mill for S^0 and SO_4^{2-} as final oxidation products (Kaplan and Rittenberg, 1964; Ivanov et al., 1976). Accordingly, the impact of this process on the isotopic record of sulfur cannot be dramatic (for possible exceptions in the Archean, see Section 4.6).

4.3.2.2 Inorganic Fractionations

The only thermodynamically controlled isotope exchange reaction quantitatively important for the crustal sulfur cycle is the equilibrium between dissolved and precipitated (evaporite) sulfate reflected by the constant

$$K = \frac{(^{32}S/^{34}S)_{\text{sulfate(aq)}}}{(^{32}S/^{34}S)_{\text{sulfate(solid)}}}.$$

Values obtained for K mostly range between 1.001 and 1.0016, corresponding to about 1–1.6‰ ^{34}S enrichment in evaporite sulfate with respect to the parent brine (Thode and Monster, 1965; Holser and Kaplan, 1966). Accordingly, the isotopic composition of sulfate in seawater is preserved in sulfate evaporites with but minor change.

Equilibrium fractionations also govern the isotope geochemistry of sulfur in hydrothermal and related high-temperature systems (Ohmoto, 1972; Ohmoto and Rye, 1979). Equilibrium fractionation between sulfate and hydrogen sulfide results in enrichment of ^{32}S in the reduced phase that increases from about 10‰ at 600°C to 74‰ at 25°C, whereas fractionations achieved in bacterial sulfate reduction at 25°C are mainly within the range of 30–45‰ (Fig. 4.6). It should be pointed out that reduction of sulfate at temperatures of 300–400°C in hydrothermal systems of mid-ocean ridges is likewise associated with an equilibrium fractionation of sulfur isotopes (Shanks et al., 1981; Kerridge et al., 1983; see also Section 4.2.3).

4.3.3 Isotope Record of Carbon and Sulfur

We now turn to a consideration of how these fractionations appear in the rock record. In accordance with the fractionations discussed in the previous section, stable isotopes of carbon and sulfur are distributed among the reservoirs by the fluxes entering and leaving each reservoir. Therefore, an inventory of light and heavy isotopes of an element in all the crustal reservoirs today gives an estimate of the isotopic composition of that element in the entire crust. If there is independent information as to what this value should be, then any significant deviation of the inventory from the expected value will suggest that one or another number or assumption is incorrect. The isotope record of the past can also be interpreted under the assumption that the isotopes of the element have only undergone redistribution while their total inventory has remained constant through geologic time.

4.3.3.1 Isotope Balance for Carbon and Sulfur

The isotopic balance for carbon, as derived by several authors, is listed in Table 4.5. $\delta^{13}C$ for total carbon of the exogenic cycle varies only from -4.4 to -5.2‰ among the estimates of the various authors. The value derived here is near the less negative end of this range and compares favorably with some estimates of the isotope composition of mantle carbon. Deines (1980a,b) found a mode of $\delta^{13}C = -5$ to -6‰ in diamonds, and Galimov and Gerasimovskiy (1978) found $\delta^{13}C = -2.5$ to -3.3‰ in Icelandic magmatic rocks.

The corresponding balance for sulfur isotopes is given in Table 4.6. The mean $\delta^{34}S$ ranges from -4.7 to 8.9‰, but the most reasonable values are in the range 1–3‰, similar to the range of values determined for what may be primary mantle sulfur: 1‰ in basalts from the Mid-Atlantic Ridge (Kanahira et al., 1973), -1.3 to 2.3‰ in ultramafics, -0.8 to 1.6‰ in basalts from the Indian Ocean (Grinenko et al., 1975), and -0.9 to -0.3‰ in unaltered basalts from the western North Atlantic (Puchelt and Hubberten, 1980).

TABLE 4.6. Balance of Sulfur Isotopes in the Exogenic Cycle

Author	Mass (10^{18} moles sedimentary S)		$\delta^{34}S$ Sediments (‰) (CDT)		
	Reduced	Oxidized	Reduced	Oxidized	Total[a]
Holser and Kaplan (1966)	84	162	-12	$+17$	$+8.9$
Li (1972)	231	149	-12	$+17$	$+1.3$
Grinenko et al. (1973)[b]	1.93	8.04	-11.7	$+13.5$	$+8.6$
Garrels and Perry (1974)	294	200	-12	$+17$	$+1.3$
Schidlowski et al. (1977)	184	198	-16	$+19$	$+3.8$
Nielsen (1978)[c] (A)	178	165	-15.2	$+17$	$+2.3$
(B)	178	41	-15.2	$+17$	-4.7
This chapter (A)	153	152	-7.9[d]	$+8.4$[d]	$+2.5$
(B)	180	126	-7.4[d]	$+12.1$[d]	$+2.9$

[a] Including about 40×10^{18} moles sulfate in seawater with $\delta^{34}S = +20$‰; except Grinenko et al. (1973).
[b] Russian platform only: rock data from Ronov; total does not include seawater.
[c] For definitions see Table 4.2.
[d] Mean values for two Caucasus geosynclines and for Russian platform (Ronov et al., 1974), each with proportions of sediment types measured by Ronov (1980) (see Tables 4.1a and 4.1b) and with proportions of tectonic types by Southam and Hay (1981) (see Tables 4.1a and 4.1b). For sulfide of oceanic sediments a mean of $\delta^{34}S = -44$‰ from Migdisov et al. (1983) and Puchelt and Hubberten (1980). Inclusion of more recent data on $\delta^{34}S$ of oceanic sediments (Scharff, 1980; Brumsack and Lew, 1982) would slightly lower $\delta^{34}S$ of reduced sediments and of total sediments.

In summary, although the presently available data do not reveal any glaring discrepancies in the isotope balance of carbon and sulfur in the exogenic cycle, many of the numbers need to be determined more precisely before this conclusion can be tied down.

4.3.3.2 Isotope Age Curves for Marine Sulfate and Carbonate

Variations of sulfur and carbon isotopes through geologic time are a measure of corresponding variations in the oxidation–reduction systems of these elements. That is, as the balance between oxidized and reduced species in a flux or in a reservoir change, the isotopic composition of the flux or the reservoir must also change. The sea is a reservoir of particular interest, and the sediments deposited from the sea exhibit variations in both sulfur and carbon δ values. For both sulfur and carbon, the oxidized inorganic deposits—sulfate in evaporites and carbonate in limestones—have been the most studied. One reason is that oxidized components are well mixed in the sea reservoir and hence reflect the composition of seawater. Both oxidized and reduced constituents provide important data, however, and the latter are also discussed in Section 4.3.3.3.

The isotope changes recorded in the gypsum and anhydrite of marine evaporites will be discussed first, because they exhibit well-documented and dramatic variations. It is a matter of observation that the geologic age of an evaporite deposit is the variable that best correlates $\delta^{34}S$ of its sulfates. Different marine evaporites of the same age have preserved the same $\delta^{34}S$, usually within $\pm 1‰$, regardless of mineralogy, facies, or tectonic setting (e.g., Holser and Kaplan, 1966; Claypool et al., 1980). At the same time, this commonality proves that the evaporites and their sulfates were of marine origin. Thus, the most important variations of $\delta^{34}S$ in marine rocks may be displayed by a *sulfur isotope age curve* that charts the variations of seawater sulfate through geologic time. Several versions of such age curves for Phanerozoic time have been published, notably by Holser and Kaplan (1966), Nielsen (1978), and Claypool et al. (1980). Figure 4.7 shows data recently compiled by Lindh (1983; see also Saltzmann et al., 1982), who used a statistical treatment of several thousand published analyses, aggregated at time intervals of 25×10^6 years. The values charted represent measured values in sulfate minerals, which as discussed in Section 4.3.2, should be about 1.3‰ more positive than the seawater from which they were deposited.

The most dramatic features of the sulfur isotope age curve are a minimum of about $\delta^{34}S = 11‰$ in the Permian and a maximum of more than 30‰ in the Cambrian. These extremes reflect substantial variation in the oxidized fraction of exogenic sulfur, although this reservoir is by no means homogeneous. The Precambrian record, although much less complete, shows a rise of $\delta^{34}S$ from near-primary values in the Archean to values in the Proterozoic that are within the same range as those for the Paleozoic, indicating that at some time between 3 and 1×10^9 years ago, the sulfur-reducing bacteria that are responsible for the isotope shift became prevalent enough to change the isotope level in the major reservoir of the sea (Section 4.6.2).

4.3 ISOTOPIC FRACTIONATIONS OF CARBON AND SULFUR

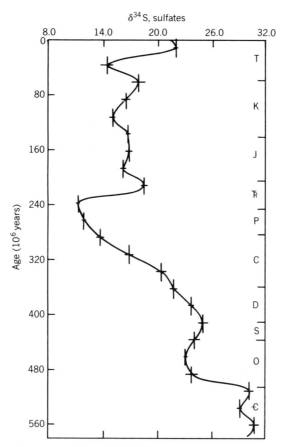

Figure 4.7 The sulfur isotope age curve for Phanerozoic time based on $\delta^{34}S$ of marine evaporite sulfate. The published single analyses were aggregated statistically at 25×10^6-year intervals. The bar indicates ± 1 standard error of the mean for each interval. Excursions to the left result from a decrease in the reduced fraction of sulfur reservoirs, to the right a corresponding increase. After Saltzman et al. (1982) and Lindh (1983).

The corresponding variations in $\delta^{13}C$ of marine carbonate with age are more subtle, as might be expected from the large size of the carbonate reservoir compared with the sulfate reservoir. In fact, reality of secular variations was in question until demonstrated conclusively by Veizer et al. (1980). Figure 4.8 shows the carbon isotope age curve for the Phanerozoic compiled on a basis similar to that of Fig. 4.7 (Lindh et al., 1981; Saltzman et al., 1982; Lindh, 1983). Note the enlarged scale of $\delta^{13}C$ relative to $\delta^{34}S$. The most prominent features of the curve are a minimum in the early Paleozoic and a high in Pennsylvanian–Permian time.

140 GEOCHEMICAL CYCLES OF CARBON AND SULFUR

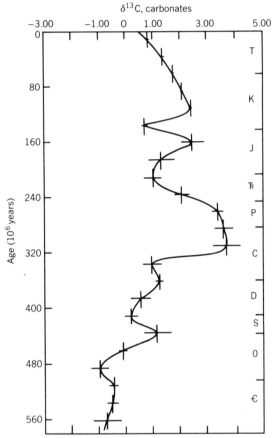

Figure 4.8 The carbon isotope age curve for Phanerozoic time $\delta^{13}C$ of marine calcite (whole rock or fossils). The published analyses and a large number of unpublished analyses were aggregated statistically at 25×10^6-year intervals. The bar indicates ± 1 standard deviation of the mean for each interval. After Lindh et al. (1981).

Fine structure of both the sulfur and carbon isotope age curves has been smoothed out by the statistical aggregation used in constructing Figs 4.7 and 4.8, but in many instances there are variations on a fine scale that have been documented by interbasin correlations. In this respect, an interesting and puzzling feature of the age curves is that they may undergo sharp excursions within a short time. A jump from $\delta^{34}S = 13‰$ to $27‰$ within the Early Triassic was labeled the *Röt chemical event* by Holser (1977) and has since been confirmed worldwide, as shown in Fig. 4.9. Most of this dramatic rise seems to have taken place within one-third of the Spathian substage, which probably lasted only $1-2 \times 10^6$ years altogether. Similar sharp changes have been recorded for $\delta^{13}C$ in the Cretaceous Period by Arthur (1979), who called them *anoxic events*,

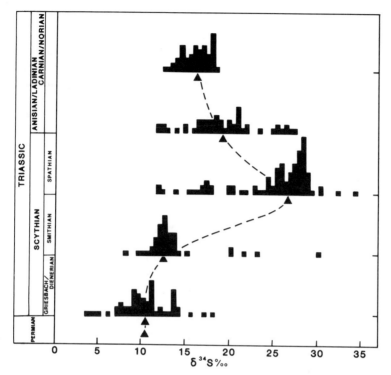

Figure 4.9 A sulfur isotope event in Early Triassic time. The dashed line connects the medians (black triangles) of the sample distributions. Earliest Triassic evaporite sulfates, from Griesbachian to Smithian substages, show a barely perceptible rise over the well-established low of $\delta^{34}S = 11‰$ in Late Permian time (Fig. 4.7). A jump to about $\delta^{34}S = 27‰$ occurred at the end of the Spathian substage—most of the values in the Spathian below 23‰ seem to precede the high insofar as can be determined by the stratigraphic correlations. The return to modern values of $\delta^{34}S = 17‰$ in Late Triassic time is more gradual. Note that the time scale for the approximate 7×10^6 years of Scythian time is magnified about seven times over the 25×10^6 years of later Triassic time. Sources of data are Sichuan: Chen et al. (1981); Israel, Germany, and Canada: Claypool et al. (1980); Alps: Cortecci et al. (1981); United States: Wilgus (1981); Greenland: Clemmensen et al. (1985).

and in the Permian Period by Magaritz (e.g., Magaritz et al., 1981, 1983). In the latter case, a rise in $\delta^{13}C$ of over 6‰ can be timed by varves to have spanned only a few thousand years.

4.3.3.3 Age Curves for Other Isotopes

Several other systems provide supplementary information that helps to understand the secular changes in the exogenic cycles of sulfur and carbon shown in Figs. 4.7 and 4.8. For example, the oxidation–reduction systems of sulfur seem to be paralleled approximately by that of cerium, as shown by variations of rare-earth patterns in fossil marine apatite (Wright et al., 1987). This parallelism suggests the ^{34}S enrichments of the sulfur isotope curve of Fig. 4.7 reflect increasing frequencies of basins with anoxic bottom water, whereas ^{34}S deficiencies reflect times in which anoxia was largely confined to interstitial water.

Age curves for the isotopes of organic carbon and sulfide are expected to parallel those for carbonate and sulfate, respectively, but at a lower level separated by a mean fractionation. The isotope geochemistry of reduced carbon was broadly surveyed by Deines (1980b); see also Section 4.6. Spotty data for $\delta^{13}C_{org}$ have been assembled into partial isotope age curves by Jackson et al. (1978), Galimov (1980), Schidlowski (1982), and Hoefs (1981). The data are apparently not yet precise enough to decide whether $\delta^{13}C_{org}$ tracks $\delta^{13}C_{carb}$ owing to variations in the relative fraction of carbon in the reduced reservoir with time (Schidlowski, 1982), or whether independent shifts of $\delta^{13}C_{org}$ result from changes in the nature of organic carbon such as the proliferation of land plants in the mid-Paleozoic (Welte et al., 1975; Galimov, 1980) and their distribution (Maynard, 1981; Yeh and Epstein, 1981). The small amount of available data for $\delta^{34}S$ sulfide range widely and seem to be so strongly dependent on the degree to which the reduction system is open and on the sedimentation rate (Ronov et al., 1974; Berner, 1978; Maynard, 1980; Scharff, 1980; Fisher and Hudson, 1987) that age trends are obscured.

Oxygen isotope ratios in sedimentary carbonates have generally been considered to reflect diagenesis or equilibration with formation waters (Veizer and Hoefs, 1976; Hudson, 1977; Hoefs, 1981); however, the clear imprint of low values from the influx of fresh water into sedimentary basins has been documented (e.g., Magaritz et al., 1981). In some cases, consistent variations of $\delta^{18}O$ of marine carbonates with age have been demonstrated (Popp et al., 1986; Veizer et al., 1986), but whether they are due to secular changes in surface temperature or to $\delta^{18}O$ of seawater is debated. No variations with age related to the exogenic cycle have been detected under this noise. Oxygen isotopes in evaporite sulfates, on the other hand, show some distinct and significant variations with age; their possible relation to the exogenic cycle of sulfur was reviewed by Holser et al. (1979).

Although the isotope ratio of strontium, $^{87}Sr/^{86}Sr$, owes its initial variations to the radioactivity of rubidium in igneous rocks, thereafter in the exogenic cycle it simulates a stable isotope. As explained by Faure (1977, p. 128), $^{87}Sr/^{86}Sr$ in marine sediments reflects its igneous provenance: the ratio approaches 0.720 for sources rich in old granitic rocks and may be as low as 0.704 for sources rich in young basaltic rocks. An age curve for marine strontium has been established from samples of conodont apatite from the Cambrian through the Triassic (Kovach, 1980) and less accurately from fossil or whole-rock carbonate during Mesozoic–Cenozoic time (e.g., Brass, 1976; Burke et al., 1982) after conodonts were extinct. As shown in Fig. 4.10, $^{87}Sr/^{86}Sr$ goes through four minima in the Paleozoic. Another minimum in Late Jurassic time is followed by a gradual rise through the Cretaceous and Cenozoic to 0.7090 in modern seawater. The

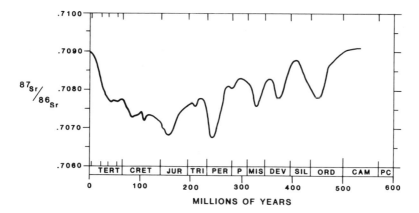

Figure 4.10 Strontium isotope age curve of Burke et al. (1982). See also Palmer and Elderfield (1986). Excursions toward lower ratios are thought to reflect an increasing influence of new, basaltic sources, whereas higher ratios reflect the influences of older granitic material (Faure, 1977).

strontium isotope age curve is of particular interest because its variations approximately correlate with the sulfur isotope age curve, although they are each thought to be controlled by different proximal processes.

4.4 RELATIONS AMONG ISOTOPE AGE CURVES

Relations among the patterns of age distribution for various isotopes may help explain the origin of the variations seen and shed light on some aspects of earth history. We first examine some observed relations, then discuss modeling of the sulfur and carbon curves.

4.4.1 Correlations

Approximate correlations of the age curves of isotopes of oxidized sulfur and carbon, as well as $^{87}Sr/^{86}Sr$ with world sea level, were discovered independently by several investigators (e.g., Hoefs, 1981). The level of this general set of relations is listed in Table 4.7 as a factor analysis matrix derived from smoothed age curves. The loadings show a strong inverse relation between carbon and sulfur, which can be attributed to their coupled oxidation–reduction systems.

Although these correlations are substantiated by long-time averaging, they break down over the short term. In constructs like Fig. 4.11, as the averaging interval is decreased from 100 to 5×10^6 years, the correlation coefficient steadily decreases and becomes less significant (Saltzman et al., 1982). With a short time constant, just about any combination of shifts can be found. The Röt event in the Early Triassic is a very sharp rise in $\delta^{34}S$ (Fig. 4.9) that does not seem to be accompanied, however, by any change in $\delta^{13}C$ (Wilgus, 1981).

144 GEOCHEMICAL CYCLES OF CARBON AND SULFUR

TABLE 4.7. Factor Analyses of the Relations Among the Age Curves of Strontium Isotopes in Carbonate, Carbon Isotopes in Carbonate, Sulfur Isotopes in Sulfate, and Sea Level for the Phanerozoic

	Factor 1	Factor 2	Factor 3	Communality
$^{87}Sr/^{86}Sr(carb)^a$	0.47	0.57*	0.32	0.65
$\delta^{13}C_{carb}^b$	−0.83*	−0.54*	0.06	0.98
$\delta^{34}S_{sulfate}^b$	0.92*	0.15	0.17	0.90
Sea-level stands^c	0.16	0.76*	0.03	0.60
Percentage of variation	81.9	15.3	2.9	
Factor interpretation	Organic redox coupling	Endogenic tectonic phenomena	Non representative data	

^{a}Data from Veizer and Compston (1974) and J. Veizer (unpublished).
^{b}Data from Veizer et al. (1980).
^{c}Data from Vail et al. (1977).
*Significant at the 0.05 level.

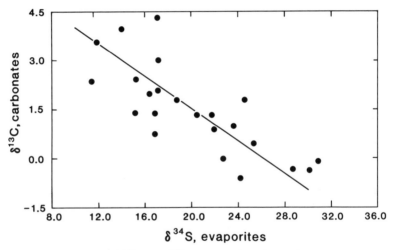

Figure 4.11 Correlation of $\delta^{13}C$ and $\delta^{34}S$. Inasmuch as the two isotope ratios are rarely determined on the same sample, and often not on the same formation or in the same basin, the mean values for 25×10^6-year intervals (Figs. 4.7 and 4.8) were treated as single points in this comparison. After Lindh (1983); see also Saltzman et al. (1982).

During the sharp rise of $\delta^{13}C$ in the Late Permian no changes of $\delta^{34}S$ can be detected, even in the same samples (Magaritz et al., 1983). In the Cretaceous, a rise in $\delta^{13}C$ is accompanied by a rise in $\delta^{34}S$ (Arthur et al., 1985).

Strontium isotopes are not influenced by bacterial oxidation–reduction processes, only by the balance of contributions from oceanic and continental crust to seawater. Thus, they provide direct evidence of past tectonic events.

Table 4.7 suggests a relation of carbon isotopes, and hence sulfur isotopes, to strontium isotopes and to sea level.

The inverse correlation between $\delta^{13}C$ and $\delta^{34}S$ shown in Fig. 4.11 seems to contradict what we know about the behavior of carbon and sulfur in modern sediments. As discussed in Section 4.2.1, reduced sulfur and reduced carbon occur today in a nearly constant molar ratio of 0.12. Thus, increases in reduced carbon should lead to increases in reduced sulfur. But Fig. 4.11 is showing the opposite behavior: times of abundant deposition of reduced sulfur (highly positive $\delta^{34}S$ in evaporites) are correlated with times of increased deposition of carbon in carbonates (negative carbonate $\delta^{13}C$). We might refer to this as the *central dilemma* of carbon–sulfur geochemistry, and most of the rest of this chapter is devoted to analyzing it.

4.4.2 Modeling Time Variations in the Carbon–Sulfur System

Garrels and Perry (1974) showed that such an inverse relation should, in fact, be true on a worldwide scale. If the O_2 content of the atmosphere has been relatively constant for the Phanerozoic Eon, as indicated by the fossil record, then an increase in $CaSO_4$ deposition should be matched by an increase in deposition of organic matter. Four reactions describe this system:

$$8Ca^{2+} + 16HCO_3^- + 8SO_4^{2-} + 16H^+ \longrightarrow 8CaSO_4$$
$$+ 16CO_2 + 16H_2O, \quad (4.4)$$

$$15CO_2 + 15H_2O \longrightarrow 15CH_2O + 15O_2, \quad (4.5)$$

$$4FeS_2 + 15O_2 + 16H_2O + 8CO_2 + 8CaCO_3 \longrightarrow 16H^+ + 8SO_4^{2-}$$
$$+ 2Fe_2O_3 + 8Ca^{2+} + 16HCO_3^-, \quad (4.6)$$

$$7MgCO_3 + 7SiO_2 \longrightarrow 7MgSiO_3 + 7CO_2. \quad (4.7)$$

Combining these to give constant O_2 yields the overall relationship

$$4FeS_2 + 7CaMg(CO_3)_2 + CaCO_3 + 7SiO_2 + 15H_2O \longrightarrow 8CaSO_4$$
$$+ 2Fe_2O_3 + 7MgSiO_3 + 15CH_2O. \quad (4.8)$$

Note that pyrite plus calcite occurs together on the left side and sulfate plus organic matter on the right side, consistent with the isotope curves.

Garrels and Lerman (1981) developed a quantitative model based on this equation for comparison with the isotope curves. As elaborated by Berner and Raiswell (1983), the model is based on conservation of the total mass of sulfur:

$$S_T = S_1 + S_2 + S_3, \quad (4.9)$$

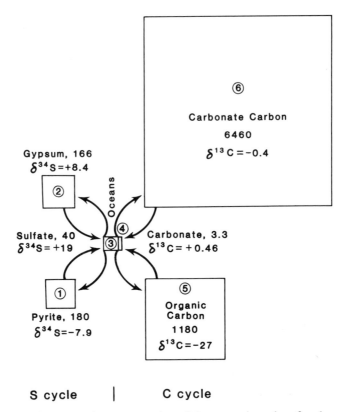

Figure 4.12 Diagrammatic representation of the exogenic cycles of carbon and sulfur. The area of each box is proportional to the size of the corresponding reservoir (in 10^{18} moles). Far more of the sulfur is dissolved in the ocean than is the case for carbon. The numbers refer to the subscripts in the equations of Section 4.4.1.

where S_T is total exogenic sulfur and S_1, S_2, and S_3 refer to pyrite sulfur, evaporite sulfur, and seawater sulfur, as labeled in Fig. 4.12. If sulfur dissolved in the ocean, S_3 is held constant, then inputs and outputs must be equal:

$$F_{1 \to 3} + F_{2 \to 3} = F_{3 \to 1} + F_{3 \to 2}, \tag{4.10}$$

where F is the flux from one reservoir to another. Also,

$$\frac{dS_1}{dt} = F_{3 \to 1} - F_{1 \to 3}, \tag{4.11}$$

$$\frac{dS_2}{dt} = F_{3 \to 2} - F_{2 \to 3}. \tag{4.12}$$

The assumption of constant seawater sulfate greatly simplifies the mathematical treatment but is perhaps too restrictive. See François and Gérard (1986) for a discussion of models with variable S_3.

Another assumption made in the Garrels–Lerman model is that the rate of weathering is a first-order function:

$$F_{1 \to 3} = k_{13} S_1, \qquad (4.13)$$

$$F_{2 \to 3} = k_{23} S_2. \qquad (4.14)$$

There must also be isotopic mass balance in the system:

$$\delta_{TST} = \delta_1 S_1 + \delta_2 S_2 + \delta_3 S_3. \qquad (4.15)$$

Finally, we wish to examine the change with time in sulfur isotopic composition of seawater (δ_3), which is given by

$$S_3 \left(\frac{d\delta_3}{dt} \right) = \delta_1 F_{1 \to 3} + \delta_2 F_{2 \to 3} - (\delta_3 - \Delta) F_{3 \to 1} - \delta_3 F_{3 \to 2}, \qquad (4.16)$$

where Δ is the average isotopic fractionation between sulfates and sulfides, assumed constant for the Phanerozoic. From present-day fluxes, k_{13} and k_{23} can be estimated; then the above equations can be solved for particular segments of the isotope age curve ($d\delta_3/dt$) by stepping backward in time from modern-day values of reservoir sizes and isotopic composition. Of particular interest are values for the fluxes, which can be used in

$$\frac{dC_5}{dt} = F_{4 \to 5} - F_{5 \to 4} = -\frac{15}{8} (F_{3 \to 1} - F_{1 \to 3}) \qquad (4.17)$$

and

$$F_{5 \to 4} = k_{54} C_5 \qquad (4.18)$$

to give estimates of the flux to reservoir 5, the burial of organic carbon (Fig. 4.12). The factor 15/8 arises from the stoichiometry of the basic equation (4.2).

Following this procedure, Berner and Raiswell (1983) obtained the pattern of fluxes shown in Fig. 4.13. These fluxes can then be converted into a C/S ratio, given by Fig. 4.14, which shows pronounced excursions from the present-day ratio during the Phanerozoic. The constant values used in such equations vary from author to author, but changing these values will only change the left-hand scale on these figures, not the patterns of variation with time. Note, however, that the model presented above uses only the isotope age curve for sulfur and neglects that for carbon.

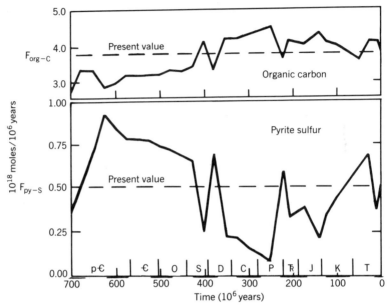

Figure 4.13 Burial rate, F, of organic carbon and pyrite sulfide as a function of time, based on modeling of sulfur isotope age distribution. After Berner and Raiswell (1983, Fig. 4).

How are these changes to be explained? What mechanism initiates the fluctuations? Examination of Eq. (4.8) indicates a number of possibilities. The most obvious nonuniform event is evaporite deposition, which might act via Ca^{2+} to change the carbonate flux. The much larger size of the carbonate reservoir argues against such a notion, nor can SO_4^{2-} deposition alone be responsible: sulfur isotopes in evaporites, unlike those in pyrite, are too close to seawater values for their deposition to change $\delta^{34}S$ of seawater. Rapid evaporite deposition could lower the SO_4^{2-} concentration of seawater worldwide, but this should not in general affect pyrite formation, because C_{org} rather than SO_4^{2-} is the limiting variable. We are then left with the balance between organic and carbonate carbon and with the silicate mineral reactions. Both are thought to be related at least indirectly to tectonics, so in Section 4.5 we examine the relation of tectonic history to the isotope curves.

4.5 TECTONICS AND ISOTOPE AGE CURVES

We have seen that the cycles of carbon and sulfur are linked by biologic oxidation and reduction processes. However, the source of the fluctuations in these two cycles is a major question. The correlations with the sea-level curve and the strontium age curve suggests that there is a tectonic cause.

4.5 TECTONICS AND ISOTOPE AGE CURVES

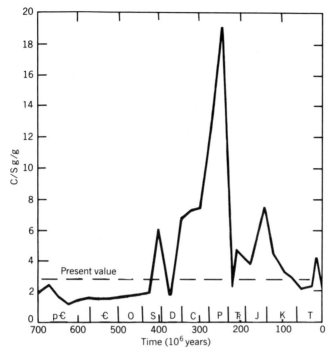

Figure 4.14 Amounts of organic carbon and sulfide sulfur in sediments as a function of time, based on modeling. After Berner and Raiswell (1983, Fig. 5).

4.5.1 Carbon and Sulfur Isotopes and Plate Tectonics

The age distribution of carbon and sulfur isotopes may be related to tectonics through the rise and fall of sea level. We will consider the known changes in sea level in the Phanerozoic, relate these to sea-floor spreading, and then discuss the possible linkage between these results and the age curves of carbon and sulfur.

4.5.1.1 The Phanerozoic Sea-Level Curve

A generalized sea-level curve for the Phanerozoic, as derived from seismic stratigraphic data by Vail et al. (1977), is shown in Fig. 4.15C. It represents the extent of onlap of coastal deposits in marine sequences and correlates well with the percentage flooding curves for North America and the Soviet Union (Turcotte and Burke, 1978; Hallam, 1984). The detailed record is more complex, but of particular interest here is the cyclic pattern of sea-level changes.

The Phanerozoic began as a time of relatively low sea level, continuing the pattern of the Late Proterozoic. Sea level rose to a maximum in Late Cambrian–early Ordovician time, then returned to a level like that of today in the late Carboniferous. This second period of relatively low sea level lasted about

150 GEOCHEMICAL CYCLES OF CARBON AND SULFUR

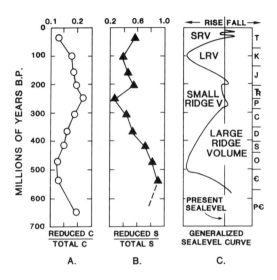

Figure 4.15 Age distribution of oxidized versus reduced carbon and sulfur, based on modeling, compared with sea level. Times of high sea level (large ridge volume) correlate with a high proportion of reduced sulfur and a low proportion of reduced carbon. Redrawn from Mackenzie and Pigott (1981).

100×10^6 years and was followed by a rise, which culminated in the major Cretaceous transgression, and then by a general lowering to the level of today.

Although other causes of shorter-term sea-level change are recognized (e.g., ice volume changes), major fluctuations are attributed to changes in the volume of mid-ocean ridges (e.g., Hays and Pitman, 1973) and, to a lesser degree, to mid-plate volcanism (Schlanger et al., 1981): large ridge volume, high sea level; small ridge volume, low sea level (Fig. 4.15C). Thus, global sea-level changes are linked indirectly to plate movements. The global sea-level minimum of the late Carboniferous through late Triassic is coincident with a time of relative stability of the supercontinent of Pangea, and thus a time of minimum ridge volume. This phase was preceded by a relatively long period of continental rifting and movement of continents from equatorial positions to a meridional distribution. The last 240×10^6 years of earth history have been mainly a time of continental breakup and rifting interrupted by the collision in the Tertiary of the Indian and Eurasian continental blocks of fragmented Pangea.

Comparison of Fig. 4.15C with the carbon and sulfur curves of Figs. 4.7 and 4.8 shows that these long-term changes in sea level are correlated with variations in the $\delta^{34}S$ and $\delta^{13}C$ isotopic composition of sedimentary rocks (Table 4.7). Periods of high ridge volume and high sea level are reflected in the isotopic distributions as times of depletion of ^{13}C in carbonate rocks and enrichment of ^{34}S in evaporites. Conversely, periods of low ridge volume and low sea level correlate with enrichment of ^{13}C in carbonates and depletion of ^{34}S in evaporites.

From the Phanerozoic sulfur and carbon isotope distributions, period averages for the masses of the sedimentary carbon and sulfur reservoirs and for the ratios of reduced carbon to total carbon and reduced sulfur to total sulfur (Fig. 4.15A,B) can be calculated, using equations of the type described in Section 4.4 (Mackenzie and Pigott 1981).

It can be seen from Fig. 4.15 that during times of progressive flooding of continental margins, carbon was transferred from the reduced carbon reservoir (mainly organic carbon in shales) to the oxidized carbon reservoir in carbonates, whereas sulfur was moving from the oxidized sulfate reservoir in evaporites to the reduced sulfide reservoir (mainly FeS_2 in shales). Regression of sea level correlates with opposite patterns of reservoir transfer for carbon and sulfur. Note, however, the anomalous period from $100-200 \times 10^6$ yr where both move in the same direction.

4.5.1.2 Tectonic States and Geochemical Responses

A tentative tectonic interpretation has been made of the observed coupling between carbon and sulfur and its relation to sea level by MacKenzie and Pigott (1981). In this interpretation, the forcing function is considered to be tectonics; that is, tectonic events are the driving force governing large-scale reservoir exchange on a long-term basis. By using the carbon–sulfur inter-reservoir transfer model of Garrels and Perry (1974), the Phanerozoic continental reconstructions and paleoclimatic interpretations of Ziegler et al. (1979), and the concept of tectonic modes developed by Sloss and Speed (1974) and Sloss (1976), a reasonable scheme for tectonic control of the carbon and sulfur cycles can be developed.

During the last $600-700 \times 10^6$ years of earth history, two cycles, each of a few hundred million years duration, can be recognized in the distributions of sedimentary mass, carbon and sulfur isotopes, and sea level (see Fischer, 1981). The cycles of carbon and sulfur and of sea level alternate between high stands of sea level and low stands. These two states correspond to the submergent and oscillatory mode of Sloss and Speed (1974), who applied them initially to the tectonic response of cratons and continental margins. The submergent mode was used to refer to a time of general submergence of the craton, whereas the oscillatory mode referred to a time of general net elevation of cratons with respect to sea level.

A state of low stand of sea level covers the time spans of Late Proterozoic to Early Cambrian, the Late Carboniferous to Early Jurassic, and much of the Cenozoic, whereas a state of high stand of sea level is characteristic of much of the Paleozoic and the time span from about Early Jurassic to Early Paleocene. The latter state was interrupted during the Paleozoic by short-lived times of emergence—the emergent mode of Sloss and Speed (1974). Details of the sea-level curve reveal these short-lived times of general emergence of cratons; it is possible that, once sufficient data are available, the distribution of carbon and sulfur isotopes in the Phanerozoic may be interpretable on the time scale of these emergent episodes (10^6-10^7 years). Whatever the case, a summary of the sedimentary reservoir transfers and reactions, and tectonic, sedimentologic, and geomorphic events for the two states are given in Tables 4.8 and 4.9. A discussion is provided below.

During a state of high-stand sea level—a time of active plate convergence, obduction and subduction at continental margins, rapid sea-floor spreading, large ridge volume, and general submergence of continental margins and interiors—the net reservoir transfer reaction during rising sea level can be expressed as the reverse of Eq. (4.8) with a gas phase included to emphasize the role of CO_2:

$$CO_{2(g)} + 2Fe_2O_3 + 8CaSO_4 + 8MgSiO_3 + 15CH_2O \longrightarrow 4FeS_2$$
$$+ 8CaMg(CO_3)_2 + 8SiO_2 + 15H_2O. \qquad (4.19)$$

That is, periods of elevated CO_2 in the atmosphere should also be periods of transfer of carbon from the reduced organic carbon to the oxidized carbonate carbon reservoir, and of sulfur from the oxidized sulfate reservoir to the reduced pyrite reservoir. Thermodynamic equilibrium for reaction 4.19 lies far to the right, implying very low P_{CO_2} values, but the earth is not a single system. Instead it is a set of relatively isolated reservoirs and reaction 4.19 represents the net transfers among those reservoirs. Thus $CO_{2(g)}$ is a reservoir (the atmosphere) and reaction 4.19 shows how its mass may have changed through time in conjunction with other reservoir transfers.

The sources of CO_2 needed to participate in reaction (4.19) are metamorphic reactions like

$$CaMg(CO_3)_2 + SiO_2 = CaCO_3 + MgSiO_3 + CO_2 \qquad (4.20)$$

and

$$5CaMg(CO_3)_2 + Al_2Si_2O_5(OH)_4 + SiO_2 + 2H_2O$$
$$= Mg_5Al_2Si_3O_{10}(OH)_8 + 5CaCO_3 + 5CO_2. \qquad (4.21)$$

These clay–carbonate mineral reactions can be a source of large volumes of CO_2 in the subsurface (Hutcheson et al., 1980) and can occur in a variety of tectonic–sedimentologic settings. One such setting would be a region of shallow igneous activity, where geothermal gradients would be high enough and pressure low enough to convert carbonate minerals to silicates with release of CO_2. These two conditions can be met around andesitic volcanic centers above subduction zones.

It has been demonstrated that CO_2-rich springs are associated mainly with young orogenic belts (e.g., circum-Pacific) and less importantly with rifting continental platforms (Irwin and Barnes, 1980). Important sources of the CO_2, as indicated by $\delta^{13}C$ analyses, are the mantle, metamorphism of carbonate rocks, and oxidation of organic matter. Thus, it would be anticipated that, as sea-floor spreading and subduction activity increase, increased volumes of CO_2 gas would be released to the surface because of increased rates of metamorphism of carbonate rocks and increased rates of CO_2 evasion from mantle-derived basaltic melts. During times of decreased global tectonic activity, the rate of release of CO_2 from the subsurface would decrease.

Reactions (4.20) and (4.21) differ from the overall equation of Garrels and Perry (1974, reaction 1) because CO_2 is not held constant in the ocean–atmosphere system but is produced by reaction of carbonates to form silicates during metamorphism. These reactions are a type of reverse weathering (Garrels, 1965; Mackenzie and Garrels, 1966a,b), in that calcium and magnesium originally derived from the weathering of silicates and deposited in the ocean as carbonates are reconstituted by reaction with SiO_2 to make silicate minerals. The material transfer associated with this buffer mechanism is small when compared to the amounts of calcium, magnesium, and carbon cycling about the earth's surface, and metamorphism of carbonates is probably varies with the rate of spreading and subduction. Thus, it is possible that the CO_2 in the atmosphere (a very small reservoir of CO_2) could vary through geologic time.

The isotopic evidence suggests that during rising sea level sulfate and organic carbon are transferred from the oxidized evaporite and reduced carbon reservoirs by erosion of sulfate minerals and old sedimentary carbon to the reduced sulfur and oxidized carbon reservoirs, respectively, by precipitation of carbonate minerals and FeS_2 in marine sediments. The terrestrial biota would be restricted during this mode because of encroachment of the sea onto continental margins and interiors, and nutrient supply to the oceans would be reduced because of decreased land area and mean continental elevations, leading to a decrease in runoff and chemical denudation rates (Meybeck, 1976, 1979a,b; Dunne, 1978). Furthermore, phosphorus may be lost to continental shelf and estuarine sediments during the initial stages of transgression of continental margins (Broecker, 1982). The effect of decreased nutrient supply to the ocean would be to decrease the amount of marine plant material produced per unit time, which in turn would lead to a decrease in the rate of accumulation of organic carbon in marine sediments.

With a fall in sea level, the above picture is reversed, culminating in the state of low-stand sea level—a time of generally convergent boundaries of oceanic and continental plates, moderate rates of sea-floor spreading, small ridge volume, elevated cratonic areas, and marked topographic relief as compared to submergent times. The net reservoir transfer reaction is opposite to that of reaction (4.19). Emergence leads to erosion of sediments deposited previously during high sea level, accompanied by oxidation of pyrite sulfide to sulfate and weathering and erosion of carbonates and silicates. Because there now would be an increase in the flux of nutrients like phosphate to the oceans from rivers,

TABLE 4.8. Submergent Mode[a], Sedimentary Reservoir Transfers and Reactions, and Sedimentologic–Geomorphic Responses[b]

Tectonic Mode	Reservoir Transfers and Reactions		Sedimentologic–Geomorphic Response
	Process	Principal Mass Transfer Reaction	
Submergent: Craton depressed below sea level; time of active plate convergence, obduction of oceanic margins on cratons; ridge volume high, sea-level high	1. High input of diagenetic and metamorphic CO_2 to ocean–atmosphere system	$CaMg(CO_3)_2 + SiO_2 \rightarrow CaCO_3 + MgSiO_3 + CO_2$	Widespread epicontinental seas; deposition of nonclastic (carbonate, cratonal evaporite) sediment important, as well as mature clastics derived from within continent; convergent plate boundaries closely coincide with cratonic margins; craton ringed by orogenic belts that locally and sporadically reach mountainous elevations. Immature sediment is derived from these highlands and progrades towards continental interiors. Small land area, low mean elevation, low stream gradients, low denudation rates. Relatively high global mean temperature because of increased atmospheric CO_2 level and perhaps decreased albedo; decreased rate of oceanic mixing; thick, widespread O_2 minimum, anoxic events important.
	2. Erosion of sulfate and organic carbon and precipitation of carbonate and pyrite	$2Fe_2O_3 + 15CH_2O + 8CaSO_4 \rightarrow 8CaCO_3 + 4FeS_2 + 7CO_2 + 15H_2O$	
	3. Restriction of terrestrial biota and decreased nutrient supply to oceans leads to increased respiration (oxidation)	$CH_2O + O_2 \rightarrow CO_2 + H_2O$	

[a] Adapted from Sloss and Speed (1974).
[b] Reservoir transfer reactions are for rising sea-level stage of this mode; for falling sea level, transfers are reversed, as shown in Table 4.9.

TABLE 4.9. Oscillatory Mode[a], Sedimentary Reservoir Transfers and Reactions, and Sedimentologic–Geomorphic Responses[b]

Tectonic Mode	Reservoir Transfers and Reactions		Sedimentologic–Geomorphic Response
	Process	Principal Mass Transfer Reaction	
Oscillatory: Craton elevated or oscillating with respect to sea level; marginal areas characterized by convergent boundaries with associated island arcs and back-arc basins; ridge volume low, sea-level low.	1. Low input of diagenetic and metamorphic CO_2 to ocean–atmosphere system	$CO_2 + CaCO_3 + MgSiO_3 \rightarrow CaMg(CO_3)_2 + SiO_2$	Cratonic interiors are far from continental margin areas; internal clastic sedimentation in intracratonal basins important; carbonates and mature clastics at craton margins are displaced by immature clastics, and in cratonic interiors immature sandstones, accompanied in places by evaporites, are deposited; large land area, high mean elevation, high denudation rates, high mechanical to dissolved load (like today). Relatively high nutrient supply to ocean from increased erosion and exposure of shelf areas. Relatively low global mean temperature because of decreased atmospheric CO_2 level and perhaps increased albedo. Tectonic, paleoclimatic, and paleogeographic conditions favor evaporite deposition after and during continental collisions (giving rise to oscillatory mode) which led to relatively long-lived basins and necessary basin hydrography.
	2. Emergence of terrestrial materials increases rate of sulfide oxidation	$4FeS_2 + 15O_2 + 8H_2O \rightarrow 2Fe_2O_3 + 8SO_4^= + 16H^+$	
	3. Erosion of carbonates	$CaCO_3 + CO_2 + H_2O \rightarrow Ca^{++} + 2HCO_3^-$	
	4. Increased photosynthesis provides O_2 for sulfide oxidation	$CO_2 + H_2O \xrightarrow{\text{nutrients}} CH_2O + O_2$	
	5. Evaporite precipitation	$Ca^{++} + SO_4^= \rightarrow CaSO_4$	

[a] Adapted from Sloss and Speed (1974).
[b] Reservoir transfer reactions are for falling sea-level stages, culminating in the oscillatory mode.

carbon would be transferred from the oxidized carbon reservoir in limestones to the reduced carbon reservoir, principally organic matter in shales.

4.5.1.3 Tectonics and S/C Ratios

We are now in a position to attempt a reconciliation of the dilemma identified in Section 4.4: Why is there an inverse correlation between reduced carbon and reduced sulfur through time, when in modern environments the two are positively correlated? In modern marine sediments with normal oxygen content, organic carbon is directly proportional to the sediment accumulation rate (Heath et al., 1977; Toth and Lerman, 1977). Therefore, we would expect that, during times of falling sea level with their faster erosion and clastic sedimentation rates and higher productivity, organic carbon and reduced sulfur accumulation rates would also increase together. Conversely, during times of rising sea level and decreasing erosion and clastic sedimentation and low productivity, organic carbon and reduced sulfur accumulation rates should decrease.

One reason for the apparent conflict is that, in Eq. (4.19), we are dealing with net global changes. Whereas the above statements may hold for local environments, it is not necessary that globally increased accumulation of organic carbon should lead to increased pyrite formation rates. Of particular importance may be oceanic anoxic events during which bottom waters contain insufficient oxygen to oxidize organic carbon at the sediment–water interface (Schlanger and Jenkyns, 1976; Arthur and Schlanger, 1979; Scholle and Arthur, 1980; Wilde and Berry, 1984). These anoxic events are also times of relatively high $\delta^{13}C$ values in Cretaceous pelagic carbonates and presumably in dissolved inorganic carbon in seawater (Schlanger et al., 1987).

Berner and Raiswell (1983) have suggested that there are in fact three main environments of reduced sulfur fixation into pyrite, each with its distinctive S/C ratio, and it is the distribution of sulfur deposition among these environments that controls the global S/C ratio. The studies of modern marine sediment cited previously were almost entirely in sediments of "normal" marine character—oxygenated bottom water with sulfate reduction developing a few centimeters below the sediment–water interface. Here the S/C weight ratio is about 0.35 (Berner and Raiswell, 1983, Fig. 1). In contrast, nonmarine environments can accumulate large amounts of carbon but, in the absence of a marine transgression over the peat swamp, do not precipitate much pyrite because of the low sulfate concentration in fresh water. The S/C ratios are between 0.035 and 0.040 (Berner and Raiswell, 1983, Fig. 2). The opposite effect is seen in euxinic marine basins, those in which sulfate reduction goes on above the sediment–water interface. The best-known modern example is the Black Sea, which has S/C ratios mostly greater than 1.0 (Berner and Raiswell, 1983, Fig. 3). The situation here is somewhat complicated by a positive sulfur intercept at zero carbon (Leventhal, 1983), so the S/C ratio increases with decreasing carbon, as shown in Fig. 4.16.

From these observations, one would predict that in times of extensive terrestrial environments, global S/C would be quite low, whereas if conditions favored

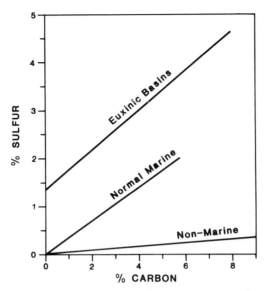

Figure 4.16 Idealized representation of the relation of pyritic sulfur to organic carbon among different environments. After Berner and Raiswell (1983, Figs. 2 and 3).

numerous euxinic basins, S/C would be much higher. Berner and Raiswell (1983, Fig. 5), from the isotope curves, identify a period of low S/C in the Late Carboniferous–Permian and one of high S/C in the Early Cretaceous (see Fig. 4.14). The first is reasonably associated with the low sea level and continental collisions of the end of the Paleozoic, the second with the high sea levels and continental separation of the Cretaceous. Higher S/C levels are also found for the Cambrian–Devonian, which may reflect the absence of an abundant terrestrial flora rather than tectonic events.

The linkage between tectonic events and the behavior of carbon and sulfur is likely through phosphate, as mentioned above. During times of low sea level, the phosphate flux is high, but much of it goes to terrestrial environments. Thus, the production of reduced carbon is high, but production of reduced sulfur is low. During times of high sea level, the phosphate flux to the oceans is restricted and so carbon production falls, but these are also times of maximum development of euxinic basins (e.g., Hallam and Bradshaw, 1979) in marine environments, and minimum development of coal swamps. Thus, times of low organic carbon became times of high reduced sulfur, explaining the inverse correlation of the isotope curves.

An alternative explanation for the inverse correlation of reduced carbon and reduced sulfur through time has been proposed by Keith (1982). He argued that the major, long-term trends in $\delta^{13}C$ of limestones are not simply related to changes in the ratio of the fluxes to the sea floor of organic carbon and carbonate carbon but that they depend on oceanic mixing states. Keith proposed that the observed patterns of decreasing $\delta^{13}C$ in limestones are indicative of changes in the ocean from a well-mixed, aerated system to a stratified, stagnant ocean. In the latter condition, vertical ocean mixing is restricted and oxidative decay of organic particles at depth gives way to anaerobic processes because of a decrease in the rate of ventilation of the deep ocean by O_2. Bacterial activity becomes more important, as well as reduced products of anaerobic bacterial decay (CH_4, H_2S, NH_3). The increased importance of anaerobic bacterial cycling of organic materials, including $\delta^{13}C$-depleted methane, and the increased utilization of bacteria depleted in ^{13}C as a food source for organisms higher in the food chain lead to sources of ^{13}C-depleted carbon for limestone production. Because of bacterial oxidation of organic matter and reduction of sulfate leading to sulfide precipitation, ocean water would become progressively enriched in ^{34}S, as would evaporitic sulfate deposits forming from this water. This would explain the observed negative correlation between limestone $\delta^{13}C$ and evaporite $\delta^{34}S$. Conditions in the well-mixed aerated ocean would be more like those of today. Changes in oceanic mixing from a stratified stagnant ocean to one like that of today presumably would result in increasing $\delta^{13}C$ of limestones and decreasing $\delta^{34}S$ of evaporitic sulfates, according to Keith's hypothesis.

4.5.2 Other Isotope Age Curves

There are some lines of evidence that support at least in part the general tectonic model of sulfur and carbon variation proposed here. These evidences come principally from the Cenozoic rock record. Although still somewhat controversial, it appears that sea-floor spreading rates and ridge volume have generally declined during the last 70×10^6 years of earth history (Fig. 4.17A). This trend can be related to trends in the $\delta^{18}O$ of invertebrates (Fig. 4.18), principally foraminifera, the Sr/Ca ratio of planktonic foraminifera, and the $^{87}Sr/^{86}Sr$ ratio of marine sedimentary precipitates (Fig. 4.17B,C).

The $\delta^{18}O$ values of a large number of samples of foraminifera show a gradual increase over the last 80×10^6 years (Fig. 4.18). Although there are sharp changes and reversals, the overall pattern is essentially unidirectional. This overall change in $\delta^{18}O$ has been interpreted as reflecting a change in the temperature of high-latitude surface seawater, and consequently low-latitude bottom water, from about 13°C at the beginning of the Tertiary to about 1°C today (e.g., Savin et al., 1975). Thus, the latitudinal temperature gradient in the late Cretaceous ocean (and atmosphere) may have been less steep than that of today. Our model linking carbon and sulfur fluxes and tectonics implies that times of high sea-floor spreading rates and mid-ocean ridge volume are times of high atmospheric CO_2 levels. If so, based on present-day climate models relating atmospheric CO_2 and temperature, the inferred latitudinal temperature gradient of the Late Cretaceous would reflect an average surface temperature of 2–3°C greater than that of today and an atmospheric CO_2 content about double that of today (Berner et al., 1983; Lasaga et al., 1985).

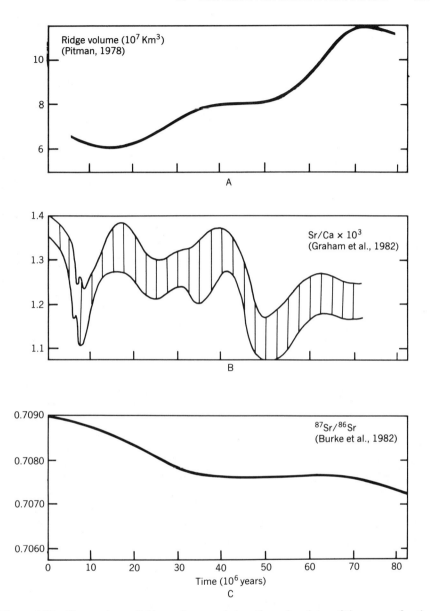

Figure 4.17 Comparison of ridge volume and strontium chemistry of the oceans for the Cenozoic. Both the Sr/Ca ratio of planktonic foraminifera and the $^{87}Sr/^{86}Sr$ ratio of marine carbonates increase with decreasing age (to the left) as ridge volume decreases, possibly reflecting an increasing contribution of continental crust to seawater chemistry compared to a more important role for oceanic crust during times of faster spreading. See also Renard (1986).

160 GEOCHEMICAL CYCLES OF CARBON AND SULFUR

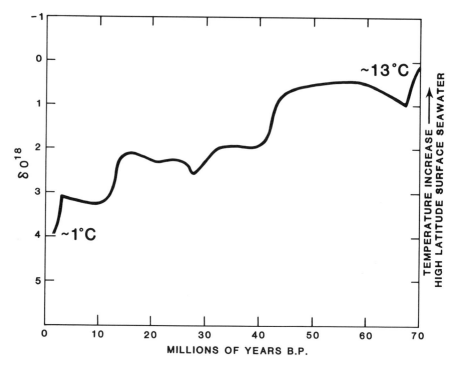

Figure 4.18 Oxygen isotopes of foraminifera through the Cenozoic. The trend probably reflects a decrease in the temperature of the oceans. This temperature decrease may in turn reflect a decline in atmospheric CO_2, as discussed in Chapter 2. See also Shackleton (1987, p. 431).

The Sr/Ca ratios of planktonic foraminifera plotted against geologic age show a poorly defined increase from the beginning of the Cenozoic to today (Fig. 4.17B). A possible interpretation of the overall trend is that during the Cenozoic, as sea-floor spreading rates and ridge volumes decreased, the input of hydrothermal fluids to the oceans slowed. Thus, the ratio of the hydrothermal flux with low Sr/Ca ratio to the river water flux with high Sr/Ca ratio decreased as tectonic activity slowed and the seas withdrew from the continents. The Sr/Ca ratios of planktonic foraminifera are then interpreted as reflecting the changes in the Sr/Ca of seawater. This interpretation of foraminiferal Sr/Ca ratios, although not the only one possible, is in accord with the tectonic model proposed in this chapter.

The $^{87}Sr/^{86}Sr$ ratio of marine sedimentary precipitates has also increased with decreasing sedimentary rock age since the peak of the Late Cretaceous high-stand sea level (Figs. 4.11, 4.17C). This trend can be interpreted as indicative of a lessening contribution of low $^{87}Sr/^{86}Sr$ fluids derived from reactions involving mafic rocks in the marine realm and an increasing contribution to the oceans of high $^{87}Sr/^{86}Sr$ river water derived from weathering of continental rocks (Section 4.3.3.3). The $^{87}Sr/^{86}Sr$ ratio of seawater apparently increased progressively through the Cenozoic with the inferred decline in tectonic activity reflected in a decrease in sea-floor spreading rates, a sea-level fall, and a land area increase.

Although the strontium isotopic curve from the past 80×10^6 years or so of earth history can be interpreted as being in accord with the tectonic model of this chapter, its interpretation during the rest of Phanerozoic time is more difficult. For example, the $^{87}Sr/^{86}Sr$ ratio of marine precipitates is low during the late Permian, a time of the stabilized mega-continent of Pangea, low tectonic activity, and low sea level. We would anticipate high $^{87}Sr/^{86}Sr$ ratios at this time because of large land area and low inputs of hydrothermal fluids to the ocean from basalt–seawater reactions, giving rise to a river-dominated, high $^{87}Sr/^{86}Sr$ strontium source for the ocean.

One final comment: a number of investigators (Sandberg, 1975, 1983, 1985; Mackenzie and Pigott, 1981; Wilkinson et al., 1985) have suggested that ooids and marine cements during Phanerozoic high stands of sea level were principally calcite in composition, whereas low stands of sea level were characterized by originally aragonitic cements and ooids. A similar relation has also been suggested for biomineralization (Wilkinson, 1979; Mackenzie and Agegian, 1987). Thus, there appears to be a long-term cyclicity in these marine carbonate precipitates corresponding to the long-term cyclicity in the sea-level curve. The relation again may be fortuitous, but it has been argued that this cyclicity in carbonate mineralogy reflects changes in the CO_2 content of the atmosphere and perhaps the saturation state of seawater with respect to calcite (e.g., Mackenzie and Pigott, 1981; Sandberg, 1983). High sea levels are times of increased tectonic activity, giving rise to high atmospheric CO_2 levels and low seawater saturation states that promote formation of calcitic cements and ooids. Conversely, low sea levels are times of decreased tectonic activity, low atmospheric CO_2 levels, and high seawater saturation states that result in oceanic environments more conductive to formation of aragonitic, as well as calcitic, precipitates. Thus, the cyclicity in marine carbonate mineralogy is in accord with the suppositions of our tectonic model.

4.6 LONG-TERM EVOLUTION OF THE BIOGEOCHEMICAL CYCLES OF CARBON AND SULFUR

In previous sections, we have seen that continuous cycling of carbon and sulfur through the ages has given rise to a system of quasi-stationary reservoirs, with the bulk of these elements concentrated in the earth's crust, most notably in sedimentary rocks (Fig. 4.12). In the case of carbon, the 10^{18} moles of carbon residing as bicarbonate, carbon dioxide, and organic carbon in the exogenic reservoirs of ocean, atmosphere, and biosphere constitute a tiny part of the huge repository of sedimentary carbon, which is on the order of 10^{21} moles. For sulfur, the relative proportions are somewhat different, with about one-tenth (10^{19} moles) of the total (10^{20} moles) stored in the ocean as dissolved sulfate.

An important problem in the evolution of the earth's carbon and sulfur cycles centers around the questions of *how* and *when* the states depicted in the box models of Figs. 4.1, 4.3, and 4.12 were established. Points of concern are (1) the evolution and possible growth through time of the total amounts of carbon and sulfur cycled through the atmosphere–ocean–crust system, and (2) the time of emergence of the biologic processes responsible for the pathways of element fluxes from the atmosphere–ocean system into sediments.

4.6.1 Evolution of the Carbon Cycle

4.6.1.1 *Source of Exogenic Carbon*

Although the mass of the earth's sedimentary shell of 2.6×10^{24} g is only $\sim 8\%$ of the mass of the crust, most crustal carbon (5 of about 8×10^{21} moles) is concentrated in sedimentary rocks (Hunt, 1972). There is agreement that the bulk of the carbon cycled through the atmosphere–ocean–sediment system during the earth's history was originally degassed from the upper mantle. Because the total carbon content of typical sediments ($\sim 3\%$) is much in excess of the carbon content of igneous rocks (~ 200 ppm) from which they were derived, most crustal carbon must stem from a gaseous precursor that is now tied up in solid phases in the sedimentary shell. Although hotly debated in the past, there seems to be fairly widespread agreement today that most of the primordial carbon was fed into the exogenic system as carbon dioxide (CO_2) rather than in a reduced form as methane (CH_4).

Assuming that the earth had differentiated into core and mantle before it started scavenging additional chondritic debris that subsequently formed the upper mantle, the ratio CO_2/CH_4 in the gas fraction released from such a mantle would reflect the general oxidation state of the chondritic precursor material. This material consisted mainly of late-stage condensates of the solar nebula formed at temperatures below 750 K, the temperature at which metallic iron is oxidized to Fe^{2+} and subsequently (below 450 K) to Fe^{3+} (cf. Grossman and Larimer, 1974; Walker, 1977, pp. 188ff). Thus, the earth's early outer veneer can be assumed to have been characterized by the absence of metallic iron and by a preponderance of Fe^{2+} with minor Fe^{3+}. Gases in equilibrium with such a system should be weakly reducing, with $CO_2 \gg CO + CH_4$, $H_2O \gg H_2$, and $N_2 \gg NH_3$, as in volcanic emanations released today from magmas of presumed mantle derivation. Hence, the Fe^{2+}/Fe^{3+} ratios of basalts and ultramafic rocks from different ages should record the oxidation state of the earth's mantle through time. The observed constancy of this ratio over the last 3.5×10^9 years gives eloquent testimony to the fact that the overall oxidation state of the mantle has not changed during this time span. Furthermore, the occurrence of sedimentary carbonates in the $\sim 3.8 \times 10^9$ year-old Isua metasedimentary series of Greenland (Schidlowski et al., 1975) provides a minimum age for an atmospheric reservoir of carbon dioxide.

4.6.1.2 *Transfer of Carbon from the Mantle to the Crust*

Given that exogenic carbon was ultimately derived from the mantle, the total amount of this element cycled through the atmosphere–ocean–crust system was primarily determined by the earth's degassing process. After a brief interlude in the atmosphere–ocean–biosphere system as CO_2, HCO_3^-, or organic compounds, virtually all carbon was transferred to sedimentary rocks, with only one-thousandth of the total remaining in the other exogenic reservoirs. Hence, an assessment of the quantitative evolution of the earth's carbon budget can be made by studying the growth of the sedimentary carbon reservoir through geologic time.

Although many details of the degassing process are still subject to debate, there appear to be two extremes for representation of the release of volatiles from the upper mantle as a function of time: the *catastrophic* and the *linear* degassing models. The first representation follows a scenario in which the bulk of the originally mantle-bound volatiles was released during or shortly after accretion of the earth's outermost chondritic layer; the second has the degassing process continuing through time (Fig. 4.19). Because reasonable arguments can be advanced for either of these alternatives (cf. Fanale, 1971; Craig et al., 1975; Alfvén and Arrhenius, 1976), the two representations are obviously not exclusive of each other. They can be combined in a model based on exponentially decreasing degassing rates that combines an initial phase of vigorous degassing with an extended, less intense period. Using such an approach based on a degassing constant of $\lambda = 1.16 \times 10^{-9}$ year^{-1} (Li, 1972), the accumulation of carbon in the sedimentary shell would follow curve 3 in Fig. 4.19. According to this curve, more than 80% of the carbon presently stored in the sedimentary shell would have been released prior to 3×10^9 years ago, and during the last 2×10^9 years the earth's exogenic carbon mass would only have increased by 5%.

4.6.1.3 *Evolution of the Sedimentary Carbon Reservoirs of the Crust*

As pointed out above, most carbon degassed from the earth's interior was bound to end up in the crust after a transient existence in the atmosphere–ocean–biosphere system. As a result of biologic activity at or near the earth's surface, a partitioning of the carbon flux into the sedimentary shell was established with oxidized carbon fixed as carbonate (C_{carb}) and reduced carbon as fossil organic matter or *kerogen* (C_{org}). Because the kinetic isotope effect inherent in biologic carbon fixation is retained during this transfer, a relative depletion of heavy carbon (^{13}C) in the organic and a corresponding ^{13}C enrichment in the inorganic reservoir resulted.

Thus, apart from possible changes through time in the carbon masses of the ocean–atmosphere–biosphere, and notably of atmospheric CO_2, the most important single event in the evolution of the carbon cycle was the establishment of a biologic control owing to the appearance of a sink for organic carbon with

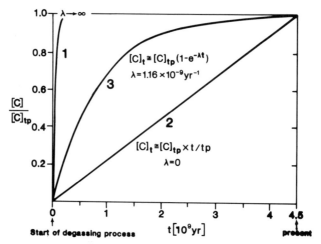

Figure 4.19 Models for the increase of the reservoir of sedimentary carbon through time, based on alternative scenarios for the earth's degassing (λ = degassing constant). Carbon reservoirs existing at time t are expressed as fractions of the present reservoir C_t/C_{tp} where tp = 4.5×10^9 years = present time. (1) Carbon evolution based on catastrophic degassing ($\lambda = \infty$) near the time of the earth's formation ($t = 0$). (2) Linear degassing model ($\lambda = 0$). (3) Exponentially decreasing degassing model ($\lambda = 1.16 \times 10^{-9}$ year^{-1}). This value of λ can be regarded as the best approximation for the time-averaged degassing constant (Li, 1972); the exponential function is likely to come closest to reality. According to this function, the carbon reservoir of the crust had reached >80% of its present size 1.5×10^9 years after the start of the degassing process. Note that reservoir growth was minimal during the last billion years and negligible during the Phanerozoic.

subsequent formation of a crustal C_{org} reservoir. It is a reasonable conjecture that, prior to the advent of life, carbon was removed from the atmosphere–ocean system mainly by precipitation in carbonate minerals because inorganic production of reduced carbon compounds on the primitive earth was unlikely to have reached geochemically significant levels. As has been pointed out previously (Broecker, 1970; Schidlowski et al., 1975), the isotopic consequences of such a prebiotic scenario are straightforward: sedimentary carbonates formed in the absence of a biologic carbon sink should have inherited the isotopic composition of the earth's primordial carbon with an average $\delta^{13}C$ value between -5 and $-6‰$ (PDB), a value similar to those for the principal forms of deep-seated carbon such as carbonatites and diamonds (Section 4.3.3.1). After the emergence of a biologic sink, light carbon would have preferentially concentrated in organic matter, thus forcing $\delta^{13}C$ values of carbonates from about $-5‰$ toward more positive values, the magnitude of the shift depending on the fraction of total carbon ending up as organic matter.

Organic substances are enriched by about 20–30‰ in light carbon as compared to oceanic bicarbonate and carbonate (Section 4.3.1). Because both

4.6 LONG TERM EVOLUTION OF THE BIOGEOCHEMICAL CYCLES

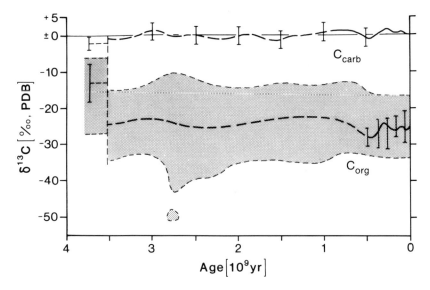

Figure 4.20 Isotopic fractionation between carbonate carbon (C_{carb}) and organic carbon (C_{org}) through geologic time. The envelope for sedimentary organic matter or kerogen is, in essence, the geochemical manifestation of the activities of the enzyme ribulose-1,5-bisphosphate carboxylase, which is principally responsible for the discrimination against heavy carbon (^{13}C) in photosynthetic carbon fixation. The extremely light kerogens at the lower fringe suggest the involvement of methane-fixing pathways based on different enzymatic requirements. Note the discontinuity at $t \approx 3.5 \times 10^9$ years, which can best be explained as a result of amphibolite-grade metamorphism of the Isua rocks with concomitant isotope shifts of both carbon species. Redrawn from Schidlowski (1987).

organic and carbonate carbon are incorporated in sediments with but minor changes in their isotopic composition (Schidlowski, 1982; Schidlowski et al., 1983), these differences are propagated into the crust, attesting to the presence of biologic carbon isotope fractionation in the geologic past. Accordingly, we may attempt to use the carbon isotope record for placing constraints on the time of emergence of a crustal C_{org} reservoir.

Figure 4.20 shows the isotopic composition of marine carbonates and sedimentary organics as a function of time. It is evident that, over the last 3.5×10^9 years, the mean for carbonate carbon is close to the zero per mill line ($\delta^{13}C_{carb} = -0.4$; Table 4.5), whereas the values of kerogen oscillate around an average of between -25 and $-27‰$, with the bulk of data points falling into the range $\delta^{13}C_{org} = -27 \pm 7‰$ (Schidlowski et al., 1983; Table 4.5). Because $\delta^{13}C_{carb}$ values are basically inherited from a bicarbonate parent with an uncertainty of usually less than 1‰, the carbonate $\delta^{13}C$ record suggests that the isotopic composition of marine bicarbonate has changed little through time. As for organic carbon, the envelope shown in Fig. 4.20 resembles the spread typical of recent organic matter (Section 4.3.1, Fig. 4.4) and hence can best be explained as representing the isotopic signature of biologic (autotrophic) carbon fixation.

The uniformity of this signature is consistent with the notion of an extreme conservatism of the principal pathways of carbon assimilation, notably Calvin cycle (C3) photosynthesis. The pronounced negative spike in the lower boundary of the envelope between 2.5 and 3.0×10^9 years ago and the other extremely negative values at the lower fringe of the band are most likely the result of the involvement of methane-fixing pathways in the formation of the respective kerogen precursors.

The only alternative to a biologic interpretation of the early $\delta^{13}C_{org}$ record would be the existence of an inorganic geochemical process able to mimic the isotope fractionation in photosynthesis with almost incredible precision. Although Fischer–Tropsch and Miller–Urey processes could, in principle, be credited with an inorganic production of organic substances, both fail to meet the constraints of the existing isotope data, with fractionations in Miller–Urey type spark-discharge syntheses only marginally overlapping the range attributed to sedimentary organics (Schidlowski, 1982; Chang et al., 1983). Furthermore, the biologic character of the observed isotope fractionations is corroborated by the recent extension of the fossil record to about 3.5×10^9 years ago (Dunlop et al., 1978; Lowe, 1980a; Walter et al., 1980; Awramik et al., 1983).

It can be argued from the above findings that the appearance of a crustal sink for organic carbon and the stabilization of the "modern" carbon cycle as depicted in the box model of Fig. 4.1 occurred early in the earth's history. From the isotope mass balance

$$\delta^{13}C_{prim} = R\delta^{13}C_{org} + (1-R)\delta^{13}C_{carb}, \qquad (4.22)$$

where $R = C_{org}/(C_{org} + C_{carb})$, the biologic sink also must have been a fairly effective one because $\delta^{13}C_{carb}$ and $\delta^{13}C_{org}$ imply that the fraction R of organic carbon within the total carbon flux must have averaged 0.2 (or 20%), using $\delta^{13}C_{prim} = -5‰$ as the value of primordial mantle carbon originally fed into the system. Hence, the ratio of fluxes of C_{org} to C_{carb} into the sedimentary rocks should have been 1:4 since 3.5×10^9 years ago.

Whether or not the biologic carbon sink was already in existence 3.8×10^9 years ago depends on interpretation of the oldest part of the record preserved in the metasediments of the Isua Supracrustal Belt, West Greenland (Moorbath et al., 1973; Allårt, 1976; Schidlowski et al., 1976). With $\delta^{13}C_{carb} = -2.3 \pm 2.2‰$ and $\delta^{13}C_{org} = -13.0 \pm 4.9‰$, the averages for both carbon species from the Isua rocks are markedly offset from the rest of the record (Fig. 4.20). There is little doubt that the observed shifts are the result of amphibolite-grade metamorphism of the Isua sediments (Schidlowski, 1982; Schidlowski et al., 1983). In fact, both carbon species have behaved under metamorphic conditions in a manner consistent with presently available thermodynamic data and observations from other metamorphic terranes (Section 4.3.1.2). If this interpretation is correct, the premetamorphic carbonate and reduced carbon of the Isua suite had isotopic compositions close to those of their geologically younger counterparts, thus justifying an extrapolation of the $\delta^{13}C_{carb}$ and $\delta^{13}C_{org}$ records back to

3.8×10^9 years. This would imply a very early origin for life on earth, a conclusion that is not implausible in view of the extension of the fossil record to 3.5×10^9 years ago and reports of suspected microfossils even from the Isua rocks (see Pflug and Jaeschke-Boyer, 1979; Bridgwater et al., 1981). It might be added that the abundance of reduced carbon in the Isua rocks does not seem to differ very much from that of other Precambrian and Phanerozoic sediments (Schidlowski, 1982, Fig. 1), with carbon contents occasionally reaching 3% in the Isua banded iron-formation.

Thus, in summary, there is reason to believe that the dual nature of the carbon flux from the atmosphere–ocean system into sedimentary rocks was established more than 3.8×10^9 years ago, with C_{org} and C_{carb} piling up in the crust ever since then in about "modern" proportions. In terms of the isotope mass balance, the small variations in the isotope record (see Fig. 4.20), would imply a correspondingly constant partitioning of about 1:4 between reduced and oxidized carbon through time, the proportion of organic carbon having been fixed close to one-fifth of the total. Unless this early stabilization of the carbon cycle can be related to the activity of some inorganic geochemical process, the conclusion seems unavoidable that biologic controls were imposed on the carbon cycle very early in the earth's history.

The conspicuous and rather constant partitioning between C_{org} and C_{carb} indicated by the record suggests the presence of an effective control mechanism. Such control could have been provided by phosphate as a limiting nutrient for primary production of organic matter (cf. Section 4.5.3). Because phosphate is made available to the biosphere by weathering processes, its concentration in the ocean–biosphere reservoir is determined by quasi-uniform turnover rates that lead to flushing of the phosphorus content of weathered rocks through the environment. With the C/P ratio of about 106:1 in living marine organic matter being largely time invariant, a reasonable case can be made for phosphate control of the size of the earth's biomass. According to this model, the growth of the organic carbon reservoir would come to an end once the phosphorus available for the synthesis of organic matter was exhausted. Though important details of this concept still await further clarification (for discussion, see Holland, 1978; pp. 215 ff), the phosphate model holds the best potential for explaining the stability of terrestrial C_{org}/C_{carb} ratios from the very onset of the sedimentary record.

Whether substantial amounts of sedimentary carbon are restored to the mantle as a result of subduction of crustal material is not quite clear, but it seems improbable in view of the fact that both carbon species would lend themselves to extensive volatilization under such conditions. Whereas carbonates undergo large-scale decarbonation during the formation of metamorphic calc-silicates, conversion of kerogen to CO_2 mainly proceeds via oxidation by water, iron oxides, or sulfate. That the carbon content of highly metamorphosed sediments such as paragneisses is found to be in the range of common igneous rocks (Hoefs, 1965) is consistent with this view, suggesting a large-scale release of carbon during high-grade metamorphism (Section 4.5.2). Hence, we may reasonably assume that the transfer of carbon between mantle and crust largely follows a one-way route. Once stored in the crust and associated minor reservoirs, the element is unlikely to be ever returned to its site of primary residence; however, some present models of the carbon cycle show circulation of carbon through the upper mantle (Arthur et al., 1985).

4.6.2 Evolution of the Sulfur Cycle

In many respects, the sulfur cycle is a mirror image of the carbon cycle, the principal differences being (1) a change in the relative proportions of reduced and oxidized element species and (2) a large oceanic reservoir that stores about one-tenth of all the sulfur circulating in the system (Figs. 4.3 and 4.12). As in the case of carbon, primordial sulfur appears to have become partitioned between an organically derived (reduced) and an inorganic (oxidized) crustal reservoir. The fluxes feeding these reservoirs are established in the ocean, which stores an oxidized phase (dissolved marine sulfate) whose subsequent reduction to sulfide by microbial sulfate reducers supplies reduced sulfur to the crust. Sulfate also leaves the ocean in the form of sulfate evaporites.

Like carbon, sulfur in sediments considerably exceeds that in average igneous rocks. We may therefore assume that the element originally entered the exogenic system mainly as hydrogen sulfide (H_2S) and sulfur dioxide (SO_2), which are the principal sulfur species encountered in present-day volcanic emanations. Subsequent hydration of SO_2 would have yielded sulfite (SO_3^{2-}), but not sulfate (SO_4^{2-}), the most abundant sulfur species in the present ocean reservoir. Conversion of sulfite to sulfate, involving a valence change from S^{4+} to S^{6+}, is contingent on an additional oxidation step. There is little doubt that the abundance of sulfate in contemporary oceans is a direct consequence of the oxygenated conditions prevailing in the earth's near-surface environments. Thus, the marine sulfate reservoir is unlikely to be a primary feature of the sulfur cycle.

Because nonbiologic oxygen sources were scant on the primitive earth (cf. Walker, 1977, pp. 220 ff), the emergence of an oceanic SO_4^{2-} reservoir almost certainly had to await a biologically mediated oxidation of primordial sulfur to the level of sulfate. At present, this oxidation is mainly achieved by free oxygen released as a by-product of water-splitting photosynthesis:

$$2H_2O^* + CO_2 \xrightarrow{h\nu} CH_2O + H_2O + O_2^*, \qquad (4.23)$$

where the asterisk (*) indicates the source of the molecular oxygen. However, because bacterial photosynthesis preceded the evolution of the H_2O-splitting reaction, we may assume that the first large-scale introduction of sulfate to the environment was caused by the activity of photosynthetic (green and purple) sulfur bacteria. These bacteria are capable of using sulfur compounds at oxidation levels lower than sulfate as sources of reducing power; for example,

$$H_2S + 2CO_2 + 2H_2O \xrightarrow{h\nu} 2CH_2O + 2H^+ + SO_4^{2-}. \qquad (4.24)$$

4.6 LONG-TERM EVOLUTION OF THE BIOGEOCHEMICAL CYCLES

The early Precambrian probably provided optimum conditions for photosynthetic prokaryotes and may have been the golden age of photosynthetic sulfur bacteria. Accordingly, sulfate as a mild oxidant should have accumulated in appreciable quantities in the ancient seas before the advent of free oxygen in the environment (cf. Broda, 1975, p. 77).

The time of appearance of an oceanic sulfate reservoir is therefore one of the principal questions pertaining to the evolution of the sulfur cycle. A second problem concerns the emergence of dissimilatory sulfate reduction. This process is responsible for the basic configuration of the present sulfur cycle (cf. Figs. 4.3 and 4.12), notably, the splitting of the sulfur flux to the crust into a sulfide branch and a sulfate branch and the ensuing partitioning of the earth's exogenic sulfur into a reduced fraction and an oxidized fraction. An important corollary of this partitioning is an isotopic disproportionation of sulfur into a *light* (sulfide) and a *heavy* (sulfate) reservoir (Section 4.3.2).

The record of sulfate evaporites might provide a convenient tool for imposing time constraints on the buildup of the oldest oceanic sulfate reservoir. Unfortunately, this record has intrinsic limitations because of preferential recycling of evaporite rocks that, owing to their solubility under postdepositional conditions, are more easily destroyed than other sediments. Thus, major beds of sulfate evaporites tend to be absent from rocks older than about 1.4×10^9 years. Older sulfates are extremely patchy and occur mostly in the form of barite ($BaSO_4$), which stands a better chance of preservation than calcium sulfates (gypsum, anhydrite). The oldest sedimentary sulfates are therefore bedded barites described from various Archean terranes ranging back in age to $\sim 3.5 \times 10^9$ years (Fig. 4.21). Although the sedimentary origin of some of these occurrences has been repeatedly questioned, there is increasing evidence that the barite was either precipitated as a chemical sediment (Heinrichs and Reimer, 1977) or originated by diagenetic replacement of a primary calcium sulfate precursor by Ba^{2+}-bearing intrastratal solutions (Dunlop, 1978). Furthermore, the host sediments usually show sedimentary features of a primary evaporite facies, and there are reports of chert or barite pseudomorphs after primary gypsum (Lowe and Knauth, 1977; Dunlop, 1978; Lambert et al., 1978; Barley et al., 1979). These results suggest that sulfate was present in ancient seas from about 3.5×10^9 years ago in concentrations sufficiently high to undergo precipitation in appropriate sedimentary environments. This statement, in turn, lends support to concepts of a conservative seawater chemistry through time (cf. Garrels and Mackenzie, 1974; Holland, 1974). However, as pointed out above, the existence of oceanic sulfate does not necessarily imply oxygenated conditions for the earth's surface because the first large-scale introduction of SO_4^{2-} to ancient environments probably resulted from bacterial photosynthesis. Furthermore, stabilization of oxidized relative to reduced sulfur occurs at low O_2 levels.

The problem of the antiquity of bacterial sulfate reduction (Monster et al., 1979; Schidlowski, 1979) is necessarily coupled to the emergence of a marine sulfate reservoir because the availability of sufficient quantities of sulfate was a

170 GEOCHEMICAL CYCLES OF CARBON AND SULFUR

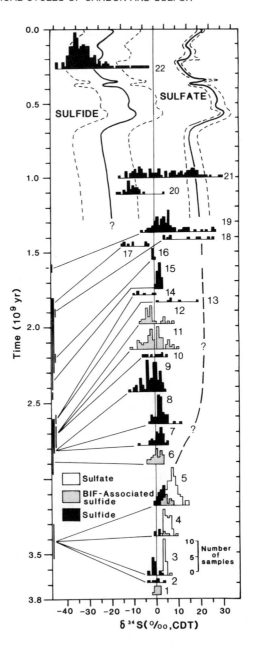

prerequisite for the evolution of sulfate respirers. Because oxidation of reduced primordial sulfur by photosynthetic sulfur bacteria was, in all probability, responsible for the first accumulation of sulfate in ancient seas, dissimilatory sulfate reduction may be conceived of as an adaptive reversal of this pathway by which H_2S and CO_2 were restored to the environment. The hypothesis that photosynthetic sulfur bacteria are the most probable ancestors of sulfate reducers is consistent with this idea (cf. Peck, 1974; Broda, 1975; Trüper, 1982). In this way, a light-powered biologic sulfur cycle was established, which in its pure form is confined in the modern oxygenated world to certain anaerobic niches (*sulfureta*) but which should have been of prime importance in the anoxic environment of the primitive earth.

Figure 4.21 Isotopic composition of sedimentary sulfide and sulfate through time. The isotope–age curve of Phanerozoic and Late Proterozoic *sulfates* is the best documented part of the record, the dashed lines enclosing the estimated uncertainty according to Claypool et al. (1980). Prior to 1.2×10^9 years ago, the record of sulfate evaporites is extremely scarce and virtually restricted to sedimentary barite ($BaSO_4$), which stands a better chance of preservation than calcium sulfates. Note that values for Early Archean barites tend to approach the zero per mill line. The corresponding age function for Phanerozoic and Late Proterozoic *sulfides* is a construct based on a difference of 40‰ between sulfide and sulfate, which is about the average fractionation observed in contemporary sedimentary environments. Considering the large spread in $\delta^{34}S$ values of sedimentary sulfides, the bulk of the measured values should fall into a range $\pm 12‰$ of this average (indicated by dashed lines on either side of the curve). The $\delta^{34}S$ pattern of the Permian Kupferschiefer (22) exemplifies a euxinic facies characterized by fractionations clearly in excess of the average. Localities: (1) Banded iron formation from Isua, West Greenland (Monster et al., 1979). (2) Swartkoppie Formation, Onverwacht Group, South Africa (Schidlowski, personal communication). (3) Warrawoona Group, Pilbara Block, Australia; sulfate as bedded barite (Lambert et al., 1978). (4) Fig Tree Group, South Africa; sulfate as bedded barite (Vinogradov et al., 1976; Lambert et al., 1978; Perry et al., (1971). (5) Iengra Series, Aldan Shield, Siberia, sulfate as barite and anhydrite (Vinogradov et al., 1976). (6) Banded iron formation from Rhodesian schist belts, mostly Sebakwian Group (Fripp et al., 1979). (7) Black shales from greenstone belts, Yilgarn Block, Australia (Donnelly et al., 1979). (8) Deer Lake greenstone belt, Minnesota, USA, (Ripley and Nicol, 1981). (9) Syngenetic "barren" sulfide deposits of Birch–Uchui greenstone belt, Superior Province, Canada (Seccombe, 1977). (10) Fortescue Group, Hamersley Basin, Australia (Schidlowski et al., 1983). (11) Michipocoten and (12) Woman River banded iron formation, Superior Province, Canada (Goodwin et al., 1976). (13) Steep Rock Lake Series, Canadian Shield (Veizer and Nielsen, personal communication). (14) Ventersdorp Supergroup, South Africa (Veizer and Nielsen, personal communication). (15) Cahill Formation, Pine Creek geosyncline, Australia; bedded sulfides (Donnelly et al., 1977). (16) Frood Series, Sudbury District, Canadian Shield (Thode et al., 1962). (17) Black shales from Outokumpu, Finland (Mäkelä, 1974 and personal communication). (18) Onwatin Slate, Sudbury Basin, Canadian Shield (Thode et al., 1962). (19) Shales, siltstones, and carbonates of McArthur Basin, Australia (Smith and Croxford, 1973, 1975). (20) Adirondack Mountains, Grenville Province, Canadian Shield; sedimentary sulfides (Buddington et al., 1969). (21) Nonesuch Shale, Superior Province, Canadian Shield (Burnie et al., 1972). (22) Kupferschiefer, Central Europe (Marowsky, 1969).

172 GEOCHEMICAL CYCLES OF CARBON AND SULFUR

Sulfate reducers can leave their mark in the geologic record because they release H_2S, which is susceptible to fixation as pyrite. Furthermore, bacterial sulfate reduction discriminates effectively against heavy sulfur (Section 4.3.2); thus, the $\delta^{34}S$ values of bacteriogenic sulfide serve as an index of biogenicity. It was through this process that the earth's primordial sulfur ($\delta^{34}S \approx 0‰$; Nielsen, 1978) was partitioned between biologically processed "light" sulfide ($\delta^{34}S \approx -28$ to $-13‰$) and "heavy" residual sulfate ($\delta^{34}S \approx 11–20‰$), with an average fractionation of 30–40‰ between the two sulfur species (Fig. 4.6). Evidence of such fractionation in ancient sediments would constitute proof of the activity of microbial sulfate reducers and thus might serve to constrain the time of emergence of this process. Ideally, the isotope data should be obtained from coeval sulfide and sulfate. This, however, is only possible for the younger (notably the Phanerozoic) record where sulfate values typically oscillate on the positive side of the δ scale and sulfide values on the negative side (Fig. 4.21). Because the Precambrian is largely devoid of evaporites, we unfortunately have to rely on the sulfide record alone for the major part of the earth's history.

Figure 4.21 shows schematically the isotope trends of sedimentary sulfide and sulfate during the Phanerozoic along with the record for the older history of the earth. The Precambrian data depicted primarily reflect the isotopic geochemistry of common sediments because major stratiform ore bodies (e.g., Broken Hill, Mount Isa) and assemblages of detrital sulfides (Witwatersrand) were excluded from this compilation. Whereas data for sulfate are largely lacking for the time span of 1.2–3.8×10^9 years ago, a certain amount of information is at least available for sedimentary sulfides. If established criteria for the identification of bacteriogenic sulfides are applied to this record, the oldest isotope patterns that might be interpreted as biogenic are those from the $\approx 2.7 \times 10^9$ year-old Michipicoten and Woman River banded iron formations (Nos. 11 and 12 of Fig. 4.21) and coeval sediments from the Birch–Uchi Greenstone Belt, Canada (No. 9 of Fig. 4.21). The most convincing of these patterns is that of the Woman River sulfides, which, although showing a fairly large spread of $\delta^{34}S$ values, is distinctly skewed toward the negative side (No. 12 of Fig. 4.21). The $\delta^{34}S$ averages of coexisting sulfide–sulfate pairs from the early Archean (3.3–3.5×10^9 years ago) differ only by a few per mill (see Nos. 3,4,5 of Fig. 4.21). These small fractionations are certainly not indicative of the presence of sulfate-reducing bacteria nor are the several Archean sulfide occurrences with $\delta^{34}S$ means close to zero per mil. However, the substantially reduced fractionation between sulfide and sulfate observed in these oldest sediments would be consistent with a process of oxidation of reduced sulfur to sulfate by photosynthetic sulfur bacteria.

Based on the record summarized in Fig. 4.21, a case can be made for the rise of bacterial sulfate reduction at or prior to 2.7×10^9 years ago. Because other evidence indicates that sulfate may have been present in some marine environments as early as 3.5×10^9 years ago, this would imply a major time lag between the availability of sulfate and its ultimate biologic utilization. However, because of lack of data, we cannot exclude the possibility that the first appearance of bacteriogenic sulfide patterns may give only a minimum age for the underlying biologic event. Bacterial sulfate reducers could have gained control of the earth's sulfur cycle perhaps as early as 3×10^9 years ago. Variations in their activity are responsible, in all likelihood, for secular imbalances of the cycle as reflected by the isotope–age curve of Phanerozoic sulfate evaporites. The effect of this variation in bacterial activity on the interconnected cycles of carbon and oxygen probably has been considerable (Schidlowski et al., 1977; Veizer et al., 1980; Schidlowski and Junge, 1981). Similar imbalances in the Precambrian are possibly lost to view in the less detailed rock record of those times.

5 THE EVOLVING EXOGENIC CYCLE

J. Veizer

5.1 INTRODUCTION

For most of the century, it has been recognized that the present-day thickness and areal extent of Phanerozoic sedimentary strata are progressively larger with their decreasing geologic age. The present day areal distributions of basement age provinces also conform to this pattern. Such a pattern has been interpreted either as reflecting an increase in the frequency of tectonic pulses—and hence in the rate of sedimentation, crustal generation, and cratonization—toward the present (Barrell, 1917; Schuchert, 1931; Stille, 1936; Holmes, 1947; Umbgrove, 1947; Hurley and Rand, 1969; Ronov, 1976; Tugarinov and Bibikova, 1976; Salop, 1977) or as resulting from better preservation of the younger part of the geologic record (Gilluly, 1949; Gregor, 1968, 1970, 1985; Garrels and Mackenzie, 1969, 1971a,b; Li, 1972; Veizer, 1976b, 1979; Veizer and Jansen, 1979, 1985).

Study of the rocks themselves led to similarly opposing conclusions. The observed secular (= age) variations in relative proportions of lithological types and in chemistry of sedimentary rocks (Daly, 1909; Vinogradov et al., 1952; Nanz, 1953; Engel, 1963; Strakhov, 1964, 1969; Ronov, 1964, 1968, 1972, 1980; Ronov and Migdisov, 1971) were mostly given an evolutionary interpretation. An opposing, uniformitarian, approach was proposed by Garrels and Mackenzie (1971a, 1972) and Garrels et al. (1972). Subsequent contributions (Van Moort, 1972, 1973; Veizer, 1973, 1978; Engel et al., 1974; Mackenzie, 1975; Schwab, 1978; Veizer and Garrett, 1978; Taylor and McLennan, 1985), although attempting to separate evolutionary from postdepositional and cyclic signatures, vacillated between these two limiting alternatives.

For isotopes, the situation is less polarized and the consensus favors deviations from the present-day steady state as the likely cause of secular trends. Such trends have been documented for $^{87}Sr/^{86}Sr$ in carbonates, $\delta^{34}S$ in sulfates, $\delta^{18}O$ in sulfates, $\delta^{13}C$ in carbonates, and perhaps $\delta^{13}C$ in organic matter (see Chapter 4 for details). Only the trends toward decreasing values of $\delta^{18}O$ with increasing age in carbonates (Clayton and Degens, 1959; Keith and Weber, 1964; Schidlowski et al., 1975, 1983; Veizer and Hoefs, 1976; Veizer et al., 1986; Popp et al., 1986), cherts (Knauth and Lowe, 1978; Perry et al., 1978; Karhu and Epstein, 1986), and phosphates (Longinelli and Nuti, 1968; Lutz et al., 1984) are still embroiled in the familiar primary versus secondary controversy.

In summary, the *evolutionists* believe that the secular trends reflect mainly variations in the nature and intensity of exogenic processes through time. The *cyclists*, on the other hand, emphasize a uniformitarian approach, with the secular variations believed to be predominantly—but not always entirely—a consequence of postdepositional phenomena or of nonrepresentative data. If stated in such a polarized form, the two interpretations are mutually exclusive. However, this is not necessarily the case. Small deviations from an apparent steady state, although within observational error on a short time scale, may result in measurable differences over a long time span. Human life represents a good homology. It is controlled by an *apparently* steady-state daily cycle. For most purposes, it can be treated as such if a short time scale is considered. On a longer time scale, however, it is obvious that human life is a unidirectional aging process and the negligible differences between daily cycles become magnified through their cumulative effect. As a consequence, the daily cycle of an octogenarian, although not much different from that of the septuagenarian, is easily distinguishable from that of the teenager. In geologic terms, the adage that the "present is a key to the past" applies to the nature of processes rather than to the rates at which they operate, although present-day rates may represent an upper or lower limit.

This chapter attempts to show that recycling and evolution are not opposing, but complementary, concepts. It will concentrate on chemical and lithological parameters reflecting the evolution of the detrital component of sediments. This restriction permits only a cursory discussion of variations in stable isotopes, silica, alkaline earth, and oxidation state of metals, all reflecting mostly the evolution of the hydrosphere–atmosphere system. Similarly, it does not deal with the evolution of mineral deposits and of life, both well amenable to the outlined conceptual treatment.

5.2 THE EARTH SYSTEM

The major outer parts of the earth are its lithosphere, hydrosphere, atmosphere, and biosphere. We are aware that rocks, water, air, and life interact, and we also assume intuitively that they are all part of the terrestrial exogenic *system*. This system should therefore be definable by the rules and approaches of the general system science theory (e.g., von Bertalanffy, 1968; Forrester, 1968; Odum, 1983) with its subsets, such as population dynamics and hierarchical structures.

5.2.1 Population Dynamics

In its abstract meaning, a *population* is a natural grouping of related constituent *units*. The terminologies used for designation of populations and units vary with the field of inquiry and some examples are given in Table 5.1. A population, however, is a dynamic phenomenon, with constituent units being *born* and *dying*. This *birth–death* cycle, or *rate of recycling*, imposes a specific age distribution

5.2 THE EARTH SYSTEM

TABLE 5.1. Examples of Populations and Their Constituent Units

Field	Population	Unit	Recycling Process
Biology	Population	People, trees	Birth/death
Astronomy	Galaxy	Stars	"Birth/death"
Geosciences	Domain reservoir	Weight or volume measures	Creation/destruction, influx/efflux
Banking	Reserve	$, DM	Debit/credit or cash flow
Insurance	Population	People, sheep	Birth/death (survivorship tables)
Industry	Standing stock	Cars	Manufacture/scrapping
Religion	?	Souls	Reincarnation

pattern on units constituting the population. If the process of recycling proceeds internally, within the population, it is termed *internal or cannibalistic* recycling. In contrast, the *external* recycling presupposes birth and death outside the population. Taking the national populations of people as an example, the internal birth–death process represents cannibalistic recycling, whereas immigration (mostly of young individuals) and emigration (of retiring people) approximates the external recycling. Most natural populations are involved in internal as well as external recycling (cf. Bailey, 1964, Chapter 8). It is only their relative importance that varies from one population to another. Both types of recycling, independently or in concert, control the age structure of the population. Taking geologic populations as an example, erosion–deposition of sediments represents almost entirely a cannibalistic recycling, whereas generation of oceanic crust from the mantle and its subduction back into the mantle is—from the exogenic perspective—almost exclusively an external recycling phenomenon.

The fundamental parameters essential for quantitative treatment of population dynamics are the population *size* (A_0) and its *recycling rate*. Because of unequal sizes of natural populations (e.g., China versus Liechtenstein), A_0 is normalized in the subsequent discussion to one population (or 100%) and the rates of recycling relate to this normalized size. Absolute rates can be established by multiplying this relative recycling rate (parameter b below) by population size (e.g., 1×10^9 inhabitants in the Chinese population).

A steady-state natural population, characterized by a continuous birth–death cycle, is usually typified by an age structure similar to that in Fig. 5.1. The proportion of progressively older constituent units, such as human individuals (see Anonymous, 1985), decreases exponentially, because *mortality*, like radioactive decay, is a first-order function of the size of the given age group. The age distribution of the population can be expressed either as a histogram or as a cumulative curve. The latter is the preferred presentation for our purposes because it defines all necessary parameters of a given population. These are its *half-life* τ_{50}, *mean age* τ_{mean}, and *oblivion age* or life expectancy τ_{max} (Fig. 5.1).

178 THE EVOLVING EXOGENIC CYCLE

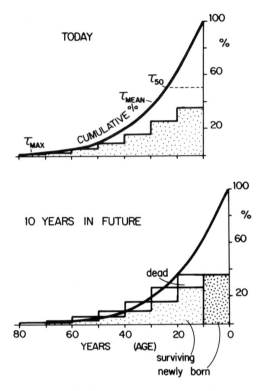

Figure 5.1 Simplified age distribution pattern for a steady-state extant population. In this case, the natality/mortality rate is 35% of the total population for a 10-year interval ($b = 35 \times 10^{-3}$ year^{-1}). τ_{max} is defined as the 5th percentile. In practice τ_{max} is the age at which the resolution of the database becomes indistinguishable from the background. The b value for the same population is inversely proportional to the available time resolution T [Eq. (5.2)]. Today's "instantaneous" rates of deposition and erosion of sediments exceed those calculated from the geological record based on time resolution of 10^6–10^7 years (cf. Sadler, 1981). It is therefore essential to stipulate the resolution T in consideration of rates. In this chapter, the resolution is specified by a superscript: b^{10} is the recycling proportionality constant. See also Gingerich (1983) and Veizer (1984, 1988). Reproduced from Veizer and Jansen (1985) by permission of the University of Chicago Press.

Complex systems, such as those structured from natural populations, correspond frequently to what system science theory describes as systems with feedback loops (cf. Forrester, 1968). For steady-state first-order (= single population) systems, the survival rate of constituent units can be expressed as

$$A_{t^*} = A_0 e^{-kt^*}, \tag{5.1}$$

where A_{t^*} is the cumulative fraction of the surviving population older than t^*, $A_0 = 1$ (one population), t^* is age (not time), and k is the rate constant for the recycling process. In the subsequent discussion, the recycling rate is considered in the form of a *recycling proportionality constant b*, which is related to the above equation through formalism:

$$b = 1 - e^{-kT}, \qquad (5.2)$$

where T is the time resolution or duration of recycling (cf. Veizer and Jansen, 1979, 1985). In general, the larger the b value—that is, the faster the rate of recycling—the steeper the slope of the cumulative curve and the shorter the τ_{50} and τ_{max} of the population. In other words, the chance of survival diminishes with increasing *mortality* rates. For a steady-state extant population, *natality* per unit time must equal combined mortality for all age groups during the same time interval. Consequently, the cumulative slope remains the same but propagates into the future (Fig. 5.1, bottom).

The above terminology is applicable to internally (cannibalistically) recycling populations. In an external type of recycling, the influx and efflux (immigration and emigration) cause a similar age structure, but the terminology differs. In this case, the average duration an individual unit resides within the population is termed the *residence time* τ_{res}. Mathematically, τ_{res} is similar to the cannibalistic τ_{mean} and it relates to the above parameters as $\tau_{max} \geqslant \tau_{res} \geqslant \tau_{50}$. It is this alternative—that is, populations interconnected by external fluxes—that is usually referred to as the familiar box model by natural scientists. Frequently, box models are nothing more than one possible arrangement for propagation of cyclic populations.

In the subsequent discussion, the terminology of the cannibalistic populations is employed. Note, however, that the age distribution patterns and the recycling rates calculated from these patterns are a consequence of both cannibalistic and external recycling. At this stage, we lack the data and the criteria for quantification of their relative significance. Nevertheless, from the point of view of preservation probability, it may be desirable, but not essential, to know whether the constituent units (geologic entities) have been created and destroyed by internal, external, or combined phenomena.

A family of interrelated populations defines a system (e.g., von Bertalanffy, 1968). Note that despite subsequent quantifications, a *quantitative* model of the *global exogenic system* is *not* being proposed. Our present understanding of interactions among constituent populations is at best semiquantitative. For a self-confined, closed system, it might be possible to derive a working dynamic model simply from the knowledge of the level variables such as population sizes (Forrester, 1968). This, however, presupposes that all fluxes—that is, rate variables—are of the external type. As discussed above, this is only one aspect of the dynamics of recycling. The development of a quantitative model must therefore await the resolution of the relative importance of internal versus external recycling rates for each constituent population. In addition, it is not yet clear on what time scales the global exogenic system can be regarded as closed and when the external inputs of energy (e.g., solar or mantle heat) must be taken into consideration. In other words, it is not clear on what time scales and to what degree the system must be treated as partially open. Note also that, for external recycling, it is implicitly assumed that all units entering a population acquire zero age. In nature, most immigrants are indeed young, but not all are babies. Thus, even for a purely externally recycling population, its mean residence time and mean age need not entirely coincide. In the geologic context, *new* sediments may contain, for example, young resedimented intrabasin grains (e.g., freshly lithified carbonates) as well as older extrabasinal detrital components. Similar problems arise, for example, with inheritance and resetting of radiometric ages and with neosom versus assimilated paleosom within plutons. While being aware of these complications, we are not yet in a position to quantify such subordinate phenomena. For all these reasons, this chapter concentrates on major patterns of intrapopulation dynamics, with no attempt to resolve interpopulation relations. The model presented here can perhaps be classified as a white box type, in contrast to the usual black box models ubiquitous, for example, in oceanography and geochemistry. The former describe the structure and interactions within boxes (populations) but do not pay much attention to interactions of boxes. The latter have the opposite attributes. It is evident that a quantitative model of the global exogenic system must be a synthesis of both approaches.

Among natural populations, two major deviations from the ideal pattern are ubiquitous (cf. Lotka, 1956; Wilson and Bossert, 1971; Pielou, 1977; Lerman, 1979). The first exception consists of populations with excessive proportions of young individuals (e.g., planktonic larval stages, third world countries) because the infant mortality is very high. The chances for survival improve considerably only with the attainment of an advanced growth stage. Such populations, if compared to the ideal ones, have shorter τ_{50} and τ_{mean}, but their τ_{max} may be

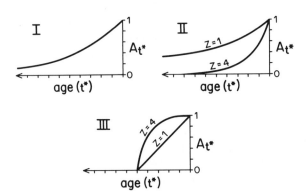

Figure 5.2 Cumulative age distribution functions for various populations. I—exponential distribution [Eq. (5.1)]; II—power law distribution approaching zero assymptotically [Eq. (5.3)]; III—age distribution attaining zero at finite age [Eq. (5.4)]. For explanation of z see the text. These three subtypes of age distribution patterns are also termed ecological, physiological, and maturational, respectively (cf. Tanner, 1978, p. 86). Modified from Lerman (1979).

similar (Fig. 5.2). The mathematical formalism for such populations (e.g., Lerman, 1979) is a power law function,

$$A_{t*} = A_0 (1 + kt^*)^{-z}, \qquad (5.3)$$

where the exponent z increases for populations with progressively larger mortality rates of young individuals.

The other common exception consists of populations with suppressed mortality of young individuals (e.g., the advanced industrial nations; Anonymous, 1985). In these instances, the mortality rates increase rapidly as the life span τ_{max} is approached (Fig. 5.2). Mathematical expression for this relation is

$$A_{t*} = A_0 [1 - (kt^*)^z], \qquad (5.4)$$

where the exponent z increases for populations with progressively larger mortality rates of old individuals.

The above relations are valid for populations of constant size, but not necessarily for growing or declining ones. For non-steady-state populations with stable age structures—that is, those with overall rates of growth or decline much slower than the rates of recycling of their constituents (e.g., generational procreation)—the age distributions approach the pattern of the constant-size populations. The calculated recycling rates are therefore identical. For populations where the overall growth (decline) approaches the rate of recycling of constituent units, independent criteria are required to differentiate recycling from the growth (decline) component. As an illustration, case II in Fig. 5.2 can be interpreted either as indicating higher mortality of young individuals within a steady-state population or as representing a population that expanded its overall size by excessive births during its latest history. Similarly, case III may be a consequence of enhanced mortality of old constituents or of fast overall growth rates in incipient stages of population development.

The above discussion assumed a quasicontinuous birth–death process for a first-order system. The earth system, however, contains a plethora of interdependent populations and is therefore a second- or higher-order entity with a multitude of feedback loops (cf. Forrester, 1968). This causes oscillations, at a hierarchy of frequencies and amplitudes, around the smooth overall patterns. Furthermore, natural processes, particularly those in geology, are usually discrete and episodic phenomena. Because of all these factors, partial intervals have birth and death rates that deviate from the smooth average rates. This appears as a scatter around the overall trends (Fig. 5.3), with the connecting tangents having either shallower or steeper slopes. The former indicates intervals when the partial mortality was suppressed or the natality enhanced, and the latter indicates the opposite situation (see also Garrels and Mackenzie, 1971a, Chapter 10). Note again that a given partial slope may reflect deviations in natality, in mortality, or in their combined effect. Usually, the problem is not resolvable, but the combined effect is the most likely alternative. As the

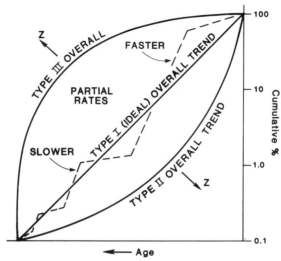

Figure 5.3 Cumulative age distribution functions for various populations plotted on a semilogarithmic scale. See Fig. 5.2 and the text for explanations.

population ages, the magnitude of this higher-order scatter diminishes to the level of uncertainties in the database (Fig. 5.3). Quantitative interpretation of such higher-order features from the preserved record is therefore possible only for a length of time roughly comparable to the life-span τ_{max} of a given population. It would be pointless to attempt quantification of oceanic spreading rates from the fragmentary record of pre-Jurassic ophiolites. Any such quantification must rely on some derivative signal, such as isotopic composition of seawater, which may be preserved in coeval sediments. In contrast to the fast cycling oceanic crust, the higher-order scatter for slowly cycling populations (e.g., continental crust) remains considerable, because it has not yet been smoothed out by the superimposed recycling. Such populations still retain vestiges of ancient episodic events.

For geologic entities (e.g., crustal segments, mineral deposits, tectonic domains, fossils), the age distribution patterns can be extracted from their stratigraphic and geochronologic assignments. At present, only major features of the record can be quantitatively interpreted, because the database is usually not of the desired reliability. It is my hope that this contribution will stimulate compilation of the required inventories and thus facilitate development of the quantitative history of the earth.

5.3 RECYCLING OF OCEANIC CRUST

The generation of the oceanic crust on mid-ocean ridges and its subsequent subduction back into the mantle is well established and can be used to demonstrate

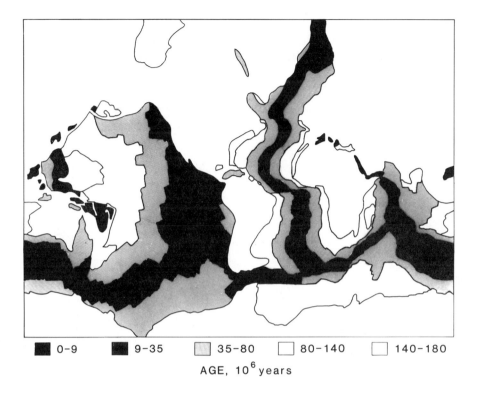

Figure 5.4 Schematic map of distribution of ages in the ocean crust. Modified from Ballard (1983).

the recycling concept. The distribution of ages in the oceanic crust is well known from magnetostratigraphic, geochronological, and paleontological information, and planimetric analyses of such maps (Fig. 5.4) yield the age distribution pattern by area, volume, or mass. This present-day age distribution indeed conforms to the theoretically predicted pattern (Fig. 5.5). The calculated rate of recycling b^{10} is $\sim 110 \pm 5 \times 10^{-10}$ year^{-1}, which multiplied by the area (or volume) translates into recycling of ~ 3.5 km^2 (or ~ 20 km^3) of oceanic crust per year. Its half-life τ_{50} is $\sim 59 \times 10^6$ years and the 95% probability of destruction (τ_{max}) should theoretically be $\sim 250 \times 10^6$ years. The actually observed τ_{max} is less ($\sim 160 \times 10^6$ years) because in detail the age distribution pattern of the oceanic crust approximates the type III rather than type I population (see Figs. 5.2 and 5.3 and Section 5.4). The generation–destruction of oceanic crust is almost entirely recycling of an external type that proceeds by generation of the crust from and resorption into the mantle. Internal recycling, such as intracrustal melting, plays only a subordinate role. Oceanic crust, however, represents only one population of the earth system. The dynamics and

184 THE EVOLVING EXOGENIC CYCLE

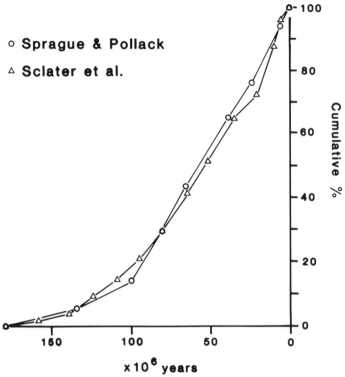

Figure 5.5 Present-day area–age cumulative distribution of oceanic crust. Modified from Veizer and Jansen (1985). Based on the data of Sprague and Pollack (1980) and Sclater et al. (1981).

evolution of this system are controlled by interactions between oceans and continents, and the working of such a machine is codified in the tenets of global plate tectonics.

5.4 GLOBAL TECTONIC REALMS AND THEIR RECYCLING RATES

The concept of global tectonics (Dietz, 1961; Hess, 1962; Morgan, 1968; Le Pichon et al., 1973) combined the earlier proposals of continental drift and sea-floor spreading into a unified theory of terrestrial dynamics. It introduced the notion of continual generation and destruction of oceanic crust and implied similar consequences for other tectonic realms. One possible classification of global tectonic realms, reflecting to some extent the restrictions of the available database, encompasses the following tectonic entities (Fig. 5.6):

1. Oceanic crust and its overlying abyssal and pelagic sedimentary basins.
2. Active margin basins.

5.4 GLOBAL TECTONIC REALMS AND THEIR RECYCLING RATES

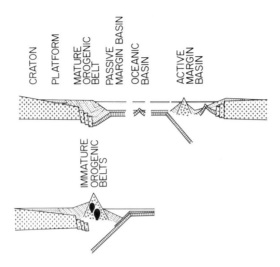

Figure 5.6 Schematic presentations of major tectonic realms in the context of the global plate tectonics. Reproduced from Veizer (1987).

3. Passive margin basins.
4. Immature orogenic belts (young mountain ranges).
5. Mature, worn-down, orogenic belts (roots of former mountains).
6. Platforms.
7. Cratons (here taken as continental crystalline and metamorphic basement).

In general, the observed mass–age or area–age relations for rocks in these tectonic realms (Fig. 5.7) correspond well with the patterns espoused in Fig. 5.2, suggesting that we are indeed dealing with populations characterized by a quasicontinuous generation and destruction of the constituent rock units. Assuming, as a first approximation, that deviations from the strictly exponential relations of Eq. (5.1) can be attributed to an unsatisfactory database, Veizer and Jansen (1985) calculated the relevant recycling proportionality constants (b values) and estimated the half-lives (τ_{50} values) for the studied tectonic realms. These results are summarized in Fig. 5.8 and Table 5.2. They are consistent with the existence of two fundamental tectonic domains: the continental and the oceanic one, with a transition at $b^{10} \approx 40 \times 10^{-10}$ year^{-1} and $\tau_{50} \approx 225 \times 10^6$ years. The *oceanic domain*, encompassing tectonic realms 1–4, represents a relatively fast cycling system with τ_{50} in the 10^7–10^8-year time range. In contrast, tectonic realms of the *continental domain*, 5–7, are being recycled on time scales of 10^8–10^9 years. The calculated parameters also indicate that mountain building (orogenesis), although a result of interaction of these two domains, is a response to the operation of the oceanic supercycle. If so, sea-floor spreading causes

186 THE EVOLVING EXOGENIC CYCLE

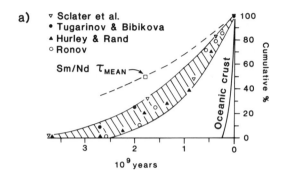

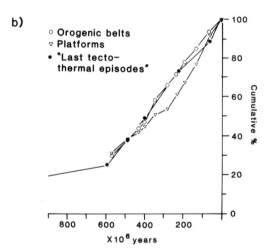

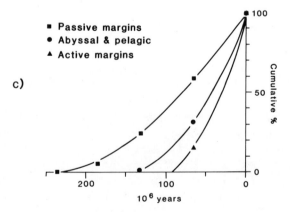

5.4 GLOBAL TECTONIC REALMS AND THEIR RECYCLING RATES

orogenesis and subsequent erosion of mountains back into the ocean. From the system science point of view, orogenesis may represent a controlling autocatalytic feedback loop in a simplified producer–consumer (P–R) model; the producer being the oceanic domain and the consumer being the continental domain.

The overal consistency of the available data with the concept of population dynamics leads to a prediction of the probable oblivion age for each tectonic realm. Theoretically, if τ_{max} is taken as the fifth percentile, the oblivion ages for the type I populations should be a factor of ~ 4.5 of the respective half-lives. Empirically, τ_{max} is usually only ~ 3.0–3.5 times τ_{50} and this discrepancy is considered below. These qualifications notwithstanding, the concept of oblivion ages dictates that the degree of tectonic diversity observed in the geologic record must be a function of time. This diversity diminishes as the given segment of solid earth ages, and the rate of memory loss is inversely proportional to recycling rates (b values) of the constituent tectonic realms. For a steady-state system, the calculated theoretical preservational probabilities are depicted in Fig. 5.9. This reasoning shows that the realms of the oceanic domain (basins of active margins to immature orogenic belts) should have only $\leqslant 5\%$ chance of survival in crustal segments older than ~ 100–300×10^6 years. This poor preservation potential, rather than any fundamental difference in the mode of coeval tectonic styles, may also have been the reason for the scarcity of, for example, glaucophane schists, paired metamorphic belts, and ophiolites in the Paleozoic and particularly the Precambrian records (cf. Ernst, 1983), because they are all manifestations of the oceanic tectonic domain. The 95% probability for obliteration of platforms and the roots of former orogenic belts is $\sim 1500 \times 10^6$ years. The implications for crystalline continental basement are discussed in Section 5.5.

If one accepts that the quality of data is sufficient for demonstration of nuances of age distributions for the tectonic realms, it appears that all these realms (Fig. 5.10) conform to the mathematical formalism (5.3) rather than (5.1) and this explains why τ_{max} is usually less than 4.5 times τ_{50}. This suggests enhanced rates of destruction ("mortality") as the constituent rock units age. Such an interpretation is in accord with our understanding of the mode of destruction (subduction and obduction) of the fast-cycling oceanic crust and its overlying sediments. Destruction is higher at the old and cold trailing edges than at the young and buoyant leading edges of oceanic plates. For slowly cycling populations, such as the continental crust, it is essential to consider also the rates of creative

---

Figure 5.7 Present-day area–age or mass–age distributions of rocks in various tectonic realms. Modified from Veizer and Jansen (1985). The curves are based on the estimates of Hurley and Rand (1969), Ronov (1976, 1982), Tugarinov and Bibikova (1976), Sclater et al. (1981) and Gregor (1985). Alternative data are available in Ronov et al. (1986a,b) and Sloss (1976). (*a*) Continental and oceanic crusts, (*b*) platforms and mature orogenic belts, (*c*) basins of passive and active margins and basins situated on oceanic crust.

TABLE 5.2. Summary of Recycling Rates and Half-Lives For Major Global Tectonic Realms[a]

Tectonic Realm	Recycling Rate (b^{10}) in units of 10^{-10} yr^{-1}		Theoretical Half-Life (τ_{50}) × 10^6 yr
	Steady-State Since 4500 × 10^6 yr Ago	"Preferred" or Logistic Growth Model[b]	
Active margin basins	223.0 ± 51.0	223.0 ± 51.0	27
Oceanic intraplate basins	126.0	126.0	51
Oceanic crust	110.0 ± 5.0	110.0 ± 5.0	59
Passive margin basins	88.0 ± 8.0	88.0 ± 5.0	75
Immature orogenic belts	85.0 ± 35.0	85.0 ± 35.0	78
Mature orogenic belts (roots)	19.3	18.4 ± 0.5	355
Platforms	19.0	18.6 ± 0.6	361
Continental basement: Probability of being affected by			
Low-grade metamorphism	10.2	8.5 ± 0.7	673
High-grade metamorphism	7.0	3.7 ± 1.3	987
Recycling into mantle	4.0	2.1 ± 1.0	1728

[a] Modified from Veizer and Jansen (1985).
[b] The "preferred" or logistic growth model assumes an initiation of generation and/or preservation of continental crust ~4000 × 10^6 years ago, fast rate of crustal growth 3–2 × 10^9 years ago, and near steady state since 2000 × 10^6 years ago. Note that the τ_{50}'s for the steady state and logistic growth models are comparable, because slower growth and recycling rates in the latter alternative yield internal age distribution patterns similar to those of the steady state.

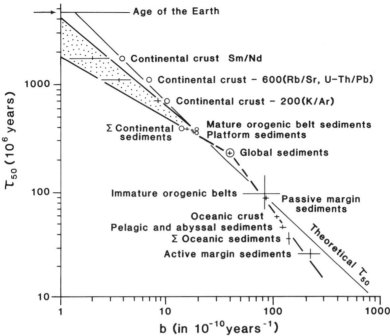

Figure 5.8 Plot of the observed half-mass and half-area ages (τ_{50}) and of recycling rates (b) for major global tectonic realms. Crosses represent solutions for the "preferred" logistic model of continental growth, assuming that continents started to grow (or be preserved) at $\sim 4000 \times 10^6$ years ago, followed ~ 3000–2000×10^6 years ago by a fast growth rate; subsequently, the continents have been of a near present-day size. The circles give results for the steady-state alternative. The two limiting solutions are identical at $b \geqslant 40 \times 10^{-10}$ year^{-1}. Continental crust -200, -600 represent solutions for the bottom and the top enveloping curves in Fig. 5.7a. Continental crust Sm/Nd is calculated for $\tau_{50} \approx 1.8 \times 10^9$ year; a value claimed as the mean age (and thus not strictly τ_{50}) of the continental crust by Jacobsen and Wasserburg (1979). Modified from Veizer and Jansen (1985).

processes ("natality") for the development of the observed type III age patterns. In general, the apparent pre-1700 $\times 10^6$-year "deficiency" of continental crust (Fig. 5.10) may have been a consequence of (1) poor data, (2) enhanced external recycling (generation *and* destruction) of continental crust via the mantle prior to 1.7 $\times 10^9$ years ago, (3) enhanced destruction of progressively older rocks at constant generation rates, (4) enhanced generation of continental crust ~ 3 to 2×10^9 years ago at constant destruction rates, and (5) any combination of the above.

5.5 GROWTH AND RECYCLING OF CONTINENTS

The observation that the ages of basement provinces decrease from central nuclei toward the periphery of continents (e.g., Fig. 5.11) is the cornerstone of

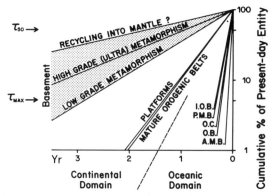

Figure 5.9 Preservation probabilities for major global tectonic realms. A.M.B. = active margin basins, O.B. = oceanic intraplate basins, O.C. = oceanic crust, P.M.B. = passive margin basins, I.O.B. = immature orogenic belts. Rates (Table 5.2) were derived on the assumption that all deviations from the ideal pattern of the type I age distributions were a consequence of the poor quality of the available database. This may not have always been the case, and the most frequently observed deviations were toward the type III age distributions, but because of z close to unity, the accepted recycling rates and half-lives are reasonable first-order approximations. Reproduced from Veizer (1988a) by permission of Pergamon Journals Inc.

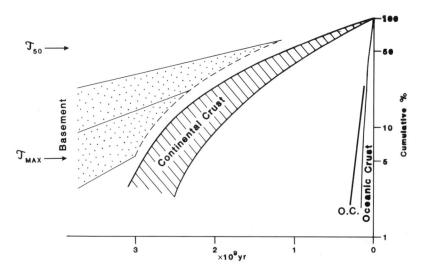

Figure 5.10 The observed present-day age distributions of continental and oceanic crusts (cf. Fig. 5.7) as compared to the probabilities of preservation based on the type I distribution patterns (cf. Figs. 5.3 and 5.9). Note the deviations toward type III distribution patterns, a feature of other tectonic realms. The envelope for the continental crust relates to the bottom half of the basement probabilities (see Fig. 5.9 for additional explanations). Modified from Veizer (1988a).

5.5 GROWTH AND RECYLING OF CONTINENTS

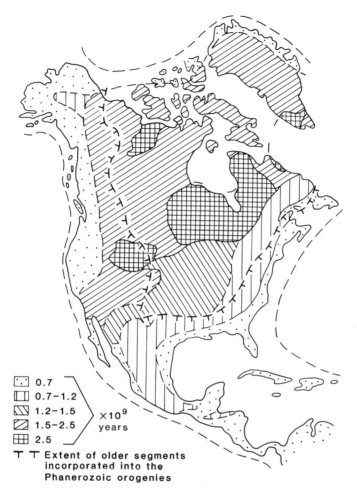

Figure 5.11 Major crustal provinces of North America, generalized and modified from Condie (1976).

the theory of continental formation by the process of lateral accretion. The available global estimates of the present-day distribution of such provinces (Figs. 5.7 and 5.10) show a band of increasing area per unit time from $\sim 3.7 \times 10^9$ years ago until the present. The lower envelope of this band is defined by the estimates of Ronov (1976) and Hurley and Rand (1969). The former paper represents the areal distribution of tectonically stabilized (cratonized) terrains, whereas the latter defines age provinces mostly from the point of view of K/Ar dates, which usually record the last thermal episodes. Thus, the lower envelope, with $\tau_{50} \approx 670 \times 10^6$ years (Table 5.2), is a reflection of intracrustal cratonization events recorded in the K/Ar systematics. The upper envelope, with $\tau_{50} \approx 990 \times 10^6$ years, is based on the areal distribution of tectonic provinces as established by the more retentive whole-rock Rb/Sr and zircon U-Th/Pb age determinations (Tugarinov and Bibikova, 1976; Sclater et al., 1981). If a similar distribution of tectonic provinces were based on the still more retentive Sm/Nd and Lu/Hf dating pairs, the resulting curves would likely be of even shallower slopes. The claimed Sm/Nd *mean age* of $\sim 1.8 \times 10^9$ years for continental crust (Jacobsen and Wasserburg, 1979) is consistent with this statement.

The conformity of the observed age distribution patterns with the theoretical predictions shows that recycling has been a major factor in generating these patterns in the continental basement. Unfortunately, the precise combination of the alternatives (1)–(4) listed in Section 5.4 is more difficult to disentangle. Nonetheless, despite the fact that the available estimates are not entirely satisfactory, it is unlikely that alternative (1), poor data, is the cause of the observed broad band or of the pre-1700 $\times 10^6$-year deficiency of continental crust. Independent estimates for mineral deposits (Veizer et al., 1988) support the first-order validity of the data.

It is my belief that crust–mantle interaction (recycling) was more vigorous on the early earth and this point is elaborated on in the subsequent text. Even so, I do not consider recycling via the mantle [alternative (2)] to be the dominant cause of the observed pattern. In the first instance, such recycling would result in simultaneous resetting of all isotopic systems. Consequently, all systematics should have identical τ_{50} and τ_{mean}, a requirement at variance with the observation. Second, an enhanced rate of crust–mantle recycling implies an overall acceleration of ancient plate tectonic processes thus shifting the whole pattern in Fig. 5.9 to the right. With such rates, only the stable metamorphic basement should be able to survive to modern times because of its *relatively* tardiest recycling rates. Yet, if compared to the Proterozoic, the Archean contains a disproportionate abundance of the perishable greenstones. The latter, regardless of their precise present-day analogue, are an expression of the ephemeral oceanic tectonic domain in the sense of Veizer and Jansen (1985). The temporal distribution of greenstones is therefore entirely opposite to that expected from continuous recycling, regardless of its actual rate (see also Veizer, 1973, 1983a). The fact that so many of them survived to this day, despite this recycling, argues for their excessive original abundance and entrainment into the precociously growing and stabilizing continents. Third, the dearth of old detrital components in Archean sediments (Section 5.8.2) argues against the existence of an old source of continental dimensions.

Mechanism (3), enhanced destruction of older rocks, appears to have been an unlikely cause, because increased tectonic maturity usually leads to more stable continental crust in direct contrast to theoretical requirements.

This elimination process leaves alternative (4), enhanced growth of continents ~ 3 to 2×10^9 years ago, as the likely cause of the observed age distribution pattern. The evolution of sediments (Sections 5.7 and 5.8.2) is in accord with such an interpretation.

The details of the growth curve for continents are presently vigorously disputed. Nevertheless, most natural systems—and solid earth is likely no

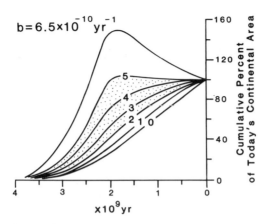

Figure 5.12 The deconvoluted growth patterns for cumulative area of the upper continental crust. The b values used for deconvolution vary from 0 to 6.5×10^{-10} year^{-1}, but the preferred (logistic) growth model falls within the 2–5×10^{-10} year^{-1} limits. Modified from Veizer and Jansen (1979).

exception—follow a *logistic* growth pattern (cf. von Bertalanffy, 1968) of the type approximated by Fig. 5.12. Mathematically, this growth rate can be expressed by the Verhulst–Pearl equation (e.g., Hendrick, 1984) as

$$\frac{dQ}{dt} = rQ\left(\frac{K - Q}{K}\right), \tag{5.5}$$

where Q is the quantity of the crust present at a given time t, K is the carrying capacity of the system (here 100% of the present-day continental crust), and r is the specific or intrinsic rate of growth, that is, the net growth possible with an unlimited extrinsic source of energy or matter.

The reality of the logistic growth for continental crust (Veizer, 1976b, 1983a; Veizer and Jansen, 1979; Taylor and McLennan, 1985) is strongly supported by recent estimates of the age distributions of continental terranes (e.g., McCulloch, 1986; Patchett and Arndt, 1986). In addition, isotopic transport models, which simulate depletion of the mantle for components residing in the present-day continents, also yield comparable growth patterns (O'Nions et al., 1979; Allègre, 1982, 1985; Allègre and Jaupart, 1985).

The generalized logistic growth curves are valid for a first-order system only. In a more realistic situation, with multiple causes and interactions (second- and higher-order systems; Forrester, 1968), growth rates oscillate, and thus appear episodic, around the mean growth values. The growth pattern that eventually emerges from a more refined data set approximates these see-saw oscillations, as reflected—after recycling—in the second-order pattern of Fig. 5.3.

The logistic growth mode tends to approach the *carrying capacity* of the system. The carrying capacity of the earth for continents has apparently been reached at $\sim 1.75 \pm 0.25 \times 10^9$ years ago. In living systems, the limits of growth are usually controlled by food supply or space (crowding). It is difficult to identify the limiting factor(s) for growth of continents. One alternative could be that the mantle–crust transport rates of "food," that is, of crustal constituents (e.g., incompatible elements), became about equal in both directions. Once the carrying capacity of the system, or steady-state, is reached the subsequent propagation is mostly via cannibalistic (in this case intracrustal) recycling, with external recycling via the mantle becoming of subordinate importance. This internal recycling, consisting of repeated episodes of erosion, metamorphism, melting, and cratonization, results in resetting and rejuvenation of radiometric ages, and it is reflected in the ubiquity of highly radiogenic $^{87}Sr/^{86}Sr$ initial ratios in coeval plutonic and batholithic rocks, a feature contrasting with the near-mantle clustering of Archean counterparts (Fig. 5.13). Such resetting is essentially a function of temperature, although fluids and other variables undoubtedly play a role. The probability that a rock will be subjected to repeated heating episodes is higher for low temperatures. Because the resistivity of isotopic pairs to resetting increases in the order K/Ar $\leqslant$ U-Th/Pb $\leqslant$ Rb/Sr $\leqslant$ Sm/Nd, it is to be expected that the low-temperature thermal episodes will affect mostly the K/Ar and to some extent the U-Th/Pb and Rb/Sr dates, but not necessarily the Sm/Nd dates. More severe conditions, or recycling via the mantle, may be required to reset all dating pairs. As a consequence, the spatial distributions of K/Ar, U-Th/Pb, Rb/Sr, and Sm/Nd terranes in the continental crust should not coincide. Therefore, the half-life (τ_{50}) of the continental crust must differ for all techniques, increasing from K/Ar to Sm/Nd. This progression is in accord with the approximate 670×10^6 (K/Ar), 990×10^6 (U-Th/Pb, Rb/Sr), and 1730×10^6-year (Sm/Nd) estimates of τ_{50} (Fig. 5.9; Table 5.2). A similar reasoning applies to their mean ages (τ_{mean}). Furthermore, these half-lives and mean ages are *not* residence times. τ_{res} represents only the external component of recycling and defines the average time an isotope will reside in the continental crust before its return into the mantle. Since the mantle $\rightarrow$ crust and particularly the reverse, crust (+ oceans, atmosphere) $\rightarrow$ mantle, transport rates are likely different for different isotopes, the respective τ_{res} must also differ. In general, τ_{res} for any isotope system should exceed its τ_{50} as well as τ_{mean}, because the latter two are a reflection of the combined external (via mantle) *and* internal (intracrustal) recycling. Much of the animated disagreement on models of crust–mantle evolution is rooted in misunderstanding of the above terms. The most common errors are the result of equating τ_{res} to τ_{mean} and of the quest for a magic, unique τ_{mean} or τ_{res} of continents.

In summary, I believe that the growth of continents approximated a logistic curve, with an inception of nucleation (and preservation) at $\sim 4 \times 10^9$ years ago, an exponential (or bandwagon) growth phase throughout the Archean, curtailment of growth rates in the early Proterozoic, and attainment of a new steady state at $\sim 1.75 \pm 0.25 \times 10^9$ years ago. The bandwagon stage coincided with high rates of external (via the mantle) recycling, while the cannibalistic intracrustal recycling became predominant in the subsequent stages.

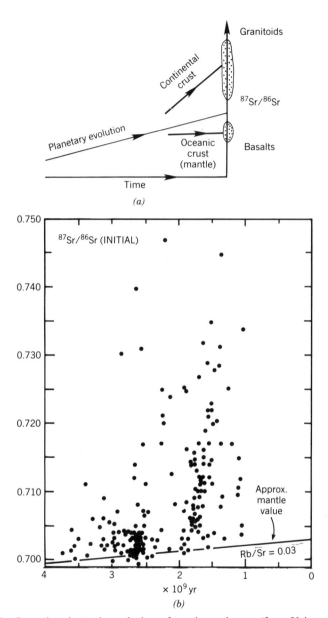

Figure 5.13 Strontium isotopic evolution of continental crust (from Veizer, 1983a). (a) Schematic representation of the evolution of $^{87}Sr/^{86}Sr$ ratios in the crust of continental and oceanic (mantle) type during geologic history. Modified from Allègre and Ben Othman (1980). (b) Initial $^{87}Sr/^{86}Sr$ ratios of granitoids, gneisses, and acid volcanic rocks plotted against their Rb/Sr isochron ages. Data are for the North Atlantic Craton, the Canadian Shield, Australia, and southern Africa. Modified from Glikson (1979). See Allègre and Ben Othman (1980) and Balashov (1985) for complementary Rb/Sr and Sm/Nd data and interpretations.

5.5.1 System Science Perspective of the Archean–Proterozoic Transition

Natural systems have a tendency to self-organize in order to maximise useful power. This is referred to as the maximum power principle or the fourth thermodynamic law (Odum, 1983, Chapters 1, 2, 7, 9). Successful systems are those able to store more energy, which can then be fed back, or depreciated (second thermodynamic law), to catalyze inflow of additional energy from an outside source. Such maximizing systems frequently follow the design of *autocatalytic modules*. These modules usually encompass an energy storage structure (continents), an autocatalytic loop or feedback (orogenic cycle), and an energy source (internal and solar heat). Additional storages may be coupled to the autocatalytic loop. The producer–consumer systems of ecology may serve as an example for a geologic system having bidirectional transport fluxes between two interconnected modules, such as those of the (upper) mantle and continents. In such analogy, the "consumer" (continents with an orogenic loop) module is coupled upstream to a "producer," that is, the Michaelis–Menten module of mantle convection. The latter (cf. Odum, 1983) consists of two or more storages, with the conserved cyclic material going from a receptive storage (mantle) to an activated storage (oceanic crust) and back. Note that the role of the autocatalytic orogenic feedback loop is to maintain elevation and thus store and maximize potential energy of the continental storage for erosion, transport, deposition, loading, subsidence, compression, metamorphism and plutonism, and renewed orogeny, via utilization of a secondary, usually dispersed (e.g., sun), source of energy.

The size and the growth rate of the storage structure (e.g., the population of continental rocks) are proportional to fluxes from external sources of energy and matter. If these were controlled (e.g., the mantle heat flow or convection rate), the growth of continents would approach asymptotically the carrying capacity of the system. Such a system is source (producer) limited. However, even if this were not the case, the internal recycle of material through the autocatalytic loop would limit the optimal size of continents. This is because the return flux of the loop is often a quadratic function of the size of the storage structure—a feature that stabilizes an otherwise unstable and ephemeral module. Equations with such quadratic feedback are called logistic equations. The quadratic term can be a consequence of several factors (Odum, 1983, Chapter 9). These may be backforce (e.g., diminished uplift with increasing elevation), quadratic drain (e.g., increased rate of erosion with higher mountains), crowding (e.g., the area available for exposed surfaces of convective cells in the mantle), conservation of recycled quantity (e.g., the incompatible elements in the crust-active mantle reservoirs), competition owing to diversity (e.g., by several tectonic regimes and modules), and other phenomena. At this stage, the role and the relative importance of such causative factors are debatable. Nevertheless, regardless of whether the global tectonic system is source limited, autocatalytic, or likely both, the resulting growth pattern should follow a *sigmoid* (or logistic) curve.

5.5 GROWTH AND RECYCLING OF CONTINENTS

The mathematical expression for sigmoid growth was given in Eq. (5.5). The regular growth rate $\dot{Q}$ for continents is thus defined by the logistic equation (Odum, 1983)

$$\dot{Q} = \frac{dQ}{dt} = rQ - \frac{r}{K}Q^2. \tag{5.6}$$

The intrinsic rate of growth can be expressed as

$$r = (k_p X - k_d), \tag{5.7}$$

where k_p is the rate of production (birth), k_d is the rate of destruction (death), and X is the constant driving force in the form of energy or material flux from an external source. With these terms of population dynamics, the equation becomes

$$\dot{Q} = k_p X Q - k_d Q - \left(\frac{k_p X - k_d}{K} Q^2\right). \tag{5.8}$$

Integrated, the time-dependent growth of continents can be expressed as

$$Q = \frac{K}{1 + C \exp^{-rt}}, \tag{5.9}$$

where C, the coefficient of efficiency, equals Q/X and is usually proportional to the volume/area ratio of the storage.

From Eq. (5.9) it is clear that the first-order logistic pattern for continental growth could be resolved if r and C were known. Unfortunately, the isotopic mantle–crust transport models do not give a unique solution with respect to this r (cf. Armstrong, 1981) and the parameter must likely be determined experimentally through expanded geochronological studies of ancient shields. The coefficient C likely depends on the nature of convecting cells in the active mantle, that is, that part of the mantle (upper versus whole) that drives plate tectonics. It is my opinion that conservation of a recycled quantity (cf. Odum, 1983, p. 170) is the most important, although not necessarily the only, factor controlling the feed into the autocatalytic loop. In such a design, the system can be donor (mantle) driven. In general, storages tend to drive pathways (e.g., the forward mantle → crust flux) in proportion to their exposed area. In the case of convective cells in the mantle, the exposure on the surface of the earth is a linear feature, as exemplified by the extant oceanic ridges. Since volume is proportional to the cube of its linear dimension,

$$C = L \sqrt[3]{\frac{V_M}{\varrho_M}} \tag{5.10}$$

where V_M is the volume of the convective cell(s) in the active mantle, ϱ_M is its density, and L is a function of the shape of the cell(s). The crucial question is therefore the depth of mantle convection directly coupled to plate tectonics. Alternatively, if the r and the shape of the logistic growth curve for continents were known, this might conceivably help to constrain the size of the convective cells in the mantle, a topic of considerable present-day controversy. It is suggested by Eqs. (5.9) and (5.10) that the rate of growth of continents, and indeed the speed of operation of the global cycle, is ultimately a function of the size of convective cells in the mantle. For example, an increase in the size of the near-surficial cells in the course of geologic history would result in decreased efficiency of the system in terms of continent production.

As a final insight, let us consider a situation where two competing autocatalytic modules operate in parallel (Odum, 1983, Chapter 12). This may be a situation with multiple tectonic regimes. For example, it is possible to envisage an essentially oceanic tectonic regime culminating in a structure akin to island arcs (e.g., greenstone belts) coexisting with a regime culminating in generation of continental cratons via an orogenic cycle (the present mode of plate tectonics). If the two autocatalytic modules were essentially analogous and differing only in coefficients (rates), the evolution would lead to suppression and competitive exclusion of the slower module (Fig. 5.14A). In the above example, this would lead to elimination of the *plate tectonic module*, a proposition at odds with reality. In a more realistic case, the two modules differ in the complexity of their

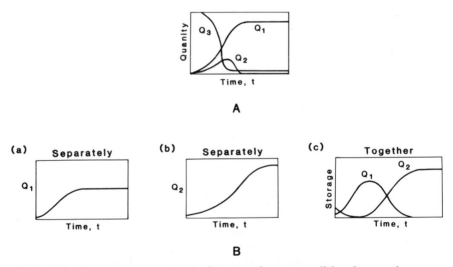

Figure 5.14 Time-dependent growth of storages for two parallel and competing autocatalytic modules. Q_1 is "greenstone" storage; Q_2 is "continents." (*A*) Two units competing for a recycling quantity in short supply (Q_3). (*B*) Competition between growth specialist and steady-state specialist: (*a*) growth specialist and growth alone; (*b*) steady-state specialist and growth alone; (*c*) growth together. Modified from Odum (1983).

structure. They both tend to maximize power in order to enhance the net growth. The greenstone module, because of its lesser structural complexity (and fewer autocatalytic subloops), is characterized by large coefficients of growth, feedback, and energy incorporation, but also depreciation, because the energy is diverted to production rather than maintenance (Odum, 1983). The plate tectonic module has the opposite attributes and can, albeit more slowly, reach a high steady-state level, because its energy is utilized for maintenance of the structure (complex autocatalytic orogenic subloops) rather than for continual replacement of the quickly decaying older units. The former is a growth or r specialist (opportunist, pioneer), whereas the latter is a steady-state or K specialist. In competition, the r specialist dominates at first but eventually is outcompeted by the K specialist (Fig. 5.14B). This process of self-organization is a common feature of natural, human, technological, or economic systems and, in my view, the Archean–Proterozoic transition is only another example of the application of the general system science theory. It is also interesting to note that populations (storages, level variables) in K modules usually have age distribution patterns of the type III curves (Wilson and Bossert, 1971, Chapter 3)—an observation in accord with the existing data for tectonic realms (Fig. 5.10).

5.6 THE GLOBAL SEDIMENTARY MASS AND ITS RECYCLING

The present-day mass of global sediments is estimated to be 2701×10^{21} g (Ronov, 1980), with 86% covering the continents and their shelves and 14% the deep ocean floor. This underscores the role of continents in generation of the global sedimentary mass. Geometric and isostatic considerations (Veizer and Jansen, 1979) suggest that the total mass of sediments on continental crust (and thus globally) grew in a fashion similar to that of their underlying basement, that is, following the advocated logistic growth. The present-day age distribution patterns for global, continental, and oceanic sediments are summarized in Fig. 5.15. The major uncertainty is in the estimate for the pre-1600×10^6-year mass, and poor data may also account for the indicated bimodality in the age distributions of global and continental sediments, with the inflection points at $\sim 570 \times 10^6$ years ago. Taking these age distributions at face value and accepting the logistic growth mode, the calculated b values are ~ 40, 16, and 142×10^{-10} year^{-1} for the global, continental, and oceanic sedimentary masses. This means that during geologic history an equivalent of 7 ± 3 global, 3.7 ± 0.5 continental, and 64 oceanic masses have been recycled. These calculations show that the rate for oceanic sediments exceeds the rate for continental sediments by a factor of ~ 9. The global rate is therefore an amalgam of the two values. In detail, however, the actual recycling of sediments within both continental and oceanic domains follows the rates typical for their constituent tectonic realms, such as passive margin basins, active margin basins, and platforms. These have been discussed in Section 5.4. Notwithstanding the above clarifications, a quantification of global sedimentary recycling rates is essential for understanding the

200 THE EVOLVING EXOGENIC CYCLE

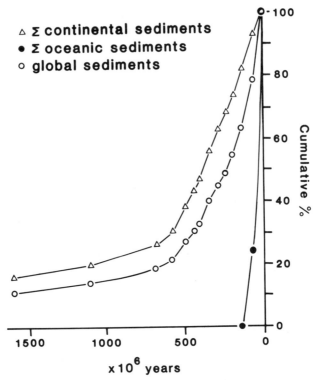

Figure 5.15 Present-day mass–age distributions of global, continental, and oceanic sediments. Modified from Veizer and Jansen (1985).

secular evolution of sediments in terms of their lithology and chemistry, a topic of subsequent discussion.

5.7 THE SEDIMENTARY SYSTEM AND ITS LITHOLOGIC EVOLUTION

As shown above, the preserved amounts of sediment per unit time decrease with increasing geologic age (Figs. 5.7 and 5.15). Furthermore, the relative proportions of different lithologies and facies within this diminishing sedimentary mass change as well (Ronov, 1964; Strakhov, 1964, 1969; Garrels and Mackenzie, 1971a, Chapter 9; Veizer, 1973; Engel et al., 1974; Mackenzie, 1975). The observed secular variations can be divided into two categories: (1) first-order variations between petrological types (lithologies) and (2) second-order variations in facies within petrologic types.

The poor quantitative data for some lithologies and facies permit only semiquantitative treatment of the subject. For this reason, no simulations or

5.7 THE SEDIMENTARY SYSTEM AND ITS LITHOLOGIC EVOLUTION

corrections for recycling are attempted. Nevertheless, even the less rigorous treatment of the subject brings out salient features in the development of the terrestrial exogenic cycle.

5.7.1 First-Order Petrologic Secular Variation

Figure 5.16 shows that the patterns of age distribution vary considerably for each lithology. In other words, the *relative* percentages of these lithologies vary with age. Younger sequences have higher absolute and relative proportions of limestones, evaporites, and phosphorites and lower relative proportions of clastics and dolostones than do older sequences (Ronov, 1964; Strakhov, 1964, 1969). Again, this pattern can be a consequence of (1) real variations in the rate of formation of different lithologies during discrete periods of geologic history (e.g., Ronov, 1964; Strakhov, 1964, 1969; Cloud, 1969); (2) decreasing survival (increasing differential recycling rates) from (bio)chemical sediments to greywackes; and (3) a combination of (1) and (2).

To explain the observed age distribution of clastic, carbonate (limestone and dolostone), and evaporite rocks (Fig. 5.16), Garrels and Mackenzie (1971a,b) introduced the concept of differential recycling rates, the respective half-lives of 600, 300, and 200×10^6 years being a function of relative solubilities. While undoubtedly important, I believe that solubilities are not the only, or even the prime, factor in sediment recycling. The rate of continental erosion is controlled essentially by relief (Garrels and Mackenzie, 1971a, Chapter 5; Hay and Southam, 1977). Consequently, the rate of mechanical erosion exceeds, by

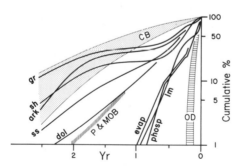

Figure 5.16 Observed cumulative mass–age distributions of major sedimentary lithological types. Carbonates and clastics after Ronov (1964, Fig. 1), evaporites from Holser (1984), and phosphorites after Cook and McElhinny (1979). gr = greywackes, sh = shales, ark = arkoses, ss = sandstones, dol = dolostones, evap = evaporites, lm = limestones, phosp = phosphorites. CB = continental basement (Fig. 5.10), P = platforms, MOB = mature orogenic belts, OD = oceanic domain (active to passive margin basins of Fig. 5.9). The database of Ronov (1964) does not include submarine sediments (Hay, 1985). If these were included, the whole pattern of sediment distribution may have been compressed somewhat toward the right-hand side of the figure, particularly for the youngest segments of the curves.

one order of magnitude, the rate of chemical erosion (Meybeck, 1979b), an observation supported by the ~5:1 ratio of suspended to dissolved load for world rivers (Martin and Meybeck, 1979; Milliman and Meade, 1983). Such mechanical erosion leads to en masse removal of all lithologies and not to preferential destruction of carbonates relative to clastics (cf. also Meybeck, 1979b). In fact, in most Phanerozoic orogenic belts, it is the carbonates and not the shales—the latter accounting for 70–80% of all clastics (Garrels and Mackenzie, 1971a; Ronov, 1980), that form the resistant lithologies. This view is exemplified by the magnificent sceneries of the Dolomites, Northern Calcareous Alps, or the Canadian Rockies. If so, the relative solubilities of different lithologies, while important for weathering and for development of stylolites and collapse breccias, likely do not control the rate of erosion. In a layer-cake stratigraphy, selective removal of carbonate or evaporite strata usually results also in the collapse and removal of the interlayered clastics. Salts can be squeezed out selectively via diapirism during deep burial and this may be a partial reason for their apparent rapid recycling (Fig. 5.16). However, limestones or phosphorites are not subject to diapirism, yet their age slopes are comparable to that for evaporites. Thus, if recycling were a cause of the observed age distribution of lithologies, such recycling must have been related to their preferential association with specific tectonic realms. The curves for evaporites, limestones, and phosphorites indicate a τ_{50} of $\sim 150 \times 10^6$ years (Fig. 5.16). This rate is about 2.5 times faster than the rate of recycling for platforms and mature (worn down) orogenic belts and ~ 2.5 times slower than the recycling rates characteristic of the oceanic domain, including its passive and active margin basins. Such present-day age distribution patterns may be a consequence of some as yet unknown partition of evaporites, limestones, and phosphorites between intracontinental and oceanic (marginal basin) depositional environments. In contrast, dolostones appear to have been associated mostly with intracontinental (platform, intracratonic basin) tectonic realms.

The age curves for clastic lithologies, however imperfect the data, indicate a type of distribution expressed mathematically by Eq. (5.3), that is, of enhanced mortality for young sediments. During the Phanerozoic, clastics appear to have been associated mostly with intracratonic settings, while older sediments display shallower slopes, comparable to those of the continental basement. Because the expected life span, τ_{max}, of intracratonic tectonic realms is $\sim 1500 \times 10^6$ years (Section 5.4), these sediments can be preserved in large quantities beyond such time limits only if incorporated into the basement. Following such incorporation, their rate of recycling is similar to that of their host tectonic realm.

The previous discussion suggests that the first-order age distribution of sedimentary lithological types is indeed a consequence of recycling, but the rates of recycling are controlled by tectonism and not by lithologies. The recycling concept does not, however, preclude simultaneous evolutionary development of first- and higher-order significance. Such evolutionary phenomena may be responsible, for example, for the observed age distribution patterns of clastic rocks. Their chemical, mineralogical, and tectonic stability increases from graywackes and arkoses to shales and sandstones, yet their apparent preservation potential is the reverse of this order (Fig. 5.16). This facies reversal is understandable if the overall tectonic evolution of the earth is taken into account. The immature graywackes and arkoses are generally associated with active tectonic settings and thus are easily dispersed mechanically and chemically. Consequently, the older crustal segments should have a deficiency, not an excess, of such immature sedimentary facies (Garrels et al., 1972; Veizer, 1973; Engel et al., 1974). The above statements lead to the conclusion that the excess of Archean graywackes and of Early Proterozoic arkoses is a relic feature of their exceptionally high original abundance. The graywacke–argillite assemblage is related to the unstable tectonic setting of the greenstone belts, the arkoses to the considerable buildup of granodioritic upper continental crust (cratonization) between 3 and 2×10^9 years ago, and the mature sandstone–shale assemblage to the ubiquity of new cratons and stable shelves (Fig. 5.12)—an explanation advanced already by Ronov (1964) and Mackenzie (1975).

5.7.2 Second-Order Secular Variations in Facies

In addition to time-dependent variation between lithologies, the variation of facies within lithologies can be summarized.

Clastic Sediments. Graywackes are the dominant sandstones of the Archean (cf. Ronov, 1964). Since then their proportion has diminished by a factor of 4. Arkoses, on the other hand, are dominant rock types of the Early and Middle Proterozoic. Orthoconglomerates and quartzites (not cherts) are scarce in the Archean but fairly abundant in Early Proterozoic sequences. Red bed siltstones and mudstones are ubiquitous sedimentary rocks since about Early Proterozoic time (Cloud, 1969; Walker et al., 1983). The relative proportion of shales is more or less constant through geologic time, although their chemical and mineralogical composition (see Section 5.8.3) is not.

Carbonates. Limestones and dolostones are rare in the Archean, although the former predominate in sedimentary greenstone sequences (cf. Strakhov, 1964; Veizer, 1973; Gimmel'farb, 1978; Sidorenko et al., 1978). Early and Middle Proterozoic sequences are dominated by epicontinental lagoonal to sublittoral early diagenetic (and primary?) dolostones. Since about 1×10^9 years ago, these dolostones have been superseded by inorganic and biochemical limestones of similar depositional milieu. In the Phanerozoic, the most important carbonate rocks are littoral or neritic organodetrital limestones and "late" diagenetic dolostones formed at their expense. In Jurassic and later deposits, the dominant carbonate sediments are deep sea limestones (Twenhofel, in Kuenen, 1950, p. 392). The deep-sea milieu prior to the Mesozoic appears to have significant deposits of graptolitic shales. Thus, the trend in carbonate lithologic facies indicates a shift in the axis of maximal sedimentation seaward with decreasing age (Veizer, 1973).

Sulfates. Except for a few localities, which may be as old as $\sim 1.3 \times 10^9$ years (see the compilations of Lefond, 1969; Cook and McElhinny, 1979), the majority of gypsum and anhydrite deposits are younger than 800×10^6 years (Ronov, 1964; Strakhov, 1964, 1969; Cloud, 1968; Holser, 1984). However, disseminated gypsum, anhydrite, and pseudomorphs after these minerals, as well as occurrences of relatively large deposits of sedimentary or exhalative barite, are found in rocks as old as $\sim 3.5 \times 10^9$ years (Heinrichs and Reimer, 1977; Dunlop, 1978; Groves et al., 1981; Walker et al., 1983).

Salts. Although pseudomorphs after halite are found in rocks of the whole Proterozoic column, economic deposits of NaCl, as well as potassium–magnesium salts are confined to the most recent 800×10^6 years (Ronov, 1964; Strakhov, 1964; Zharkhov, 1974; Holser, 1984). However, collapse and solution breccias— the possible vestiges of original evaporites—are known from sequences as old as 3.5×10^9 years (Badham and Stanworth, 1977; Lowe, 1980a).

Phosphorites. Phosphorites in economic concentrations are known only since the Middle Riphean and Middle Sinian (Strakhov, 1964; Bushinskii, 1969; Slansky, 1986) and are mostly younger than $\sim 1 \times 10^9$ years. Furthermore, the Phanerozoic is characterized by phosphorites of pelletal type, whereas the Proterozoic and the Cambrian phosphorites are usually nonpelletal (Cook and McElhinny, 1979).

Siliceous Rocks. Few long-term trends for siliceous rocks with age are known. However, the chert content of iron ores and carbonates increases with geologic age (Veizer, 1978). Cherts and massive silicification (and carbonation) zones are a remarkable phenomenon of the Archean volcanoclastic sequences (e.g., Lowe and Knauth, 1977) as well as the Phanerozoic pelagic suites (e.g., Kuenen, 1950, p. 361). Therefore, the ubiquity of siliceous sediments appears to have been associated, in part, with areas and periods of intense volcanic activity, the latter serving as an ultimate source of silica for (bio)chemical precipitation.

Organics. Formation of organics is directly related to the evolution of organic life and its decay. Carbonaceous shales are known since the Archean and no age trend has been described. Oil, gas, and asphalt accumulations are encountered principally in the Phanerozoic (cf. Tissot, 1979), with 15% of identified oil reserves in Paleozoic, 54% in Mesozoic, and 31% in Cenozoic rocks (Bois et al., 1982). Coal-like deposits are known from the early Proterozoic (Strakhov, 1964; Cloud, 1968), but the important economic accumulations date from the Late Paleozoic (e.g., Vishemirskij, 1978; Tissot, 1979; Bestougeff, 1980; Bois et al., 1982). According to Volkov (1968), there is, at least in the Soviet Union, a progressive shift in the site of coal formation from mostly paralic to terrestrial environments with decreasing age.

The evolution of organics is directly related to the evolution of life, with appearance of prokaryotes prior to 3.5×10^9 years ago (Awramik, 1981) and eukaryotes at $\sim 1.5 \times 10^9$ years ago (Schopf, 1983). This latter event was followed by colonization of continents by higher plants and animals in the Silurian–Devonian.

Consideration of the time distribution of other types of sediments, such as weathering crusts (laterites), residual deposits (placers), and of other sedimentary ores is beyond the scope of this chapter; it is discussed in detail in Veizer et al. (1988).

5.7.3 Recycling and Evolution

Pending the accumulation of more quantitative data, the relative role of evolution versus recycling in age distribution of facies can be resolved only partially. For example, the post-Archean progression of limestone types (lagoonal → shallow marine → deep sea with decreasing age) closely parallels the evolution of life (Wilson, 1975; Scholle et al., 1983). Precambrian limestones are interpreted as being mostly (bio)chemical, bacterial, and algal precipitates, frequently in partially restricted environments. Paleozoic deposits are mostly organodetrital accumulations of littoral–neritic biota, and post-Jurassic sediments are predominantly deposits of pelagic organisms such as *Globigerina*, coccoliths, and pteropods.

The preponderance of dolostones in the Proterozoic record has been suggested to be a consequence, in part, of the continuous dolomitization of limestones; the older the sequence, the higher the probability of it being dolomitized (Garrels and Mackenzie, 1974). Yet these dolostones are predominantly of an early diagenetic (and primary?) type (Veizer, 1973; Tucker, 1982) and thus were dolomitized mostly within a short time interval after deposition of their $CaCO_3$ precursor and not continually over a 10^9-year time span (see Veizer, 1978, for discussion of possible causes). This near-shore facies was complemented in the Phanerozoic by the pervasive "late" diagenetic dolomitization of shallow marine, bioclastic limestones. The latter, forming highly permeable and less compressible carbonate bodies on the flanks of subsiding, mostly shaley, basins were prime water conduits during the early diagenetic and subsequent compaction phases of the evolution of basins. This frequently led to mixing of waters of different (e.g., marine and meteoric) parentage and resulted in their ubiquitous diagenetic dolomitization (see the review of Machel and Mountjoy, 1986).

Perhaps not surprisingly, the above discussion suggests that the observed temporal variations in sediment types are a consequence of both evolution and recycling. The first *massive* appearance of many lithologies and facies is likely related to evolutionary phenomena. These may be, for example, the emergence of suitable tectonic settings (graywacke–argillite, arkoses, sandstone–shale, evaporites, dolostones), life forms (phosphorites, organics, facies of limestones), and other single or multiple factors. In many respects, and particularly in terms of tectonics, the earth appears to have reached a near present-day steady state at $\sim 1.75 \pm 0.25 \times 10^9$ years ago (Section 5.5) and the subsequent temporal distribution of sediments has increasingly been controlled by the rates of recycling of their host tectonic settings.

5.8 CHEMISTRY OF THE SEDIMENTARY SYSTEM

5.8.1 Composition of the Global Sedimentary Mass

Sedimentary accumulations are derived principally from disintegration of the upper continental crust (basement and its related cover) and the global sedimentary mass therefore should have a chemical composition comparable to this part of the crust. This is not the case, as already pointed out by Ronov (1968). Compared with the upper continental crust, sediments (Goldschmidt, 1933; Rubey, 1951; Vinogradov, 1967; Ronov, 1968; Fig. 5.17) (1) are more mafic in overall chemical composition, (2) are strongly enriched in excess volatiles, and (3) have a higher oxidation state.

Restricting this discussion to the first point, it is of interest to estimate the proportion of the excess mafic component in the global sedimentary mass. Calculations show that a source composed of $\sim 70\%$ upper continental crust (dacite) and $\sim 30\%$ oceanic crust (basalt) is required (Sibley and Wilband, 1977; Veizer, 1979). However, this mixture still does not account for the sodium and calcium chemical differences between average igneous rock and sediments (Fig. 5.17). This composition is comparable to the earth's crust as a whole ($\sim$ andesite) and not to the upper continental crust.

The anomalous enrichment in calcium (manganese, iron) and depletion in sodium perhaps can be a consequence of halmyrolitic, exhalative, and particularly hydrothermal exchange between the ocean floor and seawater, processes discussed extensively by Wolery and Sleep (Chapter 3). The general absence of clear anomalies in the normalized average composition of sediments (Fig. 5.17) supports the conclusion of these authors that exogenic $\pm$ endogenic inputs

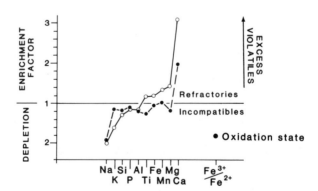

Figure 5.17 Chemical features of sediments. Full circles ($\bullet$) = average chemical composition of sediments (Ronov and Yaroshevskiy, 1969) normalized to average upper continental crust; open circles ($\circ$) = average composition of sediments normalized to the mixture of 70% upper continental crust and 30% oceanic crust. The oceanic layer II is excluded from the sedimentary mass. Its inclusion (if sedimentary) would only magnify the depletion-enrichment slope. Modified from Veizer (1979).

and sinks for most major elements—except possibly calcium and sodium—were balanced throughout the *steady-state* stage of geologic history. Thus, this process cannot be held responsible for the andesitic composition of sediments.

A direct terrigenous contribution from island arcs and ocean floor basalts can account chemically for $\sim 20\%$ of the observed basaltic admixture (Veizer, 1979) and the remainder will have to be accounted for by other phenomena. Assuming derivation of all, including nonvolcanogenic, sediments from a source more mafic than the modern upper continental crust—an alternative conceivable for the early, $\geqslant 2.5 \times 10^9$ years ago (cf. Section 5.5), earth history—the missing mafic component can perhaps be accounted for. Yet, the Archean sediments, even if derived entirely from basalts, account today for less than approximately 10% of the global sedimentary mass (Fig. 5.15).

The above discussion shows that a direct *first-cycle* clastic contribution from a mafic igneous source is probably inadequate to explain the magnitude of the observed mafic component.

5.8.2 Sedimentary Recycling

Because the continental sedimentary mass accounts for the bulk of present-day global sediments, the growth of the global sedimentary mass should be a function of the growth of the continents (see Section 5.5 and 5.6). It is feasible, however, that at least some of this mass, particularly during the early stages of the earth's history and at the beginning of each tectonomagmatic cycle, evolved on oceanic or intermediate type crust. If so, it would have to be derived from a source more mafic than the present-day uper continental crust. In an entirely cannibalistic (closed) recycling system, this composition would be perpetuated indefinitely regardless of the nature of the later continents.

Experience shows that we must be dealing with a partially open system, because some sediments are being subjected to metamorphism and melting (e.g., in subduction zones) while others are being formed at the expense of primary igneous and metamorphic rocks. Estimates based on Sm/Nd dating of sediments (Fig. 5.18) indicate that the sedimentary cycle is $\sim 90 \pm 5\%$ cannibalistic and that its t_0 is $\sim 2.5 \times 10^9$ years. The *first-order* features in Fig. 5.18 can be explained if the Archean were dominated mostly by the growth of the sedimentary mass by addition of first-cycle sediments from erosion of contemporaneous young ($\leqslant 250 \times 10^6$ years old) igneous precursors. Subsequent to a large degree of cratonization, and establishment of a substantial global sedimentary mass at $\sim 2.5 \pm 0.5 \times 10^9$ years ago, cannibalistic sediment–sediment recycling became the dominant feature of sedimentary evolution. Note the general absence of excess Sm/Nd ages, and thus of inherited old detrital components, in most Archean sediments, an observation strongly arguing against the presence of large continental landmasses prior to $\sim 3 \times 10^9$ years ago. This feature cannot be blamed on enhanced rates of crust–mantle recycling,

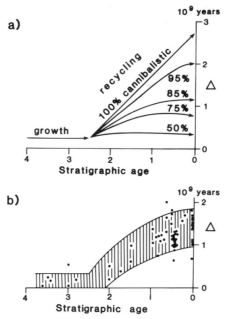

Figure 5.18 Models of Sm/Nd excess ages for sedimentary rocks. The Sm/Nd systematics dates the time of fractionation from the mantle (light REE enrichment episode) and is apparently not appreciably disturbed by later weathering, diagenetic, or even metamorphic events (McCulloch and Wasserburg, 1978). Regardless of whether most or only some of the sediments were generated during early terrestrial history, they would inherit Sm/Nd systematics from their igneous precursors. In a cannibalistic sedimentary recycling, these ancient systematics will be perpetuated and, as a consequence, Sm/Nd of all second-stage sediments should exceed their stratigraphic ages and the Δ (Sm/Nd model age minus stratigraphic age) should increase toward the present, with a 45° slope being an upper limit for a completely closed system. In order to generate the observed smaller Δ, it is necessary to add sediments formed from a young source. This can be achieved by: (1) simple addition of new sediment in a growing sedimentary mass; (2) partial loss of original old sediments (e.g., metamorphism) and their replacement by freshly derived young ones in the sedimentary mass; and (3) a combination of the above. The upper figure represents model calculations based on the assumption that prior to 2.5×10^9 years ago the sedimentary mass was growing through addition of first-cycle sediments. The post-Archean evolution assumes cannibalistic recycling of the steady-state mass, and the slopes represent the degree of cannibalism for this recycling. The bottom part is a collation of experimental data from McCulloch and Wasserburg (1978), O'Nions et al. (1983), Hamilton et al. (1983), Taylor et al. (1983), Goldstein et al. (1984), and Allègre and Rousseau (1984). Modified from Veizer and Jansen (1985).

because sediments, as opposed to oceanic crust, are difficult to subduct (cf. Patchett et al., 1984), let alone to do so wholesale. If they had existed on the surface of the early earth, they would have to be mostly recycled into younger sediments.

5.8.3 Secular Chemical and Isotopic Trends

Accepting that the primordial sedimentary mass was in part derived from a source more mafic than the present-day upper continental crust and that the sedimentary cycle is ~90% cannibalistic, the initial chemical and isotopic composition of the sedimentary mass will be altered only partly during subsequent recycling. The rate of this chemical change depends on (1) the rate of recycling, (2) the degree of open-system behavior of the cycle, and (3) the composition of the new mass, which is partially replacing, or is being added to, the old mass.

As long as the new and the old mass are not very dissimilar (as in the Archean?), no distinctive secular trends should be observed even if (1) and (2) are high. However, with the change in the source of the sediment mass in the late Archean and early Proterozoic (Fig. 5.12), its average chemical composition will start to approach the composition of the new source, with the rate of change decreasing as the chemical difference diminishes, until a new steady state is established (Fig. 5.19). Age trends in the chemical properties of rocks are indeed consistent with the predicted pattern (cf. Fig. 5.20) and they all show relatively rapid transition, of perhaps 600×10^6-year duration, around the Archean–Proterozoic boundary. Such a rapid transition from one steady state to another could not have been accomplished if a near present-day mass of sediments were extant on the pre-3×10^9-year old earth. For a 10% open system, recycling of an equivalent of more than 20 sedimentary masses is required for such a task (Fig. 5.21). With the recycling constant for global sediments of $\sim 40 \times 10^{-10}$ year^{-1}, this would require the entire 4.5×10^9 years of terrestrial history and the task would have been even more hopeless if only continental sediments were involved. This again points out that it must have been the growth of the global sedimentary mass, as a response to growing continents, that enabled the relatively sudden chemical transitions. Since the subsequent cannibalistic recycling has been almost fully closed ($\sim 90 \pm 5\%$), the chemistry of the extant sedimentary mass has been relatively insensitive to perturbations generated by variations in the composition of post-Archean sources. [This, however, does not preclude the generation of local second-order oscillations, such as those described by Michard et al.

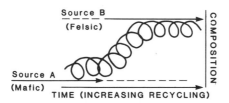

Figure 5.19 Schematic illustration of secular variations in chemical composition of sediments owing to changing sources. See the text for further explanation. Modified from Veizer and Jansen (1979).

210 THE EVOLVING EXOGENIC CYCLE

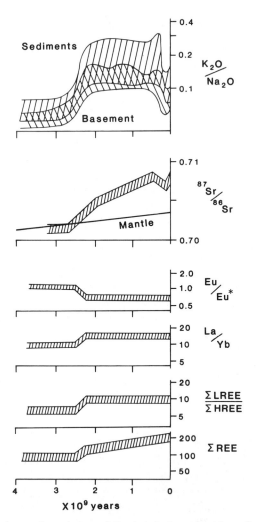

Figure 5.20 The observed secular trends in chemical composition of sedimentary rocks. The K/Na trend for undifferentiated sediments and basement from Engel et al. (1974), the $^{87}Sr/^{86}Sr$ in carbonate rocks from Veizer and Compston (1976), and the REE in argillaceous sedimentary rocks from Taylor (1979). Several other major and trace elements show similar unidirectional mafic → felsic trends (Ronov and Migdisov, 1971; Ronov, 1972; Schwab, 1978; Taylor and McLennan, 1985), but the details of the curves are lost in the coarse stratigraphic grouping. Modified from Veizer and Jansen (1979).

(1985) for French shales.] Consequently, the 30% mafic component present in the existing sedimentary mass (cf. Section 5.8.1) likely represents a relic of the original early Archean mafic sediments, not yet dispersed by subsequent recycling.

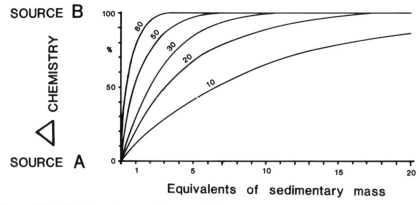

Figure 5.21 The dependency of the rate of change in chemical composition of the clastic sedimentary mass on the degree of cannibalism of the system. $\Delta_{\text{CHEMISTRY}}$ represents the difference, expressed as a percentage, in chemical composition between a source A (e.g., oceanic crust) and a source B (e.g., upper continental crust). The 10, 20, . . . over the calculated curves represent the degrees of openness of the sedimentary cycle. The range 1–20 signifies the recycling of 1–20 equivalents of a given sedimentary mass. Reproduced from Veizer (1984) with permission of VNU Science Press.

5.9 THE HYDROSPHERE AND ATMOSPHERE

The ocean and atmosphere are products of interaction of excess volatiles (Rubey, 1951) with the solid earth. The extensive debate on kinetic versus thermodynamic control of the chemistry of ocean water appears to be heading—as is frequently the case in science—toward a compromise (Chapter 1, Sections 1.1 and 1.2). The former is perhaps of greater importance for major, and the latter for some minor, elements. The multiplicity of buffer mechanisms results in a relatively conservative chemistry of ocean water during geologic history (see Chapter 1, Section 1.5) with outputs balancing the inputs. This, however, does not mean that the present-day steady-state fluxes are necessarily representative of the whole geologic past. Whereas Chapters 3 and 4 discussed possible deviations from basically cyclic developments on a $\leqslant 10^9$-year time scale, in the subsequent discussion I shall try to sketch the overall evolution of the atmosphere–hydrosphere system over the 10^8–10^9-year time interval (see also Ronov, 1968; Mackenzie, 1975; Walker, 1977; Holland, 1978, 1984), such evolution resulting from a compounding effect of shorter-term variations.

5.9.1 The Archean Oceans

The assumption of the steady-state balance of inputs and outputs can be tested if the various reservoirs have specific signatures. This is the situation, for example, for strontium, with continental crust—because of its high Rb/Sr ratio—generating considerably more radiogenic strontium than oceanic crust (Fig. 5.13). The

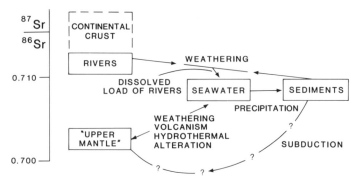

Figure 5.22 Present-day exogenic cycle of strontium. Modified from Wadleigh (1982) and discussed in Wadleigh et al. (1985).

^{87}Sr/^{86}Sr of seawater (0.709 today) therefore reflects directly the relative proportions of strontrium derived from these two sources. Because of the relatively long mean residence time ($\sim 9 \times 10^6$ years) of strontium, compared to the mixing rate of $\sim 10^3$ years for ocean water (see Chapter 3, Section 3.2) strontium in seawater is well mixed. Nevertheless, any radical change in the relative importance of a given flux will be recorded by seawater with at least a 10^7-year time scale resolution. For example, a cessation of the continental flux would, in $n \times 10^7$ years, lower the strontium isotopic ratio of seawater to that of the complementary upper mantle (= oceanic) flux (Fig. 5.22). This appears to have been the situation during the Archean (cf. Fig. 5.20). Yet toward the end of the Archean, continents attained $\approx 40\%$ of their present size (see Section 5.5) and should have generated a complementary continental flux of strontium. The *apparent* unimportance (10% of the total flux; Brevart and Allègre, 1977) of continent-derived strontium in the Archean oceanic reservoir may have been a consequence not of its absence, but rather of an exceptionally strong oceanic flux, which overwhelmed any continental contribution (Perry et al., 1978; Veizer et al., 1982).

The internal heat generation of the earth decayed exponentially with its advancing age: the global surficial heat flow 2.5×10^9 years ago being two to three times that of the present day (Birch, 1965; McKenzie and Weiss, 1975; Jessop and Lewis, 1978; Davies, 1980). Because the thickness and geothermal gradients of the Archean continental crust appear to have been comparable to their modern counterparts (Brune and Dorman, 1963; Condie, 1973; Davies, 1979), the dissipation of this additional heat must have been accomplished through a greater oceanic heat flux (Burke and Kidd, 1978). This implies a higher production of oceanic crust via faster spreading and/or more extensive ridge systems (Bickle, 1978; Hargraves, 1986), with resulting tectonic instability and large hydrothermal, exhalative, and low-temperature weathering fluxes between this crust and contemporaneous seawater. This assumption is supported by the predominantly volcanoclastic nature of the greenstone belt sequences as well as by the types and associations of Archean chemical sediments and ores. The ores show gradations from the conspicuous near-volcanic silicification and carbonation zones (with gold, copper, and zinc deposits) and sedimentary barites to Algoma type iron formations (with gold) and to volcanogenic and volcanosedimentary base metal deposits. Marine carbonates were present and cherts widespread. These assemblages, apart from differences attributable to biologic evolution, resemble the present-day spatial associations in the vicinity of oceanic spreading centers and island arcs. Furthermore, Archean calcites, if compared to their Cenozoic counterparts, are enriched in Mn^{2+} (Fig. 5.23), Fe^{2+}, depleted in $^{18}O/^{16}O$ (Veizer, 1983a,b), and contain a mantle-like ratio of $^{87}Sr/^{86}Sr$ (Fig. 5.20). Furthermore, the $^{34}S/^{32}S$ ratios in coeval sulfides and sulfates are similar (cf. Chapter 4.6). All these features are consistent with the suggestion that the oceanic flux was the dominant factor controlling the chemical and isotopic composition of the early oceans (Perry et al., 1978; Fryer et al., 1979; Veizer et al., 1982).

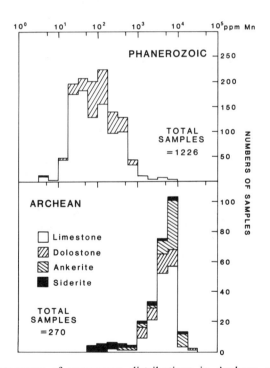

Figure 5.23 Histograms of manganese distributions in Archean and Phanerozoic carbonates. Note that these data are all for the carbonate (acid soluble) fraction of the rocks and not whole rock analyses. Reproduced from Veizer (1985), who also lists the sources of data. Similar age trends are known for iron in carbonates and for both elements in other lithologies (Ronov and Migdisov, 1971; Veizer, 1978).

5.9.2 Oxidation State of the Atmosphere–Hydrosphere System

The present-day depletion of iron and manganese in seawater is a consequence of abundant oxygen and the low solubility of Fe^{3+} and Mn^{4+} compounds. At present-day oxygen levels, the higher oceanic flux into the Archean oceans would have led to higher iron and manganese levels in clastic sediments and to more frequent formation of their ores but would not have led to higher concentrations of these elements in Archean seawater and hence in biochemically precipitated sediments. A change to Fe^{2+} and Mn^{2+} compounds would result in the desired increase in seawater iron and manganese concentrations. The advocated Fe^{2+} and Mn^{2+} enrichment in Archean calcites (Section 5.9.1 and Fig. 5.23), coupled with the observation that these elements show a unidirectional decrease in concentration in progressively younger sediments, argue for a lower Eh of the early hydrosphere–atmosphere system. The author does not advocate an oxygen-free system because the observed manganese and iron concentrations in calcites are less than would be expected for such conditions (Veizer, 1978, p. 405). The actual P_{O_2} level is highly controversial (cf. Holland, 1984; Grandstaff et al., 1986), but a decrease in this parameter could have led to an essentially anaerobic ocean below the thermocline (Drever, 1974a), while the overlying layer could have contained some free oxygen (see Veizer, 1983a). Such a proposition is somewhat analogous to the present-day ocean stratification, except that the proposed Archean P_{O_2} for both oceanic layers was probably lower and the oxygen minimum layer more pronounced than today.

5.10 SYNOPSIS

5.10.1 The Archean

The existence of granite–greenstone terranes with high-grade metamorphic belts since $\sim 3.75 \times 10^9$ years ago (Allårt, 1976; Moorbath, 1977a,b; Windley, 1984) attests to the nucleation and growth of continental crust during this period. Except for the presence of some mature sediments in the high metamorphic regions (cf. Windley, 1984; Eriksson and Donaldson, 1986), the granite–greenstone belts are characterized by immature assemblages of graywackes, paraconglomerates, and volcanoclastics, all derived from the ubiquitous and multiple (ultra)mafic to felsic magmatic cycles (Lowe and Knauth, 1977; Barley, 1978; Lowe, 1980b). These cycles are commonly capped by chemical sediments, particularly cherts and lenses of ironstones. The above assemblage indicates tectonic instability of its depositional environments; the instability being a response to a higher convective heat flux from the early mantle. This flux, dissipating preferentially through the oceanic crust, also buffered the composition of seawater (by intense oceanic crust–seawater exchange) and was ultimately responsible for the supply of silica, iron, manganese, copper, zinc, gold, barium, and sulfur for their ubiquitous deposits. The oxygen level in the hydrosphere–atmosphere system was likely lower than at present, although not entirely absent, and this enabled the deeper portions of the oceans to act as reservoirs for the exhalative–hydrothermal Fe^{2+} and Mn^{2+} (Holland, 1973b).

5.10.2 The Early Proterozoic

The transition from the Archean steady state into the Early Proterozoic one was characterized by rapid growth and cratonization of the continental crust and of the global sedimentary mass (Veizer, 1983a). This diachronous process that started $\sim 3.0 \times 10^9$ years ago in the Kaapvaal craton (the $\sim 3.0 \times 10^9$-year-old Pongola Supergroup is the oldest known cratonic sequence; Cloud, 1976; von Brunn and Mason, 1977) was largely, but not entirely, completed by $\sim 1.75 \pm 0.25 \times 10^9$ years ago. The transformation and accretion of the oceanic eugeoclinal segments to the existing cratons during the Phanerozoic may have been a continuation, at a diminishing rate, of this process. The above tectonic stabilization led to the development of the ubiquitous mature miogeoclinal and platformal sedimentary assemblages (cf. Ronov, 1964) through cyclic reworking and (mechanical as well as chemical) fractionation of their immature precursors. The mature and fractionated clastics (orthoconglomerates, quartzites, sandstones) and (bio)chemical sediments (carbonates, iron formations) were complemented by immature potassium-rich arkoses derived from the newly abundant granitic sources. The addition of this new component into the predominantly greenstone and tonalite-derived Archean precursor produced the mafic → felsic secular variations in the composition of the sedimentary mass (cf. Taylor and McLennan, 1985). The unstable tectonic (eugeoclinal) environments, with their associated sedimentary assemblages (e.g., Wopmay orogen; Hoffman and Bowring, 1984), diminished considerably in their relative importance (Kröner, 1981).

As a consequence of both the decrease in the oceanic crust–seawater flux (because of the decline in heat release from the mantle) and the concomitant increase in the river flux from the continents, seawater started to reflect a kinetic balance between these two fluxes and was not, as in the Archean, buffered mostly by the mantle. It may even be that the waning of the mantle flux (and thus the diminishing supply of reductants, such as Fe^{2+} and Mn^{2+}) rather than an increase in production of biogenic oxygen was the real reason for an increase in atmospheric P_{O_2} toward the end of this time interval (Arrhenius, 1981; Veizer et al., 1982; Veizer, 1983a). It is likely that upwelling of deep ocean waters onto the partially oxygenated shelves provided the major iron and manganese source for formation of the large Superior type iron (Holland, 1973b) and jaspilitic manganese deposits. Additional sources of Fe^{2+} and Mn^{2+}, such as continental discharge, were probably of secondary importance. However, this continental drainage under a low atmospheric P_{O_2} appears to have been essential for the formation of the fluviatile–deltaic Witwatersrand–Blind River type (gold–uranium–pyrite) conglomerate deposits.

In my view, all the processes that produced the drastic changes to the face of our planet during this late Archean–early Proterozoic transition (Fig. 5.24)

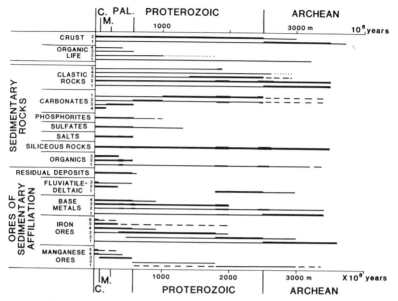

Figure 5.24 Summary of variations in the composition of the crust, biosphere, sedimentary rocks, and ores of sedimentary affiliations during geologic history. Modified from Veizer (1976a). *Crust*: 1 = smaller continental crustal nuclei; 2 = large cratonized continental crust with well-differentiated upper crust of granodioritic composition and frequent potassic granites. *Organic life*: 1 = prokaryota; 2 = eukaryota; 3 = Metazoa; 4 = continental higher plants and animals. *Clastic rocks*: 1 = greywackes; 2 = shales and slates; 3 = arkoses; 4 = orthoconglomerates and orthoquartzites; 5 = red beds. *Carbonates*: 1 = early diagenetic dolomites; 2 = (bio)chemical limestones; 3 = littoral-neritic organogenic and organodetrital limestones and their dolomitized ("late" diagenetic) equivalents; 4 = deep-sea limestones. *Organics*: 1 = carbonaceous shales; 2 = oil and gas; 3 = coal. *Residual deposits*: include iron, manganese weathering crusts (gossans), and bauxites. The so-called unconformity uranium deposits, typical mostly of the Early to Middle Proterozoic (cf. Robertson et al., 1978; Dahlkampf, 1980), not recognized as a distinct genetic group at the time of preparation (1972) of this figure, can perhaps also broadly fit into this subdivision. *Fluviatile–deltaic deposits*: 1 = conglomerate of uranium–gold–pyrite type; 2 = sandstone of uranium–vanadium–copper types; 3 = placers and paleoplacers. *Base metal deposits*: 1 = volcanogenic and volcanosedimentary type; 2 = sedimentary basinal type; 3 = sedimentary red-bed type; 4 = carbonate of lead-zinc type. *Iron ores*: 1 = Algoma type; 2 = Superior type; 3 = Clinton type; 4 = Bilbao type; 5 = Minette type; 6 = river bed, bog iron, laterite, and similar types. *Manganese ores*: 1 = jaspilitic type; 2 = volcanosedimentary and orthoquartzite–siliceous shale–manganese carbonate type; 3 = carbonate association; 4 = orthoquartzite–clay association; 5 = marsh and lake deposits.

were a consequence of the same fundamental cause, evolution of the mantle. Thus, the speedy formation and cratonization of continents, the associated development of continental shelves, the related growth and reprocessing of sedimentary mass, the change in ocean chemistry because of the waning mantle flux, the increase in P_{O_2} owing to the same reason, the formation of Superior type iron ores, and the diminishing of massive sulfides and of Algoma type iron ores are all manifestations of the same grand design. I am tempted to believe that the biota responded to this unfolding scenario orchestrated by the mantle rather than shaped it (Veizer, 1987).

5.10.3 The Middle (and Late?) Proterozoic

The transition from the preceding steady state was characterized by a diminution in the quantity of detrital pyrite and uraninite in fluviatile–deltaic deposits (Rutten, 1962; Holland, 1973b, 1984; Walker et al., 1983), by a tapering off of the Superior type iron (and manganese) deposits and by an increase in frequency of red beds, red bed copper deposits, and Clinton type iron ores (Fig. 5.24). Because all these changes are related to polyvalent metals (iron, manganese, uranium, copper), it is reasonable to suppose that this transition marked an increase in the oxidation state of the contemporaneous sedimentary environments.

There are some indications (appearance of economic phosphorites, larger evaporite accumulations, increased role of limestones) that the Late Proterozoic steady state might have been quantitatively more advanced than the Middle Proterozoic one. According to some authors (e.g., Engel et al., 1974; Kröner, 1981), this period was also a time of initial breakup of the Proterozoic supercontinent and of initiation of the modern phase of plate tectonics. To what extent these Middle to Late Proterozoic differences are real, or an artifact of postdepositional recycling phenomena, is difficult to evaluate on the basis of the available quantitative information.

5.10.4 The Early Paleozoic

Following the appearance of Metazoa, the Early Paleozoic steady state was characterized by an increased role of (bio)chemical sedimentation in the form of shallow marine limestones and their dolomitized counterparts and of phosphorites. In contrast, the high frequency of massive evaporite deposits and of hydrocarbon (oil and gas) accumulations is probably only an artifact of recycling (Section 5.7.1). Associated with these sediments are the carbonate lead–zinc (Mississippi type), iron (Bilbao type), and manganese exogenic ores and the residual deposits (bauxites). Although some of these trends are an artifact of recycling, the evolving organic life appears to have caused a real quantitative (and qualitative?) rearranging of interactions among biogenically mediated exogenic cycles of elements.

5.10.5 The Late Paleozoic–Mesozoic–Cenozoic

The last period, having many features in common with the preceding steady state, was characterized by migration of higher plants and animals onto the continents, with subsequent accumulation of organic material in terrestrial environments (e.g., coals). The high proportion of the Minette type iron ores, in spatial proximity to coal seams, and of uranium–vanadium–copper fluviatile–deltaic deposits (Colorado Plateau type), are also consistent with this phenomenon. Furthermore, the Mesozoic development (cf. Southam and Hay, 1977) of pelagic $CaCO_3$ shell-secreting biota appears to have led to a major shift in the principal carbonate sedimentation milieu from the shelves to the deep sea.

Because of space restrictions, it is not possible to discuss here the third- and higher-order secular variations, although these are—particularly during the Phanerozoic—quite pronounced. A good example is the changing biota and morphology of carbonate buildups (Wilson, 1975; Scholle et al., 1983).

In comparison to the Proterozoic, the Phanerozoic as a whole apparently contains a higher proportion of unstable tectonic settings with their associated immature (graywacke-rich) sediments (Engel et al., 1974; Fig. 5.20). This feature may be interpreted as reflecting a higher intensity of Phanerozoic plate tectonics as compared to the Proterozoic mobile belt style (Engel et al., 1974; Kröner, 1981). Alternatively, unstable tectonic realms of the oceanic domain will, in the future, be destroyed or transformed by multistage processes into cratonized continental crust, thus leaving behind a record similar to that of the Proterozoic. New advances will resolve whether this feature represents a real change in tectonic style or is an artifact of the presence of young transient segments. My preference is for the second alternative, but the time-related distribution of mineral deposits (Veizer et al., 1988) may demand an evolutionary explanation or, more likely, a combination of both factors.

5.11 CONCLUSIONS

Consideration of the terrestrial exogenic system shows that the system is fundamentally cyclic, with the rate of recycling increasing in the order: crystalline basement $\leqslant$ platforms, mature orogenic belts $\leqslant$ immature orogenic belts, passive margins $\leqslant$ oceanic crust, and its basins $\leqslant$ active margin basins $\leqslant$ hydrosphere $\leqslant$ biosphere $\leqslant$ atmosphere. From the planetary point of view, the common thread to diverse physical phenomena is the concept of population dynamics (Veizer, 1987). The populations discussed in this chapter span at least 13 orders of magnitude in mass (10^{13}–10^{26} g). The age distributions of their constituent units conform to the exponential (power law) systematics typical of populations propagated by recycling, that is, through "birth" and "death" of the constituent units. Overall, annual recycling rates appear to have been several orders of magnitude ($\sim 10^{-6\pm4}$) smaller than the population sizes. In general, the lifespans (τ_{max}) and half-lives (τ_{50}) of large populations appear to have been long and those of small populations short. The observed τ_{50} range (cf. also Lerman, 1979) from 10^{10}–10^9 years for galactic populations of stars and planets, to 10^9–10^7 years for the geologic tectonic realms, 10^8–10^2 years for the oceans,

10^7–10^{-2} years for the atmospheric constituents, and 10^2–10^{-3} years for the living systems. Viewed in this perspective, all populations are integral constituent parts of an all-encompassing entity. However, the concept of population dynamics does not require propagation of the present-day steady state. The diverse populations are in a constant intercourse on a variety of spatial and temporal scales, with some populations in apparent steady state, others growing or declining, and still others being "born" or "dying." In this scenario, the grand design appears to have been a unity, with perpetual, statistically self-regulated, internal motions.

The interdependent populations, whether living or dead (inorganic), form a nested hierarchy. In general, the larger—and thus slower cycling—populations establish the limits, and the very basis of existence, for the small populations. For subordinate populations, departures from such controlled steady states are possible, but only on time scales shorter than life spans, and thus response times, of the dominant populations. In this understanding, in contrast to the popular GAIA hypothesis (Lovelock, 1979), life is constrained by the limits imposed by the interdependent larger populations of the terrestrial exogenic system and is not a phenomenon that controls this system for its own benefit.

Superimposed on recycling are evolutionary phenomena that reflect the process of self-organization of the planet. These appear to have been a consequence of (1) the exponentially decreasing heat flux from the interior of the earth during the course of terrestrial evolution and (2) the evolution of life.

The former influenced the development of tectonic styles, and hence the buildup of the continental crust and of its associated sedimentary prism. It also exercised control—particularly in the early stages—on the chemistry of the hydrosphere and its precipitates and the atmosphere. The evolution of organic life directly caused deposition of biogenic and biochemical sediments. Equally profound may have been the indirect influence of this evolution on exogenic cycles through biogenically produced oxygen. The fact that this oxygen may not have been able to accumulate in appreciable quantities in the atmosphere as long as the mantle supply of reductants was large should not disguise the possibility that ultimately it may have been the biota that enabled the change from the early oxygen-poor, CO_2-rich atmosphere–hydrosphere into the present day oxygen-rich, CO_2-poor system.

In response to this long-term evolution, the development of the terrestrial exogenic system was characterized by a succession of evolving steady states—the fast-cycling populations adjusting particularly rapidly to the new states. As a consequence, present-day cycles in the relatively slowly cycling populations have many features in common with their early counterparts. In contrast, in the rapidly cycling systems (atmosphere, hydrosphere, biosphere), modern cycles are quantitatively, and frequently qualitatively, different from their early precursors. Although—to paraphrase James Hutton—we do not as yet see any vestiges of the beginning and end, we are starting to unravel the late youth and midlife stages of our evolving planet.

ACKNOWLEDGMENTS

The author acknowledges financial support by the Natural Sciences and Engineering Research Council of Canada, A. von Humboldt Foundation of the German Federal Republic and Canada Council (Killam Research Fellowship). J. Hayes and M. Bass typed the manuscript.

REFERENCES

Abbot, D. H., Menke, W., Hobart, M., and Anderson, R. N. (1981). Evidence for excess pore pressure in southwest Indian Ocean sediments. *JGR, J. Geophys. Res.* **86,** 813–828.

Abbott, D. H., Menke, W., and Morin, R. (1983). Constraints upon water advection in sediments of the Mariana Trough. *JGR, J. Geophys. Res.* **88,** 1075–1093.

Abbott, D. H., Menke, W. H., and Anderson, R. N. (1984). Correlated sediment thickness, temperature gradient, and excess pore pressure in a marine fault basin. *Geophys. Res. Lett.* **11,** 485–488.

Ajtay, G. L., Ketner, P., and Divigneaud, P. (1979). Terrestrial primary production and phytomass. In *The Global Carbon Cycle*, B. Bolin et al., Eds., John Wiley, New York, pp. 129–182.

Alfvén, H. and Arrhenius, G. (1976). *Evolution of the Solar System*, Washington, DC, 599 pp.

Allard, G. O. and Hurst, V. J. (1969). Brazil link supports continental drift. *Science* **163,** 528–532.

Allårt, J. H. (1976). The pre-3760 m.y. old supracrustal rocks of the Isua area, central west Greenland, and the associated occurrence of quartz-banded ironstone. In *The Early History of the Earth*, B. F. Windley, Ed., Wiley, London, pp. 177–190.

Allègre, C. J. (1982). Chemical geodynamics. *Tectonophysics* **81,** 109–132.

Allègre, C. J., Staudacher, T., Sarda, P., and Kurz, M. (1983). Constraints on evolution of the Earth's mantle from rare gas systematics. *Nature* **303,** 762–766.

Allègre, C. J. (1985). The evolving earth system. *Terra Cognita,* **5,** 5–14.

Allègre, C. J. and Ben Othman, D. (1980). Nd–Sr isotopic relationship in granitoid rocks and continental crust development: A chemical approach to orogenesis. *Nature (London)* **286,** 335–342.

Allègre, C. J. and Jaupart, C. (1985). Continental tectonics and continental kinetics. *Earth Planet. Sci. Lett.* **74,** 171–186.

Allègre, C. J. and Rousseau, D. (1984). The growth of the continents through geological time studied by Nd isotope analysis of shales. *Earth Planet. Sci. Lett.* **67,** 19–34.

Allègre, C. J., Staudacher, T., and Sarda, P. (1987). Rare gas systematics: Formation of the atmosphere, evolution and structure of the Earth's mantle. *Earth Planet. Sci. Lett.* **81,** 127–150.

Alt, J. C., Honnorez, J., Laverne, C., and Emmermann, R. (1986a). Hydrothermal alteration of a 1-km section through the upper oceanic crust, DSDP Hole 504B: Mineralogy, chemistry, and evolution of seawater-basalt interactions. *J. Geophys. Res.* **91,** 10309–10336.

Alt, J. C., Muehlenbachs, K., and Honnorez, J. (1986b). An oxygen isotope profile through the upper kilometer of the oceanic crust, DSPD Hole 504B. *Earth Planet. Sci. Lett.* **80,** 217–229.

Andersen, N. R. and Malahoff, A., Eds. (1977). *The Fat of Fossil-Fuel CO_2 in the Oceans*, Plenum, New York, 749 pp.

Anderson, A. T. (1974). Chlorine, sulfur and water in magmas and oceans. *Geol. Soc. Am. Bull.* **85,** 1485–1492.

Anderson, A. T. (1975). Some basaltic and andesitic gasses. *Rev. Geophys. Space Phys.* **13,** 37–56.

Anderson, A. T., Harris, D. M., and Ito, E. (1979). Toward geologic cycles for H_{20}, CO_2, Cl and S. *Int. Union Geod. Geophys. Abstr.* No. 17, p. 26.

Anderson, R. N. (1986). *Marine Geology: A Planet Earth Perspective*, Wiley, New York, 328 pp.

Anderson, R. N. and Hobart, M. A. (1976). The relation between heat flow and age in the southeastern Pacific. *JGR, J. Geophys. Res.* **81,** 2968–2989.

Anderson, R. N. and Zoback, M. D. (1982). Permeability, underpressure and convection in the oceanic crust near the Costa Rica Rift, eastern equatorial Pacific. *JGR, J. Geophys. Res.* **87,** 2860–2868.

Anderson, R. N., Langseth, M. G., and Sclater, J. G. (1977). The mechanisms of heat transfer through the floor of the Indian Ocean. *JGR, J. Geophys. Res.* **82,** 3391–3409.

Andrews, A. J. (1977). Low temperature fluid alteration of oceanic layer 2 basalts, DSDP leg 37. *Can. J. Earth Sci.* **14,** 911–926.

Andrews, A. J. (1979). On the effect of low-temperature seawater-basalt interaction on the distribution of sulfur in oceanic crust, layer 2. *Earth Planet. Sci. Lett.* **46,** 68–80.

Anonymous (1985). Snapshot of a changing America. *Time*, Sept. 2, pp. 16–18.

Armstrong, R. L. (1981). Radiogenic isotopes: The case for crustal recycling on a near-steady-state no-continental-growth Earth. *Philos. Trans. R. Soc. London A Ser.* **301,** 443–472.

Armstrong, R. L. and Cooper, J. (1971). Lead isotopes in island arcs. *Bull. Volcanol.* **35,** 27–37.

Arnold, M. and Shepard, S. M. F. (1981). East Pacific rise at latitude 21°N. Isotopic composition and the origin of hydrothermal sulfur. *Earth Planet. Sci. Lett.* **56,** 148–156.

Arrhenius, G. (1981). Interaction of ocean-atmosphere with planetary interior. *Adv. Space Res.* **1,** 37–48.

Arthur, M. A. (1979). Paleoceanographic events—recognition, resolution, and reconsideration *Rev. Geophys. Space Phys.* **17,** 1484–1494.

Arthur, M. A. and Schlanger, S. O. (1979). Cretaceous 'oceanic anoxic events' as causal factors in development of reef-reservoired giant oil fields. *Am. Assoc. Pet. Geol. Bull.* **63,** 870–885.

Arthur, M. A., Dean, W. A., and Schlanger, S. O. (1985). Variations in the global carbon cycle during the Cretaceous related to climate, volcanism, and changes in atmospheric CO_2. *Geophys. Monogr., Am. Geophys. Union* **32,** 504–529.

Averitt, P. (1975). Coal resources of the United States. *Geol. Surv. Bull. (U.S.)* **1412,** 131 pp.

Awramik, S. M. (1981). The pre-Phanerozoic biosphere: Three billion years of crises and opportunities. In *Biotic Crises in Ecological and Evolutionary Time*, M. H. Nitecki, Ed., Academic Press, New York, pp. 83–102.

Awramik, S. M., Schopf, J. W., and Walter, M. R. (1983). Filamentous fossil bacteria from the Archaean of Western Australia. *Precambrian Res.* **20,** 357–374.

Baas-Becking, L. G. M. (1925). Studies on the sulfur bacteria. *Ann. Bot. (London)* **39,** 613–650.

Bach, W. (coordinator) (1980). The carbon dioxide problem, and interdisciplinary survey. *Experientia* **36,** 767–812.

Bach, W., Pankrath, J., and Kellog, W., Eds. (1979). *Man's Impact on Climate*, Elsevier, Amsterdam, 327 pp.

Bachinski, D. J. (1977). Alteration associated with metamorphosed ophiolitic cupriferous iron sulfide deposits: Whalesback Mine, Notre Dame Bay, Newfoundland. *Mineral. Deposita* **12,** 48–63.

Badham, J. P. N. and Stanworth, C. W. (1977). Evaporites from the Lower Proterozoic of the east arm, Great Slave Lake. *Nature (London)* **268,** 516–517.

Badiozamani, K. (1973). The dorag dolomitization model—Application to the Middle Ordovician of Wisconsin. *J. Sediment. Petrol.* **43,** 965–984.

Bailey, Sir E. B. (1967). *James Hutton, The Founder of Modern Geology*, Elsevier, Amsterdam, 161 pp.

Bailey, N. T. S. (1964). *The Elements of Stochastic Processes with Application to the Natural Sciences*, Wiley, New York, 384 pp.

Balashov, Yu. A. (1985). Intensity of mantle volcanism and rates of continent's growth. *Geokhimiya* **12,** 1683–1985 (in Russian).

Ballard, R. D. (1983). *Exploring our Living Planet*, Nat. Geogr. Soc., Washington, DC, 366 pp.

Barley, M. E. (1978). Shallow-water sedimentation during deposition of the Archaean Warrawoona Group, eastern Pilbara Block, Western Australia. *Publ. Geol. Dep., Ext. Serv., Univ. West Aust.* **2,** 22–29.

Barrell, J. (1917). Rhythms and the measurement of geological time. *Geol. Soc. Am. Bull.* **28,** 745–904.

Barron, E. J., Sloan, J. L., II, and Harrison, C. G. A. (1980). Potential significance of land-sea distribution and surface albedo variations as a climatic forcing factor, 180 m.y. to the present. *Palaeogeogr., Palaeoclimatol. Palaeoecol.* **30,** 17–40.

Barth, T. W. F. (1961a). Abundance of the elements, areal averages, and geochemical cycles. *Geochim. Cosmochim. Acta* **23,** 1–8.

Barth, T. W. F. (1961b). Ideas on the interrelation between igneous and sedimentary rocks. *Bull. Comm. Geol. Finl.* **196,** 321–326.

Basset, C. A. (1815). *Explication de Playfair sur la théorie de la terre de Hutton, et Examen comparatif des systèmes géologiques fondés sur le feu et sur l'eau, par J. Murray, en réponse à l'Explication de Playfair* (translation), Paris and London, xxxvi + 424 + 196.

Beaty, R. D. and Manuel, O. K. (1973). Tellurium in rocks. *Chem. Geol.* **12,** 155–159.

Beauford, W., Barber, J., and Barringer, A. R. (1977). Release of particles containing metals from vegetation into the atmosphere. *Science* **195,** 571–573.

Becker, G. F. (1910). The age of the earth. *Smithson, Misc. Collect.* **56,** 6–28.

Behrens, E. W. and Frishman, S. A. (1971). Stable carbon isotopes in blue-green algal mats. *J. Geol.* **79,** 94–100.

Belyaev, S. S., Wolkin, R., Kenealy, W. R., DeNiro, M. J., Epstein, S., and Zeikus, G. (1983). Methanogenic bacteria from the Bondyuzhskoe oil field: General characterization and analysis of stable carbon isotopic fractionation. *Appl. Environ. Microbiol.* **45**, 691–697.

Benedict, C. R. (1978). The fractionation of stable carbon isotopes in photosynthesis. *What's New Plant Physiol.* **9**, 13–16.

Berger, W. H. (1976). Biogenous deep sea sediments: Production, preservation, and interpretation. In *Chemical Oceanography*, 2nd ed., Vol. 5, J. P. Riley and R. Chester, Eds., Academic Press, New York, pp. 265–388.

Berner, R. A. (1971). Worldwide sulfur pollution of rivers. *J. Geophys. Res.* **76**, 6597–6600.

Berner, R. A. (1972). Sulfate reduction, pyrite formation, and the oceanic sulfur budget. In *The Changing Chemistry of the Oceans*, Nobel Symp. D. Dryssen and D. Jagner, Eds., Wiley (Interscience), New York, pp. 347–361.

Berner, R. A. (1975). The role of magnesium in the crystal growth of calcite and aragonite from sea water. *Geochim. Cosmochim. Acta* **39**, 489–504.

Berner, R. A. (1977). Sedimentation and dissolution of pteropods in the ocean. In *The Fate of Fossil Fuel CO_2 in the Oceans*, N. R. Andersen and A. Malahoff, Eds., Plenum, New York, pp. 243–260.

Berner, R. A. (1978). Sulfate reduction and the rate of deposition of marine sediments. *Earth Planet. Sci. Lett.* **37**, 492–498.

Berner, R. A. (1982). Burial of organic carbon and pyrite sulfur in the modern ocean: Its geochemical and environmental significance. *Am. J. Sci.* **282**, 451–473.

Berner, R. A. (1984). Sedimentary pyrite formation: An update, *Geochim. Cosmochim. Acta* **48**, 605–615.

Berner, R. A. (1987). Models for carbon and sulfur cycles and atmospheric oxygen: Application to Paleozoic geologic history. *Amer. J. Sci.* **287**, 177–196.

Berner, R. A. and Holdren, G. R., Jr. (1978). Sulfur distribution in holes 380 and 380A of leg 42B. *Initial Rep. Deep Sea Drill. Proj.* **42**, Pt. 2, 625–626.

Berner, R. A. and Raiswell, R. (1983). Burial of organic carbon and pyrite sulfur in sediments over Phanerozoic time: A new theory. *Geochim. Cosmochim. Acta* **47**, 855–862.

Bestougeff, M. A. (1980). Summary of mondial coal resources and reserves: Geographic and geologic repartition. *Rev. Inst. Fr. Pet.* **35**, 353–366.

Bickle, M. J. (1978). Heat loss from the earth: A constraint on Archean tectonics from the relation between geothermal gradients and the rate of plate production. *Earth Planet. Sci. Lett.* **40**, 301–315.

Birch, F. (1965). Speculations on the earth's thermal history. *Geol. Soc. Am. Bull.* **76**, 4377–4388.

Bischoff, J. L. (1980). Geothermal system at 21°N, East Pacific Rise, physical limits on geothermal fluid and role of adiabatic expansion. *Science* **207**, 1465–1469.

Bischoff, J. L. and Dickson, F. W. (1975). Seawater-basalt interaction at 200°C and 500 bars: Implications for origin of seafloor heavy metal deposits and regulation of seawater chemistry. *Earth Planet. Sci. Lett.* **25**, 385–397.

Bischoff, J. L. and Rosenbauer, R. J. (1984). The critical point and two-phase boundary of seawater, 200°–500°C. *Earth Planet. Sci. Lett.* **68**, 172–180.

Bischoff, J. L. and Seyfried, W. E. (1978). Hydrothermal chemistry of seawater from 25° to 350°C. *Am. J. Sci.* **278**, 838–860.

Bischoff, J. L., Rosenbauer, R. J., Aruscavage, P. J., Baedecker, P. A., and Crock, J. G., (1983). Sea-floor massive sulfide deposits from 21°N, East Pacific Rise, Juan de Fuca Ridge, and Galapagos Rift, bulk Chemical composition and economic implications. *Econ. Geol.* **78**, 1711–1720.

Bjornsson, S., Arnorsson, S., and Tomasson, J. (1972). Economic evaluation of the Reykjanes thermal brine area, Iceland. *Am. Assoc. Pet. Geol. Bull.* **56**, 2380–2391.

Bois, C., Bouche, P., and Pelet, R. (1982). Global geologic history and hydrocarbon reserves. *Am. Assoc. Pet. Geol. Bull.* **66**, 1248–1270.

Bolin, B., Degens, E. T., Kempe, S., and Ketner, P., Eds. (1979). *The Global Carbon Cycle*, Wiley, New York, 491 pp.

Bostrom, K. (1976). Particulate and dissolved matter as sources for pelagic sediments. *Stockholm Contrib. Geol.* **30**, 1–79.

Bottinga, Y. (1969). Calculated fractionation factors for carbon and hydrogen isotope exchange in the system calcite-carbon dioxide-graphite-methane-hydrogen-water vapor. *Geochim. Cosmochim. Acta* **33**, 49–64.

Bowers, T. S. and Taylor, H. P., Jr. (1985). An integrated chemical and stable-isotope model of the origin of midocean ridge hot spring systems. *JGR, J. Geophys. Res.* **90**, 12583–12606.

Boyle, E. A., Collier, R., Dengler, A. T., Edmond, J. M., Ng, A. C., and Stallard, R. P. (1974). On the chemical mass balance in estuaries. *Geochim. Cosmochim. Acta* **38**, 1719–1728.

Boyle, E. A., Sclater, F., and Edmond, J. M. (1976). On the marine geochemistry of cadmium. *Nature (London)* **263**, 42–44.

Brass, G. W. (1976). The variation of the marine $^{87}Sr/^{86}Sr$ ratio during Phanerozoic time: Interpretation using a flux model. *Geochim. Cosmochim. Acta* **40**, 721–730.

Brevart, O. and Allègre, J. C. (1977). Strontium isotopic ratios in limestone through geologic time as a memory of geodynamic regimes. *Bull. Soc. Geol. Fr.* **19**, 1253–1257.

Bridgwater, D., Allårt, J. H., Schopf, J. W., Klein, C., Walter, M. R., Barghoorn, E. S., Strother, P., Knoll, A. H., and Gorman, B. E. (1981). Microfossil-like objects from the Archaen of Greenland: A cautionary note. *Nature (London)* **289**, 51–53.

Broda, E. (1975). *The Evolution of the Bioenergetic Processes*, Pergamon, Oxford, 220 pp.

Broecker, W. S. (1970). A boundary condition on the evolution of atmospheric oxygen. *J. Geophys. Res.* **75**, 3553–3557.

Broecker, W. S. (1971a). A kinetic model for the chemical composition of seawater. *Quat. Res. (N.Y.)* **1**, 188–207.

Broecker, W. S. (1971b). Calcite accumulation rates and glacial to interglacial changes in oceanic mixing. In *Late Cenozoic Glacial Ages*, K. K. Turekian, Ed., Yale University Press, New Haven, CT, pp. 239–265.

Broecker, W. S. (1974). *Chemical Oceanography*, Harcourt Brace Jovanovich, New York.

Broecker, W. S. (1982). Glacial to interglacial changes in ocean chemistry. *Prog. Ocean OGR* **7**, 151–197.

Broecker, W. S. and Peng, T.-H. (1982). *Tracers in the Sea*, Lamont-Doherty Geological Observatory, Palisades, NY, 690 pp.

Broecker, W. S., Takahashi, T., Simpson, H. J., and Peng, T. H. (1979). Fate of fossil fuel carbon dioxide and the global carbon budget. *Science* **206**, 409–418.

Brotzen, O. (1966). The average igneous rock and the geochemical balance. *Geochim. Cosmochim. Acta* **30**, 863–868.

Brown, L., Klein, J., Middleton, R., Sacks, I. S., and Tera, F. (1983). Beryllium-10: Island-arc volcanoes. *Year Book—Carnegie Inst. Washington* **82**, 455–456.

Brumsack, H. J. and Lew, M. (1982). Inorganic geochemistry of Atlantic Ocean sediments with special reference to Cretaceous black shales. In *Geology of Northwest African Continental Margin*, U. Von Rad, Ed., Springer-Verlag, Berlin, pp. 661–685.

Brune, J. and Dorman, J. (1963). Seismic waves and earth structure in the Canadian shield. *Bull. Seismol. Soc. Am.* **53**, 167–210.

Bryan, W. B., and Moore, J. B. (1977). Compositional variations of young basalts in the Mid-Atlantic Ridge rift valley near lat. 36.49'N. *Geol. Soc. Am. Bull.* **88**, 556–570.

Buat-Menard, P. (1979). Influence de la retombée atmosphérique sur la chimie des métaux en trace dans la matière en suspension de l'Atlantique nord. Ph.D. thesis, University of Paris, 434 pp.

Buckle, T. H. (1861). *A History of Civilization in England*, London.

Buddington, A. F., Jensen, M. L., and Mauyer, R. L. (1969). Sulfur isotopes and origin of northwest Adirondack sulfide deposits. *Geol. Soc. Am. Mem.* **115**, 423–451.

Buffon, Comte de (1807). *Epoques de la Nature*, Paris.

Burchfield, J. D. (1975). *Lord Kelvina and the Age of the Earth*, Sci. Hist. Pub., New York, 260 pp.

Burke, K. and Kidd, W. S. F. (1978). Were Archean continental geothermal gradients much steeper than those of today? *Nature (London)* **272**, 240–241.

Burke, W. H., Denison, R. F., Hetherington, E. A., Koepnick, R. F., Nelson, H. F., and Otto, J. B. (1982). Variation of seawater $^{87}Sr/^{86}Sr$ throughout Phanerozoic time. *Geology* **10**, 516–519.

Burnie, S. W., Schwartz, H. P., and Crocket, J. H. (1972). A sulfur isotopic study of the White Pine mine, Michigan. *Econ. Geol.* **67**, 895–914.

Burns, L. E. (1985). The border Ranges ultramafic and mafic complex, south-central Alaska: cumulate fractionates of island-arc volcanics. *Can. J. Earth Sci.* **22**, 1020–1038.

Burns, R. C. and Hardy, R. W. F. (1975). *Nitrogen Fixation in Bacteria and Higher Plants*, Springer-Verlag, New York, 225 pp.

Bushinskii, G. I. (1969). *Old Phosphorites of Asia and their Genesis*, Acad. Sci. USSR (Geol. Inst. Transl., 149, Israel Program for Scientific Translations, Jerusalem, 266 pp.).

Byers, C. W. (1976). Bioturbation and the origin of the metazoans: Evidence from the Belt Supergroup, Montana. *Geology* **4**, 565–567.

Cadle, R. D. (1980). A comparison of volcanic with other fluxes of atmospheric trace gas constituents. *Rev. Geophys. Space Phys.* **18**, 746–752.

Cadle, R. D., Lazrus, A. L., Heubert, B. J., Heidt, L. E., Rose, W. I., Jr., Woods, D. C., Chuan, R. L., Stoiber, R. E., Smith, D. B., and Zielinski, R. (1979). Atmospheric implications of studies of Central American volcanic eruptions, *JGR, J. Geophys. Res.* **84**, 6961–6968.

Calder, J. A. and Parker, P. L. (1973). Geochemical implications of induced changes in ^{13}C fractionation by blue-green algae. *Geochim. Cosmochim. Acta* **37**, 133–140.

Calvert, S. E. (1976). The mineralogy and chemistry of near-shore sediments. In *Chemical Oceanography*, 2nd ed. Vol. 6, J. P. Riley and R. Skirrow, Eds., Academic Press, New York, pp. 187–280.

Canuto, V. M., Levine, J. S., Augustsson, T. R., and Imhoff, C. L. (1982). UV radiation from the young sun and oxygen and ozone levels in the prebiological palaeoatmosphere. *Nature (London)* **296,** 816–820.

Carstens, H. (1949). Et nytt prinsipp ved geochemiske beregninger. *Nor. Geol. Tidsskr.* **28,** 47–50.

Casadevall, T. J., Johnson, D. A., Harris, D. M., Rose, W. I., Jr., Malinconico, L. L., Stoiber, R. E., Bornhorst, R. J., Williams, S. N., Woodruff, L., and Thompson, J. M. (1981). SO_2 emission rates at Mount St. Helens from March 29 through December, 1980. *Geol. Surv. Prof. Pap. (U.S.)* **1250,** 193–200.

Chambers, L. A. (1980). Sulfur isotope fractionation in intertidal sediments, Spencer Gulf, South Australia. *Biogeochem. Ancient Mod. Environ., Int. Symp., 4th Proc., 1979,* p. 25.

Chambers, L. A. and Trüdinger, P. A. (1979). Microbiological fractionation of stable sulfur isotopes: A review and critique. *Geomicrobiol. J.* **1,** 249–293.

Chameides, W. L., Stedman, D. H., Dickerson, R. R., Rusch, D. W., and Cicerone, R. J. (1977). NO_x production of lightning. *J. Atmos. Sci.* **34,** 134–139.

Chameides, W. L. and Walker, J. C. G. (1981). Rates of fixation by lightning of carbon and nitrogen in possible primitive atmospheres. *Origins Life* **11,** 291–302.

Chang, S., Des Marais, D., Mack, R., Miller, S. R., and Strathearn, G. E. (1983). Prebiotic organic synthesis and the origin of life. In *The Earth's Earliest Biosphere: Its Origin and Evolution*, J. W. Schopf, Ed., Princeton Univ. Press, Princeton, NJ, pp. 53–92.

Chen, J., Zhao, R., Huo, W., Yao, Y., Pan, S., Shao, M., and Hai, C. (1981). Sulfur isotopes of some marine gypsum. *Sci. Geol. Sin.* pp. 273–278.

Chester, R. and Aston, S. R. (1976). The geochemistry of deep-sea sediments. In *Chemical Oceanography*, 2nd ed., Vol. 6, J. P. Riley and G. Skirrow, Eds., Academic Press, New York, pp. 281–391.

Clarke, F. W. (1908). The data of geochemistry. *Geol. Survey Bull. (U.S.)* **330,** 716 pp.

Claypool, G. E., Holser, W. T., Kaplan, I. R., Sakai, H., and Zak, I. (1980). The age curves of sulfur and oxygen isotopes in marine sulfate and their mutual interpretation. *Chem. Geol.* **28,** 199–260.

Clayton, R. N. and Degens, E. T. (1959). Use of C isotope analyses for differentiating fresh-water and marine sediments. *Am. Assoc. Pet. Geol. Bull.* **43,** 890–897.

Clemmensen, L., Holser, W. T., and Winter, D. (1985). Stable isotope study through the Permian-Triassic boundary in East Greenland. *Bull. Geol. Soc. Den.* **33,** 253–260.

Cloud, P. E. (1968). Pre-metazoan evolution and the origin of metazoa. In *Evolution and Environment*, E. T. Drake, Ed., Yale Univ. Press, New Haven, CN, pp. 1–72.

Cloud, P. E. (1969). Pre-Paleozoic sediments and their significance for organic geochemistry. In *Organic Geochemistry*, G. Eglington and M. T. J. Murphy, Eds., Springer-Verlag, Berlin, pp. 772–786.

Cloud, P. E. (1976). Major features of crustal evolution. *Trans. Geol. Soc. S. Afr.* **79,** 1–32.

Coats, R. R. (1962). Magma type and crustal structure in the Aleutian Arc, in The Crust of the Pacific Basin. *Am. Geophys. Union, Mon.* **6,** 92–109.

Coleman, R. G. (1977). *Ophiolites*, Springer-Verlag, New York, 229 pp.

Condie, K. C. (1973). Archean magmatism and crustal thickening. *Geol. Soc. Am. Bull.* **84,** 2981–2992.

Condie, K. C. (1976). *Plate Tectonics and Crustal Evolution*, Pergamon, New York, 288 pp.

Converse, D. R., Holland, H. D., and Edmond, J. M. (1984). Hydrothermal flow rates in axial hot springs of the East Pacific Rise (21°N). Implications for the EPR heat budget. *Earth Planet. Sci. Lett.* **69,** 159–175.

Conway, E. J. (1943). The chemical evolution of the oceans. *Irish Acad. Proc.* **48B,** 161–212.

Cook, P. J. and McElhinny, M. W. (1979). A reevaluation of the spatial and temporal distribution of sedimentary phosphate deposits in the light of plate tectonics. *Econ. Geol.* **74,** 315–330.

Cope, J. J. and Chaloner, W. G. (1980). Fossil charcoal as evidence of past atmospheric composition. *Nature (London)* **283,** 647–649.

Corliss, J. B. (1971). The origin of metal-bearing submarine hydrothermal solutions. *J. Geophys. Res.* **76,** 8128–8138.

Corliss, J. B., Lyle, M., and Dymond, J. (1978). The chemistry of hydrothermal mounds near the Galapagos rift. *Earth Planet. Sci. Lett.* **40,** 12–24.

Corliss, J. B., Dymond, J., Gordon, L. I., Edmund, J. M., von Herzen, R. P., Ballard, R. D., Green, K. L., Williams, D., Brainbridge, A. L., Crane, K., and van Andel, T. H. (1979a). Submarine thermal springs on the Galapagos rift. *Science* **203,** 1073–1083.

Corliss, J. B., Gordon, L. E., and Edmond, J. M. (1979b). Some implications of heat/mass ratios in Galapagos rift hydrothermal fluids for models of seawater rock interaction and the formation of oceanic crust. In *Deep Drilling Results in the Atlantic Ocean: Ocean Crust*, M. Talwani, C. G. Harrison, and D. E. Hayes, Eds., Am. Geophys. Union, Washington, DC, pp. 391–402.

Cortecci, G., Reyes, E., Berti, G., and Casati, P. (1981). Sulfur and oxygen isotopes in Italian marine sulfates of Permian and Triassic ages. *Chem. Geol.* **34,** 65–79.

Cox, A., Ed. (1973). *Plate Tectonics and Geomagnetic Reversals*, Freeman, San Francisco, CA, 702 pp.

Craig, H. L. (1953). The geochemistry of stable carbon isotopes. *Geochim. Cosmochim. Acta* **3,** 53–92.

Craig, H. L., Clarke, W. B., and Beg, M. A. (1975). Excess ^{3}He in deep water on the East Pacific Rise. *Earth Planet. Sci. Lett.* **26,** 125–132.

Craig, H. L. and Lupton, J. E. (1981). Helium-3 and mantle volatiles in the ocean and the oceanic crust. In *The Sea*, 7, Wiley, New York, pp. 391–428.

Craig, H. L., Lupton, J. E., and Haribe, H. (1978). A mantle helium component in circum-Pacific volcanic gases. Hakone, the Marianas and Mt. Lassen. *Adv. Earth Planet. Sci.* **3,** 897–900.

Crocket, J. H. and Kuo, H. Y. (1979). Sources for gold, palladium and iridium in deep sea sediments. *Geochim. Cosmochim. Acta* **43,** 831–842.

Crocket, J. H., MacDougall, J. D., and Harriss, R. C. (1973). Gold, palladium and iridium in marine sediments. *Geochim. Cosmochim. Acta* **37,** 2547–2556.

Croll, J. (1871). On a method of determining the mean thickness of the sedimentary rocks of the globe. *Geol. Mag.* **8,** 97–102, 285–287.

Cuong, N. B., Bonsang, B., and Lambert, G. (1974). The atmospheric concentration of sulfur dioxide and sulfate aerosols over Antarctic, sub-Antarctic areas and oceans. *Tellus* **26,** 241–249.

Dahlkamp, F.-J. (1980). The time related occurrence of uranium deposits. *Mineral. Deposita* **5,** 69–79.

Daly, R. A. (1909). First calcareous fossils and evolution of limestones. *Geol. Soc. Am. Bull.* **20,** 153–170.

Dasch, E. J. (1969). Strontium isotopes in weathering profiles, deep-sea sediments, and sedimentary rocks. *Geochim. Cosmochim. Acta* **33,** 1521–1552.

Davies, S. F. (1979). Thickness and thermal history of continental crust and root zones. *Earth Planet. Sci. Lett.* **44,** 231–238.

Davies, S. F. (1980). Thermal histories of convective Earth models and constraints on radiogenic heat production in the Earth. *JGR, J. Geophys. Res.* **85,** 2517–2530.

Davis, J. B. and Kirkland, D. W. (1978). Bioepigentaic sulfur deposits. *Econ. Geol.* **74,** 462–468.

Davis, J. B. and Yarbrough, H. D. (1966). Anaerobic oxidation of hydrocarbons by *Desulfovibrio desulfuricans*. *Chem. Geol.* **1,** 137–144.

Deffeyes, J. S. (1970). The axial valley: A steady-state feature of the terrain. In *The Megatectonics of Continents and Oceans*, Chap. 9, H. Johnson and B. L. Smith, Eds., Rutgers Univ. Press, New Brunswick, NJ.

Degens, E. T. (1982). Riverine carbon—An overview. *Mitt. Geol. Palaeontol. Inst. Univ. Hamburg* **52,** 1–12.

Degens, E. T. and Ittekkot, V. (1987). The carbon cycle—Tracking the path of organic particles from sea to sediment. In *Marine Petroleum Source Rocks*, J. Brooks and A. J. Fleet, Eds., Geol. Soc. Lond. Spec. Pub. 26, pp. 121–135.

Deines, P. (1980a). The carbon isotopic composition of diamonds: Relationship to diamond shape, color, occurrence and vapor composition. *Geochim. Cosmochim. Acta* **44,** 943–961.

Deines, P. (1980b). The isotopic composition of reduced organic carbon. In *Handbook of Environment Isotope Geochemistry, Vol. 1, The Terrestrial Environment*, P. Fritz and J. C. Fontes, Eds., Elsevier, Amsterdam, pp. 329–406.

Deines, P. and Gold, D. P. (1973). The isotopic composition of carbonatite and kimberlite carbonates and their bearing on the isotopic composition of deep-seated carbon. *Geochim. Cosmochim. Acta* **37,** 1709–1733.

Delaney, J. R., Muenow, D. W., and Graham, D. G. (1978). Abundance and distribution of water, carbon, and sulfur in the glassy rinds of submarine pillow basalts. *Geochim. Cosmochim. Acta* **42,** 581–594.

Delany, J. M. and Helgeson, H. C. (1978). Calculation of thermodynamic consequences of dehydration in subduction oceanic crust to 100 km and 800°C. *Am. J. Sci.* **278,** 638–686.

De Luc, J. A. (1790–1791). Letters, Monthly Review (cited in Geikie, A., 1905, *Founders of Geology*, MacMillan, NY, p. 296).

de Maillet, B. (1748). *Telliamed*, 2 vols., Amsterdam.

DeMaster, D. J. (1981). The supply and accumulation of silica in the marine environment. *Geochim. Cosmochim. Acta* **45**, 1715–1732.

DePaolo, D. J. (1980a). Sources of continental crust: Neodymium isotope evidence from the Sierra Nevada and Peninsular Ranges. *Science* **209**, 684–687.

DePaolo, D. J. (1980b). Crustal growth and mantle evolution: Inferences from models of element transport and Nd and Sr isotopes. *Geochim. Cosmochim. Acta* **44**, 1185–1196.

DePaolo, D. J. and Johnson, R. M. (1979). Magma genesis in the New Britain island-arc: Constraints from Nd and Sr isotopes and trace-element patterns. *Contrib. Mineral. Petrol.* **70**, 367–379.

DePaolo, D. J. and Wasserburg, G. J. (1977). The sources of island arcs as indicated by Nd and Sr isotopic studies, *Geophys. Res. Lett.* **1**, 465–468.

Des Marais, D. J. and Moore, J. G. (1984). Carbon and its isotopes in mid-oceanic basaltic glasses. *Earth Planet. Sci. Lett.* **69**, 43–57, 1984.

Desmarest, N. (1774). Mémoire sur l'origine et la nature du basalte, à grandes colonnes polygones, derminées par l'histoire naturelle de cette pierre, observée en Auvergne. *Mèm. Acad. Sci.* for 1771, 706–708.

Detrick, R. S., Buhl, P., Vera, E., Mutter, J., Orcutt, J., Madsen, J., and Brocher, T. (1987). Multi-channel seismic imaging of a crustal magma chamber along the East Pacific Rise. *Nature (London)* **326**, 35–41.

Deuser, W. G. and Degens, E. T. (1967). Carbon isotope fractionation in the system CO_2 (gas)-CO_2(aqueous)-HCO_3. *Nature (London)* **215**, 1033–1035.

DeVooys, C. G. N. (1979). Primary production in aquatic environments. In *The Global Carbon Cycle*, B. Bolin et al., Eds., Wiley, New York, pp. 259–292.

Dietz, R. S. (1961). Continent and ocean basin evolution by spreading of the sea floor. *Nature (London)* **190**, 854–857.

Dimroth, E. and Kimberley, M. M. (1976). Precambrian atmospheric oxygen: Evidence in the sedimentary distributions of carbon, sulfur, uranium, and iron. *Can. J. Earth Sci.* **13**, 1161–1185.

Donnelly, T. H. and Ferguson, J. (1980). A stable isotope study of three deposits in the Alligator Rivers uranium field, N.T. In *Uranium in the Pine Creek Geosyncline*, J. Ferguson and A. B. Coleby, Eds., Int. Atomic Energy Comm., Vienna, pp. 397–407.

Donnelly, T. H., Lambert, I. B., Oehler, D. Z., Hallberg, J. A., Hudson, D. R., Smith, J. W., Bavington, O. A., and Golding, L. (1977). A reconnaissance study of stable isotope ratios in Archean rocks from the Yilgarn Block, Western Australia, *J. Geol. Soc. Austr.* **24**, 409–420.

Donnelly, T. W., Thompson, G., and Robinson, P. T. (1979). Very low-temperature alteration of the oceanic crust and the problem of fluxes of potassium and magnesium. In *Deep Drilling Results in the Atlantic Ocean: Ocean Crust*, M. Talwani, C. G. Harrison, and D. E. Hayes, Eds., Am. Geophys. Union, Washington, DC, pp. 369–381.

Dostalek, M. and Kvet, R. (1963). Utilization of the osmotolerance of sulphate reducing bacteria in study of the genesis of subterranean waters. *Folia Microbiol. Prague* **9**, 103–114.

Drever, J. I. (1971). Early diagenesis of clay minerals, Rio Ameca Basin, Mexico. *J. Sediment. Petrol.* **41**, 982–994.

Drever, J. I. (1974a). Geochemical model for the origin of Precambrian banded iron formations. *Geol. Soc. Am. Bull.* **85,** 1099–1106.

Drever, J. I. (1974b). The magnesium problem. In *The Sea*, Vol. 5, E. D. Goldberg, Eds., Wiley (Interscience), New York, pp. 337–357.

Duce, R. A. and Duursma, E. K. (1977). Inputs of organic matter to the ocean. *Mar. Chem.* **5,** 319–339.

Duce, R. A. and Hoffman, E. J. (1977). Chemical fractionation at the air/sea interface. *Ann. Rev. Earth Planet. Sci.* **4,** 187–228.

Dunlop, J. S. R. (1978). Shallow water sedimentation at North Pole, Pillbara Block, Western Australia. In *Archean Cherty Metasediments: Their Sedimentology, Micropaleontology, Biogeochemistry and Significance to Mineralization*, Vol. 2, J. E. Glover and D. I. Growes, Eds., Geol. Dep. and Ext. Serv., West Aust., Perth, 30–38.

Dunlop, J. S. R., Muir, M. O., Milne, V. A., and Groves, D. I. (1978). A new microfossil assemblage from the Archean of Western Australia. *Nature (London)* **274,** 676–678.

Dunne, T. (1978). Rates of chemical denudation of silicate rocks in tropical catchments. *Nature (London)* **274,** 244–246.

Dymond, J. and Hogan, L. (1973). Noble gas abundance patterns in deep sea basalts Primordial gases from the mantle. *Earth Planet. Sci. Lett.* **20,** 131–139.

Dymond, J. and Hogan, L. (1978). Factors controlling the noble gas abundance patterns of deep-sea basalts. *Earth Planet. Sci. Lett.* **38,** 117–128.

East Pacific Rise Study Group (1981). Crustal processes of the mid-ocean ridge. *Science* **213,** 40.

Edmond, J. M., Corliss, J. B., and Gordon, L. I. (1979a). Ridge crest-hydrothermal metamorphism at the Galapagos spreading center and reverse weathering. In *Deep Drilling Results in the Atlantic Ocean: Ocean Crust*, M. Talwani, C. G. Harrison, and D. E. Hayes, Eds., Am. Geophys. Union, Washington, DC, pp. 383–390.

Edmond, J. M., Measures, C., McDuff, R. E., Chan, L. H., Collier, R., Grant, B., Gordon, L. I., and Corliss, J. B. (1979b). Ridge crest hydrothermal activity and the balances of the major and minor elements in the ocean: the Galapagos data. *Earth Planet. Sci. Lett.* **46,** 1–18.

Edmond, J. M., Measures, C., Mangam, B., Grant, B., Sclater, F. R., Collier, R., Hudson, A., Gordon, L. I., and Corliss, J. B. (1979c). On the formation of metal-rich deposits at ridge crests. *Earth Planet. Sci. Lett.* **46,** 19–30.

Edmond, J. M., VonDamm, K. L., McDuff, R. E., and Measures, C. I. (1982). Chemistry of hot springs on the East Pacific Rise and their effluent dispersal. *Nature (London)* **297,** 187–191.

Elderfield, H. (1976). Manganese fluxes to the ocean. *Mar. Chem.* **4,** 103–132.

El Wakeel, S. K. and Riley, J. R. (1961). Chemical and mineralogical studies of deep sea sediments. *Geochim. Cosmochim. Acta* **25,** 110–146.

Emery, K. O. (1960). *The Sea off Southern California*, Wiley, New York.

Engel, A. E. J. (1963). Geologic evolution of North America, *Science* **140,** 143–152.

Engel, A. E. J., Itson, S. P., Engel, C. G., Stickney, D. M., and Cray, E. J. (1974). Crustal evolution and global tectonics, a petrogenic view. *Bull. Geol. Soc. Am.* **85,** 843–858.

Eriksson, E. (1960). The yearly circulation of chloride and sulfur in nature; meteorological, geochemical and pedologic implications. Part II. *Tellus* **12,** 63–71.

Erikkson, K. A. and Donaldson, J. A. (1986). Basinal and shelf sedimentation in relation to the Archean-Proterozoic boundary. *Precambr. Res.* **33**, 103–121.

Ernst, W. G. (1983). A summary of Precambrian crustal evolution. In *Revolution in the Earth Sciences*, S. G. Boardman, Ed., Kendall-Hunt, Dubuque IA, pp. 36–55.

Estep, M. F., Tabita, F. R., Parker, P. L., and Van Baalen, C. (1978). Carbon isotope fractionation by ribulose-1.5-biphosphate carboxylase from various organisms. *Plant Physiol.* **61**, 680–687.

Fanale, F. P. (1971). A case for catastrophic early degassing of the Earth. *Chem. Geol.* **8**, 79–105.

Fanning, K. A. and Pilson, M. E. Q. (1973). The lack of inorganic removal of dissolved silica during river–ocean mixing. *Geochim. Cosmochim. Acta* **37**, 2405–2415.

Faure, G. and Powell, J. L. (1972). *Strontium Isotope Geology*, Springer-Verlag, Heidelberg, 188 pp.

Faure, G. (1977). *Isotope Geology*, Wiley, New York, 464 pp.

Fenchel, T. and Blackburn, T. H. (1979). *Bacteria and Mineral Cycling*, Academic Press, London, 225 pp.

Fischer, A. G. (1981). Climatic oscillations in the biosphere. In *Biotic Crises in Ecological and Evolutionary Time*, M. Nitecki, Ed., Academic, New York, pp. 103–131.

Fisher, D. E. (1979). Helium and xenon in deep-sea basalts as a measure of magnetic differences. *Nature* **282**, 825–827.

Fisher, D. E. (1986). Rare gas abundances in MORB. *Geochim. Cosmochim. Acta* **50**, 2531–2541.

Fisher, I. St. J. and Hudson, J. D. (1987). Pyrite formation in Jurassic shales of contrasting biofacies. In *Marine Petroleum Source Rocks*, J. Brooks and A. J. Fleet, Eds., Geol. Soc. Lond. Spec. Pub. 26, pp. 69–78.

Fisher, O. (1900). An estimate of the geological age of the earth, By J. Joly, M. A. etc., *Geol. Mag., Decade* **4** (New Series) **7** 124–132.

Fitton, W. H. (1803). A comparative view of the Huttonian and Neptunian systems of geology. *Edinburgh Rev.* **2**, 337–348 (Published anonymously).

Fitzgerald, W. F. (1982). Evidence for anthropogenic atmospheric mercury input to the oceans. *Trans. Am. Geophys. Union* **63**, 245.

Fleming, N. C. and Roberts, D. G. (1973). Tectonic-eustatic changes in sea level and seafloor spreading. *Nature* **243**, 19–22.

Folk, R. and Land, L. S. (1975). Mg/Ca ratio and salinity: Two controls over crystallization of dolomite. *Amer. Assoc. Petrol. Geol. Bull.* **59**, 60–68.

Forrester, J. W. (1968). *Principles of Systems*, Wright-Allen Press, Cambridge, MA, 420 pp.

François, L. M., and Gerard, J-C. (1986). A numerical model of the evolution of ocean sulfate and sedimentary sulfur during the last 800 million years. *Geochim. Cosmochim. Acta* **50**, 2289–2302.

Frey, R. W. and Seilacher, A. (1980). Uniformity in marine invertebrate ichnology. *Lethaia* **13**, 183–207.

Friend, J. P. (1973). The global sulfur cycle. In *Chemistry of the Lower Atmosphere*, S. I. Rasool, Ed., Plenum, New York, pp. 177–201.

Fripp, R. E. P., Donnelly, T. H., and Lambert, I. B. (1979). Sulphur isotope results from Archean banded iron formations, Rhodesia. *Geol. Soc. S. Afr. Spec. Pub.* **5**, 205–208.

Froelich, P. N., Hambrick, G. A., Andreae, M. O., Mortlock, R. A., and Edmond, J. M. (1985). The geochemistry of inorganic germanium in natural waters. *J. Geophys. Res.* **90**, 1133–1141.

Fry, B., Scanlan, R. S., and Parker, P. L. (1977). Stable carbon isotope evidence for two sources of organic matter in coastal sediments: Seagrasses and plankton. *Geochim. Cosmochim. Acta* **41**, 1875–1877.

Fryer, B. J., Fyfe, W. S., and Kerrich, R. (1979). Archean volcanogenic oceans. *Chem. Geol.* **24**, 25–33.

Fuchs, G., Thauer, R., Ziegler, H., and Stichler, W. (1979) Carbon isotope fractionation by *Methanobacterium thermoautotrophicum*. *Arch. Microbiol.* **120**, 135–139.

Fyfe, W. S. (1974). Archean tectonics. *Nature (London)* **249**, 338.

Galimov, E. M. and Gerasimovskiy, V. I. (1978). Isotopic composition of carbon in Icelandic magmatic rocks. *Geokhimiya* **11**, 1615–1621; *Geochem. Int.* **15** (6, 1–69.).

Galimov, E. M. (1980). C^{13}/C^{12} in kerogen. In *Kerogen*, B. Durand, Ed., Technip. Editions, Paris, pp. 271–299.

Galimov, E. M., Migdisov, A. A., and Ronov, A. B. (1975). Variation in the isotopic composition of carbonate and organic carbon in sedimentary rocks during Earth's history. *Geokhimiya* **11**, 323–342; *Geochem. Int.* **12** (2) pp. 1–19.

Garcia, M. O., Liu, N. W. K., and Menow, D. W. (1979). Volatiles in submarine volcanic rocks from the Mariana Island Arc and Trough. *Geochim. Cosmochim. Acta* **43**, 305–312.

Gardner, W. S. and Menzel, D. W. (1974). Phenolic aldehydes as indicators of terrestrially derived organic matter in the sea. *Geochem. Comochim. Acta* **38**, 813–822.

Garrels, R. M. (1965). Silica: Role in the buffering of natural waters. *Science* **148**, 69.

Garrels, R. M. and Lerman, A. (1981). Phanerozoic cycles of sedimentary carbon and sulfur. *Proc. Nat. Acad. Sci. U.S.A.* **78**, 4652–4656.

Garrels, R. M. and Mackenzie, F. T. (1969). Sedimentary rock types: relative proportions as a function of geological time. *Science* **163**, 570–571.

Garrels, R. M. and Mackenzie, F. T. (1971a). *Evolution of Sedimentary Rocks*, Norton, New York, 397 pp.

Garrels, R. M. and Mackenzie, F. T. (1971b). Gregor's denudation of the continents. *Nature (London)* **231**, 382–383.

Garrels, R. M. and Mackenzie, F. T. (1972). A quantitative model for the sedimentary rock cycle. *Mar. Chem.* **1**, 27–41.

Garrels, R. M. and Mackenzie, F. T. (1974). Chemical history of the oceans deduced from postdepositional changes in sedimentary rocks. In *Spec. Publ.—Soc. Econ. Paleont.* **20**, 193–204.

Garrels, R. M. and Perry, E. A., Jr. (1974). Cycling of carbon, sulfur and oxygen through geologic time. In *The Sea*, Vol. 5, E. D. Goldberg, Ed., Wiley-Interscience, New York, pp. 303–336.

Garrels, R. M., Mackenzie, F. T., and Siever, R. (1972). Sedimentary cycling in relation to the history of continents and oceans. In *The Nature of the Solid Earth*, E. C. Robertson, Ed., McGraw-Hill, New York, pp. 93–121.

Garrels, R. M., Mackenzie, F. T., and Hunt, C. (1975). *Chemical Cycles and the Global Environment*, William Kaufmann, San Francisco, CA, 206 pp.

Garrels, R. M., Lerman, A., and Mackenzie, F. T. (1976). Controls of atmospheric O_2 and CO_2: Past, present and future. *Am. Sci.* **64**, 306–315.

Geikie, Sir A., Ed. (1899). *Theory of the Earth, with Proofs and Illustrations*, Vol. 3, W. Greech, London, 278 pp.

Gilluly, J. (1949). Distribution of mountain building in geologic time, *Geol. Soc. Am. Bull.* **60**, 561–590. **120**, 135–139.

Gimmel'farb, G. B. (1978). General features of Precambrian carbonate deposition based on the example of the Aldan and Baltic Shields. *Problems of the Precambrian Sedimentary Geology*, Vol. 4, Part I, Nauka, Moscow, pp. 167–180. (in Russian).

Gingerich, P. D. (1983). Rates of evolution: Effects of time and temporal scaling. *Science* **222**, 159–161.

Glikson, A. Y. (1979). Early Precambrian tonalite-trondhjemite sialic nuclei. *Earth Sci. Rev.* **15**, 1–73.

Goldberg, E. D. (1976). Rock volatility and aerosol composition. *Nature (London)* **260**, 128–129.

Goldberg, E. D., Broecker, W. S., Gross, M. G., and Turekian, K. K. (1971). Marine chemistry. In *Radioactivity in the Marine Environment*, U.S. Nat. Sci. Found., Washington, DC, pp. 137–146, 272.

Goldhaber, M. B. and Kaplan, I. R. (1974). Mechanisms of sulfur incorporation and isotope fractionation during early diagenesis in sediments of the Gulf of California. *Mar. Chem.* **9**, 95–143.

Goldhaber, M. B. and Kaplan, I. R. (1975). Controls and consequences of sulfate reduction rates in recent marine sediments. *Soil Sci.* **119**, 42–55.

Goldhaber, M. B. and Kaplan, I. R. (1980). Mechanisms of sulfur incorporation and isotope fractionation during early diagenesis in sediments of the Gulf of California. *Mar. Chem.* **9**, 95–143.

Goldschmidt, V. M. (1933). Grundlagen der quantitativen Geochemie. *Fortschr. Mineral., Kristallogr. Petrogr.* **17**, 1–112.

Goldstein, S. D., O'Nions, R. K., and Hamilton, P. J. (1984). A Sm-Nd isotopic study of atmospheric dust and particulates from major river systems. *Earth Planet. Sci. Lett.* **70**, 221–236.

Goodwin, A. M., Monster, J., and Thode, H. G. (1976). Carbon and sulfur isotope abundances in Archaean iron formations and early Precambrian life. *Econ. Geol.* **71**, 870–891.

Goody, R. M. and Walker, J. C. G. (1972). *Atmospheres*, Prentice-Hall, Englewood Cliffs, NJ, 160 pp.

Gordeyev, V. V. and Lisitsyn, A. J. (1978). The average chemical composition of suspensions in the world's rivers and the supply of sediments to the ocean by streams. *Dokl. Akad. Nauk SSSR* **238**, 150–153.

Graedel, T. T. (1978). *Chemical Compounds in the Atmosphere*, Academic Press, New York, 440 pp.

Graham, D. W. (1982). Strontium–calcium ratios in Cenozoic planktonic foraminifera. *Geochim. Cosmochim. Acta* **46**, 1281–1292.

Granat, L., Rohde, H. and Hallberg, R. (1976). The global sulfur cycle. In *Nitrogen, Phosphorus, and Sulfur—Global Cycles*, B. H. Svensson and R. Soderlund, Eds., SCOPE Report of the Royal Swedish Academy of Sciences, Orsundsobro, pp. 90–110.

Grandstaff, D. E., Edelman, M. J., Foster, R. W., Zbinden, E., and Kimberley, M. M. (1986). Chemistry and mineralogy of Precambrian paleosols at the base of the Dominion and Pongola Groups (Transvaal, South Africa). *Precambrian Res.* **32**, 97–131.

Gray, J. and Boucot, A. J. (1978). The advent of land plant life. *Geology* **6**, 489–492.

Gregor, C. B. (1967). The geochemical behavior of sodium. *Verh. K. Ned. Akad. Wet.* **24**, (2), 1–67.

Gregor, C. B. (1968). The rate of denudation in post-Algonkian time. *Proc. K. Ned. Akad. Wet. Ser B: Phys. Sci.* **71**, 22–30.

Gregor, C. B. (1970). Denudation of the continents. *Nature (London)* **228**, 273–275.

Gregor, C. B. (1980). Weathering rates of sedimentary and crystalline rocks. *Proc. K. Ned. Akad. Wet. Ser. B., Phys. Sci.* **83**, 173–181.

Gregor, C. B. (1985). The mass-age distribution of Phanerozoic sediments. In *Geochronology and the Geologic Record*, Mem. No. 10, N. J. Snelling, Ed., Geol. Soc., London, pp. 284–289.

Grinenko, V. A., Migdisov, A. A., and Barskaya, N. V. (1973). Sulfur isotopes in the sedimentary mantle of the Russian platform. *Dokl. Akad. Nuak SSSR* **210**, 445–448; *Dokl. Acad. Sci. USSR, Earth Sci. Sect. (Engl. Transl.)* **210**, 224–227.

Grinenko, V. A., Dmitriev, L. V., Migdisov, A. A., and Sharas'kin, A. Ya. (1975). Sulfur contents and isotope compositions for igneous and metamorphic rocks from mid-ocean ridges. *Geokhimiya*, (2) 199–206; *Geochem. Int.* **12** (1) 132–136.

Grossman, L. and Larimer, J. S. (1974). Early chemical history of the solar system. *Rev. Geophys. Space Phys.* **12**, 71–101.

Groves, D. I., Dunlop, J. S. R., and Buick, R. (1981). An early habitat of life. *Sci. Am.* **245**, 64–73.

Guettard, J. E. (1761). Mémoire sur certaines montagnes qui ont été des volcans. *Mém. Acad. Sci.* for 1756, p. 27.

Hajash, A. (1975). Hydrothermal processes along mid-ocean ridges: An experimental investigation. *Contrib. Mineral. Petrol.* 53, 205–226.

Hajash, A. (1980). Altered rocks from Deep Sea Drilling Project, Leg 59. *Init. Rep. Deep-Sea Drill. Proj.*, **59**, 735–736.

Hajash, A. and Archer, P. (1980). Experimental seawater/basalt interactions: Effects of cooling. *Contrib. Mineral, Petrol.* **75**, 1–13.

Hajash, A. and Chandler, G. W. (1981). An experimental investigation of high-temperature interactions between seawater and rhyolite, andesite, basalt, and periodotite. *Contrib. Mineral. Petrol.* **78**, 240–254.

Hale, L. D., Morton, C. J., and Sleep, N. H. (1982). Reinterpretation of seismic reflection data over the East Pacific Rise. *JGR, J. Geophys. Res.* **87**, 7707–7718.

Hallam, A. (1984). Pre-quaternary sea-level changes. *Annu. Rev. Earth Planet. Sci.* **12**, 205–243.

Hallam, A., and Bradshaw, M. J. (1979). Bituminous shales and oolitic ironstones as indicators of transgressions and regressions. *JGR, J. Geol. Soc.* **13**, 157–164.

Hamilton, E. I., Watson, P. G., and Cleary, J. J. (1979). The geochemistry of recent sediments of the Bristol Channel-Severn Estuary system. *Mar. Geol.* **31,** 139–182.

Hamilton, P. J., O'Nions, R. K., Bridgwater, D., and Nutman, A. (1983). Sm–Nd studies of Archean metasediments and metavolcanics from west Greenland and their implications for the Earth's early history. *Earth Planet. Sci. Lett.* **62,** 263–272.

Hampicke, U. (1979). Net transfer of carbon between the land biota and the atmosphere, induced by man. In *The Global Carbon Cycle*, B. Bolin et al., Eds., Wiley, New York, pp. 219–236.

Handa, N. (1977). Land sources of marine organic matter. *Mar. Chem.* **5,** 341–359.

Hargraves, R. B. (1986). Faster spreading or greater ridge length in the Archean? *Geology* **14,** 750–752.

Harper, H. E. and Knoll, A. H. (1975). Silica, diatoms, and Cenozoic radiolarian evolution, *Geology* **3,** 175–177.

Harris, D. M. (1981). The concentration of CO_2 in submarine tholeiitic basalts. *J. Geol.* **89,** 689–701.

Harrison, A. G. and Thode, H. G. (1958). Mechanism of bacterial reduction of sulphate from isotope fractionation studies. *Trans. Faraday Soc.* **54,** 84–92.

Hart, R. (1973). A model for chemical exchange in the basalt-seawater system of oceanic layer. II. *Can. J. Earth Sci.* **10,** 799–816.

Hart, R. (1976). Progressive alteration of the oceanic crust. *Initial Rep. Deep Sea Drill. Proj.* **34,** 433–438.

Hart, S. R. and Staudigel, H. (1978). Ocean crust: Age of hydrothermal alteration. *Geophys. Res. Lett.* **5,** 1009–1012.

Hart, S. R. and Straudigel, H. (1982). The control of alkalies and uranium in seawater by ocean crust alteration. *Earth Planet. Sci. Lett.* **58,** 202–212.

Hartmann, M. and Nielsen, H. (1969). δ^{34}S-Werte in rezenten Meeressedimenten und ihre Deutung am beispiel einiger Sedimentprofile aus der westlichen Ostsee, *Geol. Rundsch.* **58,** 621–655.

Haughton, R. A., Boane, R. D., Fruci, J. R., Hobbie, J. E., Melillo, J. M., Palm, C. A., Peterson, B. J., Shaver, G. R., Woodwell, G. M., Moore, B., Skole, D. L., and Myers, M. (1987). The flux of carbon from terrestrial ecosystems to the atmosphere in 1980 due to changes in land use: geographic distribution of the global flux. *Tellus* **39B,** 122–139.

Hawkesworth, C. J., O'Nions, R. K., and Arculus, R. J. (1979). Nd and Sr isotope geochemistry of island arc volcanics, Grenada, Lesser Antilles. *Earth Planet. Sci. Lett.* **45,** 237–248.

Hay, W. W. (1985). Potential errors in estimates of carbonate rock accumulating through time. *Geophys. Monogr., Am. Geophys. Union* **32,** 573–583.

Hay, W. W. and Southam, J. R. (1975). Calcareous plankton and loss of CaO from the continents. *Geol. Soc. Am. Abstr. Programs* **7,** 1105.

Hay, W. W. and Southam, J. R. (1977). Modulation of marine sedimentation by the continental shelves. In *The Fat of Fossil Fuel CO_2 in The Oceans*, N. R. Andersen and A. Malahoff, Eds., Plenum, New York, pp. 569–604.

Hays, J. D. and Pitman, W. C. (1973). Lithospheric plate motion, sea level changes and climatic and ecological consequences. *Nature (London)* **246,** 18–22.

Heath, G. R., Moore, T. C., Jr., and Dauphin, J. P. (1977). Organic carbon in deep-sea sediment. In *The Fate of Fossil Fuel CO_2 in the Oceans*, N. R. Andersen and A. Malahoff, Eds., Plenum, New York, pp. 605–625.

Hedges, J. I. and Parker, P. L. (1976). Land-derived organic matter in surface sediments from the Gulf of Mexico. *Geochim. Cosmochim. Acta* **40,** 1019–1029.

Heinrichs, T. K. and Reimer, T. O. (1977). A sedimentary barite deposit from the Archean Fig Tree Group of the Barberton Mountain Land (South Africa). *Econ. Geol.* **72,** 1426–1441.

Helgeson, H. C. (1968). Evaluation of irreversible reactions in geochemical processes involving minerals and aqueous solutions. I. Thermodynamic relations. *Geochim. Cosmochim. Acta* **3,** 853–857.

Helgeson, H. C. and MacKenzie, F. T. (1970). Silicate-seawater equilibria in the ocean system. *Deep-Sea Res.* **17,** 877–892.

Helgeson, H. C., Garrels, R. M., and Mackenzie, F. T. (1969). Evaluation of irreversible reactions in geochemical processes involving minerals and aqueous solutions. II. Applications. *Geochim. Cosmochim. Acta* **33,** 455–481.

Henderson-Sellers, A. and Meadows, A. J. (1979). A simplified model for deriving planetary surface temperatures as a function of atmospheric chemical composition. *Planet. Space Sci.* **27,** 1095–1099.

Hendrick, R. W. (1984). *Population Biology*, Jones and Bartlett, Boston, MA, 445 pp.

Herron, T. L., Ludwig, W. J., Stoffa, P. L., Kan, T. K., and Buhl, P. (1978). Structure of the East Pacific Rise crest from multichannel seismic data. *JGR, J. Geophys. Res.* **83,** 798–804.

Hess, H. H. (1962). History of ocean basins. In *Petrologic Studies*, A. E. J. Engel, H. L. James, and B. F. Leonard, Eds., Geol. Soc. Am., Boulder, CO, pp. 599–620.

Hitchcock, D. R. (1976). Microbial contributions to the atmospheric load of particulate sulfate. In *Environ. Biogeochem., Proc. Int. Symp., 2nd, 1975*, Vol. 1, pp. 351–367.

Hoefs, J. L. (1965). Ein Beitrag zur Geochemie des Kohlenstoffs in magmatischen und metamorphen Gesteinen. *Geochim. Cosmochim. Acta* **29,** 399–428.

Hoefs, J. L. (1969). Carbon. In *Handbook of Geochemistry*, Chap. 6, K. H. Wedepohl, Ed., Springer-Verlag, Berlin.

Hoefs, J. (1981). Isotopic composition of the ocean atmospheric system in the geologic past. *Am. Geophys. Union, Geodyn. Ser.* **5,** 110–118.

Hoffman, P. F. and Bowring, S. A. (1984). Short-lived 1.9 Ga continental margin and its destruction, Wopmay orogen, northwest Canada. *Geology* **12,** 68–72.

Holeman, J. N. (1968). The sediment yield of major rivers of the world. *Water Resour. Res.* **4,** 737–747.

Holland, H. D. (1965). The history of ocean water and its effect on the chemistry of the atmosphere. *Proc. U.S. Natl. Acad. Sci. A.* **53,** 1173–1182.

Holland, H. D. (1968). The abundance of CO_2 in the earth's atmosphere through geotime. In *Origin and Distribution of the Elements*, L. H. Ahrens, Ed., Pergamon, New York, pp. 949–954.

Holland, H. D. (1972). The geologic history of sea water—an attempt to solve the problem. *Geochim. Cosmochim. Acta* **36,** 637–657.

Holland, H. D. (1973a). Ocean water, nutrients, and atmospheric oxygen. *Proc. Symp. Hydrogeochem. Biogeochem. 1970*, pp. 68–81.

Holland, H. D. (1973b). The oceans: A possible source of iron in iron formations. *Econ. Geol.* **68,** 1169–1172.

Holland, H. D. (1974). Marine evaporites and the composition of sea water during the Phanerozoic. *Spec. Publ.—Soc. Econ. Paleontol. Mineral.* **20,** 187–192.

Holland, H. D. (1978). *The Chemistry of the Atmosphere and Oceans*, Wiley, New York, 351 pp.

Holland, H. D. (1984). *The Chemical Evolution of the Atmosphere and Oceans*, Princeton University Press, Princeton, 582 pp.

Holland, H. D. and Schidlowski, M., Eds. (1982). *Mineral Deposits and the Evolution of the Biosphere*, Springer-Verlag, Berlin, 332 pp.

Holland, H. D., Lazar, B., and McCaffrey, M. (1986). Evolution of the atmosphere and oceans. *Nature (London)* **320,** 27–33.

Holmes, A. (1947). The construction of the geological time scale. *Trans. Geol. Soc. Glasgow* **21,** 117–152.

Holmes, A. (1965). *Principles of Physical Geology*. Nelson, Edinburgh, 1025 pp.

Holser, W. T. (1979). Mineralogy of evaporites. *Mineral. Soc. Am. Rev. Mineral.* **6,** 211–294.

Holser, W. T. (1984). Gradual and abrupt shifts in ocean chemistry during Phanerozoic time. In *Patterns of Change in Earth Evolution*, H. D. Holland and A. F. Trendall, Eds., Springer-Verlag, Berlin, pp. 123–143.

Holser, W. T. and Kaplan, I. R. (1966). Isotope geochemistry of sedimentary sulfates. *Chem. Geol.* **1,** 93–135.

Holser, W. T., Kaplan, I. R., Sakai, H., and Zak, I. (1979. Isotope geochemistry of oxygen in the sedimentary sulfate cycle. *Chem. Geol.* **25,** 1–17.

Holser, W. T., Hay, W. W., Jory, D. E., and O'Connel, W. J. (1980). A census of evaporites and its implications for oceanic geochemistry. *Geol. Soc. Am., Abstr. Programs* **12,** 449.

Honnorez, J. (1981). The aging of the oceanic crust at low temperature. In *The Sea*, Vol. 7, C. Emiliani, Ed., Wiley, New York, pp. 525–587.

Horn, M. K. and Adams, J. A. S. (1966). Computer-derived geochemical balances and element abundances. *Geochim. Cosmochim. Acta* **30,** 279–297.

Hower, J., Eslinger, M. E., Hower, M. E., and Perry, E. A. (1976). Mechanism of burial metamorphism of argillaceous sediment: I. Mineralogical and chemical evidence. *Geol. Soc. Am. Bull.* **87,** 725–737.

Hsu, K. J., Montadert, L., Bernoulli, D., Cita, M. B., Ericson, A., Garrison, R., Kidd, R. B., Melieres, F., Muller, C., and Wright, R. (1977). History of the Mediterranean salinity crisis. *Nature (London)* **267,** 299–403.

Hudson, J. D. (1977). Stable isotopes and limestone lithification. *J. Geol. Soc., London* **133,** 637–660.

Humphris, S. E. and Thompson, G. (1978a). Hydrothermal alteration of oceanic basalts by seawater. *Geochim. Cosmochim. Acta* **42,** 107–125.

Humphris, S. E. and Thompson, G. (1978b). Trace element mobility during hydrothermal alteration of oceanic basalts. *Geochim. Cosmochim. Acta* **42,** 127–136.

Hunt, J. M. (1972). Distribution of carbon in crust of Earth. *Am. Assoc. Pet. Geol. Bull.* **56,** 2273–2277.

Hunt, T. S. (1875). *Chemical and Geological Essays*, Boston and London. Cited in Brock, W. H. (1978) Chemical Geology or geological chemistry? In *Images of the Earth*, L. J. Jordanova and R. S. Porters, Eds., British Soc. History Science Bucks., pp. 147–170.

Hurley, P. M. and Rand, J. R. (1969). Pre-drift continental nuclei. *Science* **164,** 1229–1242.

Hutcheson, I., Oldershaw, A., and Ghent, E. D. (1980). Diagenesis of Cretaceous sandstones of the Kootenay Formation at Elk Valley (southeastern British Columbia) and Mt. Allan (southeastern Alberta). *Geochim. Cosmochim. Acta* **44,** 1425–1435.

Hutton, J. (1785). Abstract of a dissertation read in the Royal Society of Edinburgh upon the seventh of March and fourth of April, 1785, concerning the system of the earth, its duration and stability, Edinburgh, 30 pp.

Hutton, J. (1788). Theory of the earth, etc. *Trans. R. Soc. Edinburgh* **1,** Pt. 2, 209–304.

Hutton, J. (1795). *Theory of the Earth, with Proofs and Illustrations*, Vols. 1 and 2, W. Greech, Edinburgh and London, 620 pp. and 567 pp. resp.

Ibach, L. E. J. (1982). Relationship between sedimentation rate and total organic carbon content in ancient marine sediments. *Am. Assoc. Pet. Geol. Bull.* **66,** 170–188.

Irving, E. (1977). Drift of the major continental blocks since the Devonian. *Nature (London)* **270,** 304–309.

Irwin, W. P. and Barnes, I. (1980). Tectonic relations of carbon dioxide discharges and earthquakes. *JGR, J. Geophys. Res.* **85,** 3115–3121.

Ito, E., Harris, D. M., and Anderson, A. T. (1983). Alteration of oceanic crust and the geological cycling of chlorine and water. *Geochim. Cosmochim. Acta* **47,** 1613–1624.

Ivanov, M. V. (1981). The global biogeochemical sulphur cycle. In *Some Perspectives of the Major Biogeochemical Cycles*, G. E. Likens, Ed., Wiley, New York, pp. 61–80.

Ivanov, M. V., Gogotova, G. I., Matrosov, A. G., and Zyakun, A. M. (1976). Fractionation of sulfur isotopes by phototrophic sulfur bacteria, *Ectothiorhodospira shaposhnikovii*. *Microbiology (Eng. Transl.)* **45,** 655–659.

Jackson, T. A., Fritz, P., and Drimmie, R. (1978). Stable carbon isotope ratios and chemical properties of kerogen and extractable organic matter in pre-Phanerozoic and Phanerozoic sediments—their interrelations and possible paleobiological significance. *Chem. Geol.* **21,** 335–350.

Jacobsen, S. B. and Wasserburg, G. J. (1979). The mean age of mantle and crustal reservoirs, *JGR, J. Geophys. Res.* **84,** 7411–7427.

Jaeschke, W., Claude, H., and Herrmann, J. (1980). Sources and sinks of atmospheric H_2S. *JGR, J. Geophys. Res.* **85,** 5639–5644.

Jenkins, N. J., Edmond, J. M., and Corliss, J. B. (1978). Excess ^{3}He and ^{4}He in Galapagos submarine hydrothermal waters. *Nature (London)* **272,** 156–158.

Jessop, A. M. and Lewis, T. (1978). Heat flow and heat generation in the Superior province of the Canadian Shield. *Tectonophysics* **50,** 55–57.

Johnston D. A. (1980). Contribution of volcanic gas to the atmosphere during explosive eruptions—estimate based upon magmatic gas composition. *EOS, Trans. Am. Geophys. Union* **61,** 67.

Joly, J. (1899). An estimate of the geological age of the earth, *Sci. Trans. R. Dublin Soc.* [2] **7**, 23–66.

Jørgensen, B. B., Hanse, M. H., and Ingvorsen, K. (1978). Sulfate reduction in coastal sediments and the release of H_2S to the atmosphere. *Environ. Biogeochem. Geomicrobiol. Proc. Int. Symp., 3rd, 1977,* 1, pp. 245–253.

Junge, C. E. (1963). *Atmospheric Chemistry and Radioactivity*, Academic Press, New York, 382 pp.

Kaleska, M., Rao, K. S., and Somayajulu, B. L. K. (1980). Deposition rates in the Godavari Delta. *Mar. Geol.,* **34**, M57–M66.

Kanahira, K. L., Yui, S., Sakai, H., and Sasaki, A. (1973). Sulphide globules and sulphur isotope ratios in the abyssal tholeiite from the Mid-Atlantic Ridge near 30°N latitude. *Geochem. J.* **7**, 89–96.

Kaplan, I. R. and Rittenberg, S. C. (1964). Microbiological fractionation of sulfur isotopes. *J. Gen. Microbiol.* **34**, 195–212.

Kaplan, I. R., Emery, K. O., and Rittenberg, S. C. (1963). The distribution and isotopic abundance of sulfur in recent marine sediments off southern California. *Geochim. Cosmochim. Acta* **27**, 297–331.

Karhu, J. and Epstein, S. (1986). The implication of the oxygen isotope records in coexisting cherts and phosphates. *Geochim. Cosmochim. Acta* **50**, 1745–1756.

Karig, D. E. and Sharman, G. F., III (1975). Subduction and accretion in trenches. *Geol. Soc. Am. Bull.* **86**, 377–389.

Kasting, J. F. and Richardson, S. M. (1985). Seafloor hydrothermal activity and spreading rates: the Eocene carbon dioxide greenhouse revisited. *Geochim. Cosmochim. Acta,* **49**, 2541–2544.

Kasting, J. F. and Walker, J. C. G. (1981). Limits on oxygen concentration in the prebiological atmosphere and the rate of abiotic fixation of nitrogen. *JGR, J. Geophys. Res.* **86**, 1141–1158.

Kasting, J. F., Liu, S. C., and Donahue, T. M. (1979). Oxygen levels in the prebiological atmosphere. *JGR, J. Geophys. Res.* **84**, 3097–3107.

Kasting, J. F., Zahnle, K. J., and Walker, J. C. G. (1983). Photochemistry of methane in the Earth's early atmosphere. *Precambrian Res.* **20**, 121–148.

Kay, R. W. (1980). Volcanic arc magmas: Implications of a melting-mixing model for element recycling in the crust-upper mantle system. *J. Geology* **88**, 497–522.

Kay, R., Sun, S.-S., and Lee-Hu, C. N. (1978). Pb and Sr isotopes in volcanic rocks from the Aleutian Islands and Pribilof Islands, Alaska. *Geochim. Cosmochim. Acta* **42**, 263–273.

Keays, R. R. and Scott, R. B. (1976). Precious metals in ocean-ridge basalts: Implications for basalts as source rocks for gold mineralization. *Econ. Geol.* **71**, 705–720.

Keith, M. L. (1982). Violent volcanism, stagnant oceans and some inferences regarding petroleum, strata-bound ores and mass extinctions. *Geochim. Cosmochim. Acta* **28**, 1787–1816.

Keith, M. L. and Weber, J. N. (1964). Carbon and oxygen istopic composition of selected limestones and fossils. *Geochim. Cosmochim. Acta* **28**, 1787–1816.

Kellogg, W. W., Cadle, R. D., Allen, E. R., Lazarus, A. L., and Martell, E. A. (1972). The sulfur cycle. *Science* **175**, 587–596.

Kemp, A. L. W. and Thode, H. G. (1968). The mechanism of bacterial reduction of sulphate and sulphite from isotope fractionation studies. *Geochim. Cosmochim. Acta* **32**, 71–91.

Kempe, S. (1979a). Carbon in the freshwater cycle. In *The Global Carbon Cycle*, B. Bolin et al., Eds., Wiley, New York, pp. 317–342.

Kempe, S. (1979b). Carbon in the rock cycle. In *The Global Carbon Cycle*, B. Bolin et al., Eds., Wiley, New York, pp. 343–377.

Kerridge, J. F., Haymon, R. M., and Kastner, M. (1983). Sulfur isotope systematics at the 21°N site, East Pacific Rise. *Earth Planet. Sci. Lett.* **66**, 91–100.

Kinsman, D. J. J. (1974). Evaporite deposits of continental margins. *Symp. Salt, 4th, 1973*, Vol. 2, pp. 255–259.

Kirwan, R. (1793). Examination of the supposed origin of stony substances. *Trans. R. Ir. Acad.* **5**, 1–51.

Knauth, L. P. and Lowe, D. R. (1978). Oxygen isotope geochemistry of cherts from the Onverwacht Group (3.4 billion years), Transvaal, South Africa, with implications for secular variations in the isotopic composition of cherts. *Earth Planet. Sci. Lett.* **41**, 209–222.

Knoll, A. H. (1979). Archean photoautotrophy: Some alternatives and limits. *Origins Life* **9**, 313–327.

Kohlmaier, G. H., Fischbach, U., Kratz, G., Brohl, H., and Schunck, W. (1980). The carbon cycle: Sources and sinks of atmospheric CO_2. *Experientia* **36**, 776–796.

Kovach, J. (1980). Variations in the strontium isotopic composition of seawater during Paleozoic time determined by analysis of conodonts. *Geol. Soc. Am. Abs.* **12**, 465.

Kröner, A. (1981). Precambrian plate tectonics. In *Precambrian Plate Tectonics*, A. Kröner, Ed., Elsevier, Amsterdam, pp. 57–90.

Kroopnick, P. (1977). The SO_4/Cl ratio in oceanic rainwater. *Pac. Sci.* **31**, 91–106.

Ku, T. L. (1966). Uranium series disequilibrium in deep sea sediments. Ph.D. thesis, Columbia University, New York, 157 pp.

Ku, T. L., Broecker, W. S., and Opdyke, N. (1968). Comparison of sedimentation rates measured by paleomagnetic and the ionium methods of age determination. *Earth Planet. Sci. Lett.* **4**, 1–16.

Kuenen, Ph.H. (1950). *Marine Geology*, Wiley, New York, 568 pp.

Kump, L. R. and Garrels, R. M. (1986). Modeling atmospheric O_2 in the global sedimentary redox cycle. *Am. J. Sci.* **286**, 337–360.

Kurz, M. D. and Jenkins, W. J. (1979). The distribution of helium in oceanic basalt glasses. *Earth Planet. Sci. Lett.* **53**, 41–54.

Lambert, I. B., Donnelly, T. H., Dunlop, J. S. R., and Groves, D. I. (1978). Stable isotopic compositions of Early Archaean sulphate deposits of probable evaporitic and volcanogenic origins. *Nature (London)* **276**, 80–81.

Langmuir, C., Bender, J. L., Bence, A., Hanson, G., and Taylor, S. (1977). Petrogenesis of basalts from the FAMOUS area, mid-Atlantic ridge. *Earth Planet. Sci. Lett.* **36**, 133–156.

Lasaga, A. C. (1980). The kinetic treatment of geochemical cycles. *Geochim. Cosmochim. Acta* **44**, 815–828.

Lasaga, A. C., Holland, H. D., and Dwyer, M. J. (1971). Primordial oil slick. *Science* **174**, 53–55.

Lasaga, A. C., Berner, R. A., and Garrels, R. M. (1985). An improved geochemical model of atmospheric CO_2 fluctuations over the past 100 million years. *Am. Geophys. Monogr., Am. Geophys. Union* **32,** 397–411.

Lawrence, J. R., Gieskes, J. M., and Broecker, W. S. (1975). Oxygen isotope and cation composition of DSDP pore waters and the alteration of Layer II basalts. *Earth Planet. Sci. Lett.* **27,** 1–10.

Lefond, S. J. (1969). *Handbook of World Salt Resources*, Plenum, New York, 384 pp.

Le Pichon, X., Francheteau, J., and Bonnin, J. (1973). *Plate Tectonics*, Elsevier, Amsterdam, 300 pp.

Lerman, A. (1979). *Geochemical Processes: Water and Sediment Environments*, Wiley, New York, 481 pp.

Leventhal, J. S. (1979). The relationship between organic carbon and sulfide sulfur in recent and ancient marine and euxinic sediments. *EOS, Trans. Am. Geophys. Union* **60,** 282.

Leventhal, J. S. (1983). An interpretation of carbon and sulfur relationships in Black Sea sediments as indicators of environments of deposition. *Geochim. Cosmochim. Acta* **47,** 133–137.

Leventhal, J. S. (1987). Carbon and sulfur relationships in Devonian shales from the Applachian Basin as an indicator of environment of deposition. *Am. J. Sci.* **287,** 33–49.

Lewis, B. T. R. (1982). Constraints on the structure of the East Pacific Rise from gravity. *JGR, J. Geophys. Res.* **87,** 8491–8500.

Li, Y.-H. (1972). Geochemical mass balance among lithosphere, hydrosphere and atmosphere. *Am. J. Sci.* **272,** 119–137.

Li, Y.-H. (1981). Geochemical cycles of elements and human perturbation. *Geochim. Cosmochim. Acta* **45,** 2073–2084.

Libby, W. F. (1971). Terrestrial and meteorite carbon appear to have the same isotopic composition. *Proc. Natl. Acad. Sci. U.S.A.* **68,** 377.

Lindh, T. B. (1983). Temporal variations in ^{13}C, ^{34}S and global sedimentation during the Phanerozoic. Master's thesis, University of Miami, Miami, FL, 98 pp.

Lindh, T. B., Saltzman, E. S., Sloan, J. L., II, Mattes, B. W., and Holser, W. T. (1981). A revised $\delta^{13}C$-age curve. *Geol. Soc. Am. Abstr. Programs* **13,** 498.

Livingstone, D. A. (1963). Chemical composition of rivers and lakes. *Geol. Surv. Prof. Pap. (U.S.)* **440-G,** 64 pp.

Longinelli, A. and Nuti, S. (1968). Oxygen isotope composition of phosphorites from marine formations. *Earth Planet. Sci. Lett.* **5,** 13–16.

Lotka, A. J. (1956). *Elements of Mathematical Biology*, Dover, New York, 465 pp.

Lovelock, J. (1979). *Gaia, A New Look at Life on Earth*. Oxford University Press, London and New York, 157 pp.

Lowe, D. R. (1980a). Stromatolites: 3,400-Myr old from the Archean of Australia. *Nature (London)* **284,** 441–443.

Lowe, D. R. (1980b). Archean sedimentation. *Annu. Rev. Earth Planet. Sci.* **8,** 145–167.

Lowe, D. R. and Knauth, L. P. (1977). Sedimentology of the Onverwacht Group (3.4 billion years), Transvaal, South Africa, and its bearing on the characteristics and evolution of the Earth. *J. Geol.* **85,** 699–723.

Lowenstam, H. A. (1961). Mineralogy, O^{18}/O^{16} ratios and strontium and magnesium contents of Recent and fossil brachiopods and their bearing on the history of oceans. *J. Geol.* **69**, 241–260.

Lupton, J. E. (1983a). Terrestrial rare gases: Isotope tracer studies and clues to primordial components in the mantle. *Annu. Rev. Earth Planet. Sci.* **11**, 371–414.

Lupton, J. E. (1983b). Fluxes of helium-3 and heat from submarine hydrothermal systems: Guaymas Basin Versus 21°N EPR. *EOS. Trans. Am. Geophys. Union* **64**, 723.

Lupton, J. E. and Craig, H. (1978). ^{3}He in the Pacific Ocean: Injection at active spreading centers and applications to deep circulation studies. *Trans. Am. Geophys. Union* **59**, 1105.

Lutz, B., Kolodny, Y., and Kovach, J. (1984). Oxygen isotope variations in phosphate of biogenic apatites. III. Conodonts. *Earth Planet. Sci. Lett.* **69**, 255–262.

Lyle, H. (1976). Estimation of hydrothermal manganese input to the oceans. *Geology* **4**, 733–736.

Lyons, W. B. and Gaudette, H. E. (1979). Sulfate reduction and the nature of organic matter in estuarine sediments. *Org. Geochem.* **1**, 151–155.

Macdonald, K. C., Becker, K., Spiess, F. H., and Ballard, R. D. (1980). Hydrothermal heat flux of the "black smoker" vents on the East Pacific rise. *Earth Planet. Sci. Lett.* **48**, 1–7.

Machel, H. G. and Mountjoy, E. W. (1986). Chemistry and environments of dolomitization: A reappraisal. *Earth Sci. Rev.* **23**, 175–222.

Mackenzie, F. T. (1975). Sedimentary cycling and the evolution of sea water. In *Chemical Oceanography* 2nd ed., Vol. 1, J. P. Riley and G. Skirrow, Eds., Academic Press, London, pp. 309–364.

Mackenzie, F. T. and Agegian, C. (1987). Biomineralization and tentative links to plate tectonics. In *Origin, Evolution, and Modern Aspects of Biomineralization in Plants and Animals*, Rex E. Crick, Ed., Plenum, N.Y. (in press).

Mackenzie, F. T. and Garrels, R. M. (1966a). Chemical mass balance between rivers and oceans. *Am. J. Sci.* **264**, 507–525.

Mackenzie, F. T. and Garrels, R. M. (1966b). Silica-bicarbonate balance in the ocean and early diagenesis. *J. Sediment Petrol.* **36**, 1075–1084.

Mackenzie, F. T. and Pigott, J. D. (1981). Tectonics control of Phanerozoic sedimentary rock cycling. *J. Geol. Soc., London* **138**, 183–196.

Mackenzie, F. T. and Wollast, R. (1977). Thermodynamic and kinetic controls of global chemical cycles of the elements. In *Global Chemical Cycles and Their Alterations by Man*, W. Stumm, Ed., Dahlem Konferenzem, Berlin, pp. 45–60.

Madsen, J. A., Forsyth, D. W., and Detrick, R. S. (1984). A new isostatic model for the East Pacific Rise crest. *JGR, J. Geophys. Res.* **B89**, 9997–10015.

Magaritz, M. D., Whitford, D. J., and James, D. E. (1978). Oxygen isotopes and the origin of high $^{87}Sr/^{86}Sr$ andesites. *Earth Planet. Sci. Lett.* **40**, 220–230.

Magaritz, M. D., Turner, P., and Kading, K. C. (1981). Carbon isotopic change at the base of the Upper Permian Zechstein sequence. *Geol. J.* **16**, 243–254.

Magaritz, M. D., Anderson, R. Y., Holser, W. T., Saltzman, E. S., and Garber, J. (1983). Isotope shifts in the Late Permian of the Delaware Basin, Texas, precisely timed by varved sediments. *Earth Planet. Sci. Lett.* **66**, 111–124.

Mäkelä, M. (1973). A study of sulfur isotopes in the Outokumpu ore deposit, Finland. *Finl. Comm. Geol. Bull.* **267,** 45 pp.

Mangini, A., Sonntag, C., Bertsch, G., and Muller, E. (1979). Evidence for a higher natural uranium content in world rivers. *Nature (London)* **278,** 337–339.

Manheim, F. T. (1976). Interstital waters of marine sediments. In *Chemical Oceanography*, J. P. Riley and R. Chester, Eds., Acad. Press, New York, 2nd ed., Vol. 6, pp. 115–186.

Manuel, O. K. (1978). A comparison of terrestrial and meteoritic noble gases. *Adv. Earth Planet. Sci.* **3,** 85–91.

Marowsky, G. (1969). Schwefel, Kohlenstoff- und Sauerstoff-Isotopenuntersuchungen am Kupferschiefer also Beitrag zur genetischen Deutung. *Contrib. Mineral. Petrol.* **22,** 290–334.

Marowsky, G. and Wedepohl, K. H. (1971). General trends in the behavior of Cd, Mg, Tl and Bi in some major rock forming processes. *Geochim. Cosmochim. Acta* **35,** 1255–1267.

Martin, J. M. and Meybeck, M. (1978). Major elements in river dissolved and particulate loads. In *Biogeochemistry of Estuarine Sediments*, E. E. Goldberg, Ed., UNESCO, New York, pp. 95–110.

Martin, J.-M. and Meybeck, M. (1979). Elemental mass-balance of material carried by major world rivers. *Mar. Chem.* **7,** 173–206.

Marvin, U. B. (1973). *Continental Drift: The Evolution of a Concept*, Smithsonian, Washington, DC, 239 pp.

Mathez, E. A. (1976). Sulfur solubility and magmatic sulfides in submarine basalt glass. *JGR, J. Geophys. Res.* **81,** 4269–4275.

Maynard, J. B. (1976). The long-term buffering of the oceans. *Geochim. Cosmochim. Acta* **40,** 1523–1532.

Maynard, J. B. (1980). Sulfur isotopes of iron sulfides in Devonian-Mississippian shales of the Appalachian Basin: Control by rate of sedimentation. *Am. J. Sci.* **280,** 772–786.

Maynard, J. B. (1981). Carbon isotopes as indicators of dispersal patterns in Devonian-Mississippian shales of the Appalachian Basin. *Geology* **9,** 262–265.

Maynard, J. B. (1983). *Geochemistry of Sedimentary Ore Deposits*. Springer-Verlag, New York, 305 pp.

McAlester, A. L. (1970) Animal extinctions, oxygen consumption and atmospheric history. *J. Paleontol.* **44,** 405–409.

McCready, R. G. L. and Krouse, H. R. (1980). Sulfur isotope fractionation by *Desulfovibrio vulgaris* during metabolism of $BaSO_4$ *Geomicrobiol. J.* **2,** 55–62.

McCulloch, M. T. (1986). Sm-Nd isotopic constraints on Proterozoic crustal evolution in the Australian continent. *Terra Cognita* **6,** 126.

McCulloch, M. T. and Wasserburg, G. J. (1978). Sm-Nd and Rb–Sr chronology of continental crust formation. *Science* **200,** 1003–1011.

McDuff, R. E. (1981). Major cation gradients in DSDP interstitial waters. The role of diffusive exchange between seawater and upper oceanic crust. *Geochim. Cosmochim. Acta* **45,** 1705–1713.

McDuff, R. E. and Edmond, J. M. (1982). On the fate of sulfate during hydrothermal circulation at mid-oceanic ridges. *Earth Planet. Sci. Lett.* **57,** 117–132.

McKenzie, D. P. and Weiss, N. O. (1975). Speculations on the thermal and tectonic history of the earth. *Geophys. J. R. Astron. Soc.* **42,** 131–174.

McPhee, J. (1981). *Basin and Range*, Farrar, Strauss and Giroux, New York, 216 pp.

Mead, W. J. (1914). The average igneous rock. *J. Geol.* **22,** 772–782.

Meade, R. H., Nordin, C. F., Curtis, W. F., Rodrigues, F. M. C., Do Vale, C. M., and Edmond, J. M. (1979). Sediment loads in the Amazon River. *Nature (London)* **278,** 161–163.

Mekhtieva, V. L. (1971). Isotope composition of sulfur of plants and animals from reservoirs of different salinity. *Geokhimiya* **6,** 725–730.

Mel'nik, Y. P. (1982). *Precambrian Banded Iron-Formations.* Elsevier, Amsterdam, 310 pp.

Menard, H. W. and Smith, S. M. (1966). Hypsometry of ocean basin provinces. *J. Geophys. Res.* **71,** 4305–4325.

Meybeck, M. (1976). Total mineral dissolved transport by world major rivers. *Bull. Hydrol. Sci.* **21,** 265–284.

Meybeck, M. (1977). Dissolved and suspended matter carried by rivers: Composition, time, and space variations, and world balance. In *Interaction Between Sediments and Fresh Waters,* H. L. Golterman, Ed., Junk and Pudoc, Amsterdam, pp. 25–32.

Meybeck, M. (1979a). Concentrations des eaux fluviales en éléments majeurs et apports en solution aux océans. *Rev. de Geol. Dyn. Geogr. Phys.* **21,** 215–246.

Meybeck, M. (1979b). Pathways of major elements from land to ocean through rivers. In *Review and Workshop on River Inputs to Ocean Systems,* FAO, Rome, pp. 26–30.

Meyer, B. (1977). *Sulfur, Energy and Environment,* Elsevier, Amsterdam, 448 pp.

Michard, A., Gurriet, P., Soudant, M., and Albarede, F. (1985). Nd isotopes in French Phanerozoic shales: External vs. internal aspects of crustal evolution. *Geochim. Cosmochim. Acta* **49,** 601–610.

Michard, G., Abarede, F., Michard, A., Minster, J.-F., Charlous, J.-L., and Tan, N. (1984). Chemistry of solutions from the 13°N East Pacific Rise hydrothermal site. *Earth Planet. Sci. Lett.* **67,** 297–307.

Migdisov, A. A., Ronov, A. B., and Grinenko, V. A. (1983). The sulphur cycle in the lithosphere. Part I. Reservoirs. In *The Global Biogeochemical Sulphur Cycle,* M. V. Ivanov and J. R. Freney, Eds., Wiley, New York, pp. 25–94.

Milliman, J. D. and Boyle, E. (1975). Biological upstake of dissolved silica in the Amazon River estuary. *Science* **189,** 995–997.

Milliman, J. D. and Meade, R. H. (1983). Worldwide delivery of river sediment to the oceans. *J. Geol.* **91,** 1–21.

Miyake, Y. and Tsunogai, S. (1963). Evaporation of iodine from the ocean. *J. Geophys. Res.* **68,** 3989–3993.

Monster, J., Appel, P. W. U., Thode, H. G., Schidlowski, M., Carmichael, C. M., and Bridgewater, B. (1979). Sulfur isotope studies in Early Archean sediments from Isua, West Greenland. Implications for the antiquity of bacterial sulfate reduction. *Geochim. Cosmochim. Acta* **43,** 405–413.

Mook, W. G., Bommerson, J. C., and Staverman, W. H. (1974). Carbon isotope fractionation between dissolved bicarbonate and gaseous carbon dioxide. *Earth Planet. Sci. Lett.* **22,** 169–176.

Moorbath, S. (1977a). The oldest rocks and the growth of the continents. *Sci. Am.* **236,** 92–104.

Moorbath, S. (1977b). Ages, isotopes and evolution of Precambrian continental crust. *Chem. Geol.* **20,** 151–187.

Moorbath, S., O'Nions, R. K., and Pankhurst, R. J. (1973). Early Archean age for the Isua iron-formation, West Greenland. *Nature (London)* **245,** 138–139.

Moore, C. B., and Lewis, C. (1976). Total sulfur contents of basalts from DSDP leg 34. *Init. Rep. Deep Sea Drill. Proj.* **34,** 375.

Moore, J. G. and Fabbi, B. P. (1971). An estimate of the juvenile sulfur content of basalt. *Contrib. Mineral. Petrol.* **33,** 118–127.

Moore, P. D. (1983). Plants and the palaeoatmosphere. *J. Geol. Soc., London* **140,** 13–25.

Moore, W. S. (1967). Amazon and Mississippi River concentrations of U, Th and Ra isotopes. *Earth Planet. Sci. Lett.* **2,** 231–234.

Mopper, K. and Degens, E. T. (1979). Organic carbon in the ocean: Nature and cycling. In *The Global Carbon Cycle*. B. Bolin et al., Eds., Wiley, New York, pp. 293–316.

Morgan, W. J. (1968). Rises, trenches, great faults, and crustal blocks. *J. Geophys. Res.* **73,** 1959–1982.

Morton, J. L. and Sleep, N. H. (1985a). Seismic reflections from a Lau Basin magma chamber. In *Geology and Offshore Resources of Pacific Island arcs-Tonga Regions*, Earth Sci. Ser., Vol. 2, D. W. Scholl, and T. L. Vallier, Eds., Circum-Pacific Council for Energy and Mineral Resources, Houston, TX, pp. 441–453.

Morton, J. L. and Sleep, N. H. (1985b). A mid-ocean ridge thermal model. Constraints on the volume of axial hydrothermal heat flux. *JGR, J. Geophys. Res.* **90,** 11345–11353.

Morton, J. L., Sleep, N. H., Normark, W. R., and Tompkins, D. H. (1987). Structure of the southern Juan de Fuca ridge from seismic reflection records. *JGR, J. Geophys. Res.* **92** (in press).

Mottl, M. J. (1976). Chemical exchange between sea water and basalt during hydrothermal alteration of the oceanic crust. Ph.D. dissertation, Harvard University, Cambridge, MA.

Mottl, M. J. (1983). Metabasalts, axial hot springs, and the structure of hydrothermal systems at mid-oceanic ridges. *Geol. Soc. Am. Bull.* **94,** 161–180.

Mottl, M. J. and Holland, H. D. (1978). Chemical exchange during hydrothermal alteration of basalt by seawater. I. Experimental results for major and minor components of seawater. *Geochim. Cosmochim. Acta* **42,** 1103–1115.

Mottl, M. J., Holland, H. D., and Corr, R. F. (1979). Chemical exchange during hydrothermal alteration of basalt by seawater. II. Experimental results for Fe, Mn, and sulfur species. *Geochim. Cosmochim. Acta* **43,** 869–884.

Moyers, J. L. and Duce, R. A. (1972). Gaseous and particulate iodine in the marine atmosphere. *J. Geophys. Res.* **77,** 5227–5238.

Mroz, E. J. and Zoller, W. H. (1975). Composition of atmospheric particulate matter from the eruption of Neimaey, Iceland. *Science* **190,** 461–469.

Muehlenbachs, K. (1977a). Oxygen isotope geochemistry of rocks from DSDP Leg 37. *Can. J. Earth Sci.* **14,** 771–776.

Muehlenbachs, K. (1977b). Low temperature oxygen isotope exchange between the ocean crust and sea water. In *Rock-Water Interactions*, Vol. 1, *Int. Assoc. Geochem. Cosmochem.* (UNESCO), Strasbourg, France, 317–326.

Muehlenbachs, K. (1980). The alteration and aging of the basaltic layer of sea floor, oxygen isotopic evidence from DSDP/IPOD legs 51, 52, and 53. *Initial Rep. Deep-Sea Drill. Proj.* **51,** 1159–1167.

Muehlenbachs, K. and Clayton, R. (1976). Oxygen isotope composition of the oceanic crust and its bearing on seawater. *JGR, J. Geophys. Res.* **81,** 4365–4369.

Muenow, D. W., Lui, N. W. K., Garcia, M. O., and Saunders, A. D. (1980). Volatiles in submarine volcanic rocks from the spreading axis of the East Scotia Sea back-arc basin. *Earth Planet. Sci. Lett.* **47,** 272–278.

Murray, J. (1803). *A Comparative View of the Huttonian and Neptunian Systems of Geology.* In *Answer to the Illustrations of the Huttonian Theory of the Earth by Professor Playfair,* Ross and Blackwood, Edinburgh.

Naldrett, A. J., Goodwin, A. M., and Fisher, T. L. (1977). Sulfur in Leg 37 basalts. *Initial Rep. Deep Sea Drill. Proj.* **37,** 561–562.

Naldrett, A. J., Goodwin, A. M., Fisher, T. L., and Ridler, R. H. (1978). The sulfur content of Archean volcanic rocks and a comparison with coean floor basalts. *Can. J. Earth Sci.* **15,** 715–728.

Nanz, R. H., Jr. (1953). Chemical composition of Precambrian slates with notes on the geochemical evolution of lutites. *J. Geol.* **61,** pp. 51–64.

Newmark, R. L., Anderson, R. N., Moos, D., and Zoback, M. D. (1984). Sonic and ultrasonic logging in hole 504B and its implications for the structure, porosity, and stress regime of the upper 1 km of the oceanic crust. *Initial Rep. Deep Sea Drill. Proj.* **83,** 479–510.

Nielsen, H. (1978). Sulfur isotopes in nature. In *Handbook of Geochemistry,* Part 16B, K. H. Wedepohl, Ed., Springer-Verlag, Berlin.

Nieuwenkamp, W. (1948a). Geochemistry of sodium. *Proc. Int. Geol. Congr., 18th Sess.,* Pt. 2, pp. 96–100.

Nieuwenkamp, W. (1948b). *Actualism and catastrophism,* Inaugural Lecture, University of Utrecht.

Nieuwenkamp, W. (1956a). Geochimie classique et transformiste. *Bull. Soc. Geol. Fr.,* Ser. 6, **7,** 407–429.

Nieuwenkamp, W. (1956b). Korrelation von Sediment und Eruptivgesteine. *Trans. R. Neth. Geol. Mineral. Soc.* **16,** 309–316.

Nieuwenkamp, W. (1968). Oceanic and continental basalts in the geochemical cycle. *Geol. Rundsch.* **57,** 362–372.

Odum, H. T. (1983). *Systems Ecology: An Introduction,* Wiley, New York, 644 pp.

Ohmoto, H. (1972). Systematics of sulfur and carbon isotopes in hydrothermal ore deposits. *Econ. Geol.* **67,** 551–578.

Ohmoto, H. and Rye, R. O. (1979). Isotopes of sulfur and carbon. In *Geochemistry of Hydrothermal Ore Deposits,* 2nd ed., H. L. Barnes, Ed., Wiley, New York, pp. 509–567.

O'Leary, M. H. (1981). Carbon isotope fractionation in plants. *Phytochemistry* **20,** 553–567.

Olson, J. M. (1970). The evolution of photosynthesis. *Science* **168,** 438–446.

O'Nions, R. K., Evensen, N. M., and Hamilton, P. J. (1979). Geochemical modeling of mantle differentiation and crustal growth. *JGR, J. Geophys. Res.* **84,** 6091–6101.

O'Nions, R. K., Hamilton, P. J., and Hooker, P. J. (1983). A Nd isotope investigation of sediments related to crustal development in the British Isles. *Earth Planet. Sci. Lett.* **63,** 329–338.

Orr, W. (1974). Sulfur biogeochemistry. Sec. 16-L. In *Handbook of Geochemistry,* K. H. Wedepohl, Ed., Springer-Verlag, Berlin.

Ospovat, A. M. (1971). Translation and reproduction of A. G. Werner, *Short Classification and Description of the Various Rocks*, Hafner, New York, 194 pp.

Oversby, V. and Ewart, A. (1972). Lead isotopic composition of Tonga-Kermedec volcanics and their petrogenetic significance. *Contrib. Mineral. Petrol.* **37**, 181–210.

Owen, R. M. and Rea, D. K. (1984). Sea floor hydrothermal activity links climate to tectonics: The Eocene carbon dioxide greenhouse. *Science* **227**, 166–169.

Palmer, M. R. and Elderfield, H. (1986). Sr-isotope composition of seawater over the past 75 Myr. *Nature* **314**, 526–528.

Pardue, J. W., Scalan, R. S., Van Baalen, C., and Parker, P. L. (1976). Maximum carbon isotope fractionation in photosynthesis by bluegreen algae and a green alga. *Geochim. Cosmochim. Acta* **40**, 309–312.

Park, R. and Epstein, S. (1980). Carbon isotope fractionation during photosynthesis. *Geochim. Cosmochim. Acta* **21**, 110–126.

Patchett, P. J. and Arndt, N. T. (1986). Nd isotopes and tectonics of 1.9–1.7 Ga crustal genesis. *Earth Planet. Sci. Lett.* **78**, 329–338.

Patchett, P. J., White, W. M., Feldmann, H., Keilinczuk, S., and Hofmann, A. W. (1984). Hafnium/rare earth element fractionation in the sedimentary system and crustal recycling into the Earth's mantle. *Earth Planet. Sci. Lett.* **69**, 365–378.

Peck, H. D. (1974). The evolutionary significance of inorganic sulfur metabolism. *Symp. Soc. Gen. Microbiol.* **24**, 241–262.

Perry, E. C., Monster, J., and Reimer, T. (1971). Sulfur isotopes in Swaziland System barites and the evolution of the Earth's atmosphere. *Science* **171**, 1015–1016.

Perry, E. C., Ahmad, S. N., and Swulius, T. M. (1978). The oxygen isotope composition of 3,800 m.y. old metamorphosed chert and iron formation from Isukasia west Greenland. *J. Geol.* **86**, 223–239.

Petrenchuk, O. P. (1980). On the budget of sea salts and sulphur in the atmosphere. *JGR, J. Geophys. Res.* **85**, 7439–7444.

Pflug, H. D. and Jaeschke-Boyer, H. (1979). Combined structural and chemical analysis of 3,800-Myr-old microfossils. *Nature (London)* **280**, 483–486.

Pielou, E. C. (1977). *Mathematical Ecology*, Wiley, New York, 384 pp.

Pigott, J. D. and MacKenzie, F. T. (1979). Phanerozoic ooid diagenesis: A signature of paleo-ocean and -atmospheric chemistry. *Geol. Soc. Am. Abstracts* **11**, 495–496.

Piper, D. Z. (1974). Rare earth elements in Ferromanganese nodules and other marine phases. *Geochim. Cosmochim. Acta* **38**, 1007–1022.

Piper, D. Z., Long, K., and Cannon, W. F. (1979). Manganese nodule and surface sediment compositions: Domes sites A, B and C. In *Marine Geology and Oceanography of the Pacific Manganese Nodule Province*. J. L. Bischoff and D. Z. Piper, Eds., Plenum, New York, 309–348.

Pitman, W. C., Jr. (1978). Relationship between eustacy and stratigraphic sequences of passive margins. *Geol. Soc. Am. Bull.* **89**, 1389–1403.

Playfair, J. (1802). *Illustrations of the Huttonian Theory of the Earth*. W. Greech, Edinburgh, 528 pp. (reprinted 1956 by Univ. of Illinois Press, Urbana).

Poldervaart, Arie (1955). Chemistry of the Earth's Crust. *Spec. Pap. Geol.—Soc. Am.* **62**, 119–144.

Popp, B. N., Anderson, F. T., and Sandberg, P. A. (1986). Brachiopods as indicators of original isotopic compositions in Paleozoic limestones. *Geol. Soc. Am. Bull.* **97,** 1262–1269.

Porrenga, D. H. (1967). Clay mineralogy and geochemistry of Recent marine sediments in typical areas as exemplified by the Niger delta, the Orinoco shelf, and the shelf off Sarawak. Amsterdam Univ. Fys.-Geogr. Lab. Publ. No. 9, 143 pp.

Postgate, J. R. (1979). *The Sulphate-Reducing Bacteria,* Cambridge Univ. Press, London and New York, 2 pp.

Pressler, J. W. (1978). Gypsum. *Miner. Yearb.* pp. 627–637.

Puchelt, H. and Hubberten, H. W. (1980). Preliminary results of sulfur isotope investigations on deep sea drilling project cores from Legs 52 and 53. *Initial Rep. Deep Sea Drill. Proj.* **51/52/53,** Pt. 2, 1145–1148.

Pytkowicz, R. M. (1975). Some trends in marine chemistry and geochemistry. *Earth-Sci. Rev.* **11,** 1–46.

Rahn, K. S. (1976). *The Chemical Composition of the Atmospheric Aerosol,* Tech. Rep., University of Rhode Island, Kingston.

Raiswell, R. and Berner, R. (1985). Pyrite formation in euxinic and semi-euxinic sediments. *Am. J. Sci.* **285,** 710–724.

Raiswell, R. and Berner, R. A. (1986). Pyrite and organic matter in Phanerozoic normal marine shales. *Geochim. Cosmochim. Act* **50,** 1967–1976.

Randle, K. (1974). Some trace element data and their interpretation for several new reference samples obtained by neutron activation analysis. *Chem. Geol.* **13,** 237–256.

Reade, T. M. (1879). *Chemical Denudation in Relation to Geological Time,* David Bogue, London.

Reade, T. M. (1893). Measurement of geological time. *Geol. Mag.* [3] **10,** pp. 97–100.

Redfield, A. C., Ketchum, B. H., and Richards, F. A. (1963). The influence of organisms on the composition of seawater. In *The Sea,* M. N. Hill, Ed., Interscience, New York, Vol. 2, 26–77.

Rees, C. E. (1970). The sulphur isotope balance of the ocean: An improved model. *Earth Planet. Sci. Lett.* **7,** 366–370.

Rees, C. E. (1973). A steady state model for sulphur isotope fractionation in bacterial reduction processes. *Geochim. Cosmochim. Acta* **37,** 1141–1162.

Reibach, P. H., and Benedict, C. R. (1977). Fractionation of stable carbon isotopes by phosphoenolpyruvate carboxylase from C^4 plants. *Plant. Physiol.* **59,** 564–568.

Renard, M. (1986). Pelagic carbonate chemostratigraphy (Sr, Mg, ^{18}O, ^{13}C). *Marine Micropaleo.* **10,** 117–164.

Ringwood, A. E. (1975). *Composition and Petrology of the Earth's Mantle,* McGraw-Hill, New York, 618 pp.

Ripley, E. M. and Nicol, D. M. (1981). Sulfur isotope studies of Archean slate and graywacke from northern Minnesota: Evidence of sulfate-reducing bacteria. *Geochim. Cosmochim. Acta* **45,** 839–846.

Robertson, D. S., Tilsley, J. E., and Hogg, G. M. (1978). The time-bound character of uranium deposits. *Econ. Geol.* **73,** 1409–1419.

Rodgers, C. D. and Walshaw, C. D. (1966). The computation of infrared cooling rates in planetary atmospheres. *Q. J. R. Meteorol. Soc.* **92,** 67–92.

Rona, P. A., Bostrom, K., Laubier, L., and Smith, K. L., Eds. (1983). *Hydrothermal Processes at Seafloor Spreading Centers*, Plenum, New York, 796 pp.

Rona, P. A., Klinkhammer, G., Nelsen, T. A., Trefry, J. H., and Elderfield, H. (1986). Black smokers, massive sulphides, and vent biota at the mid-Atlantic Ridge. *Nature* **321**, 33–37.

Ronov, A. B. (1959). On the post-Precambrian geochemical history of the atmosphere and hydrosphere. *Geokhimiya* **5**, 493–506.

Ronov, A. B. (1964). Common tendencies in the chemical evolution of the earth's crust, ocean and atmosphere. *Geochem. Int.* **1**, 713–737.

Ronov, A. B. (1968). Probable changes in the composition of sea water during the course of geological time. *Sedimentology* **10**, 25–43.

Ronov, A. B. (1972). Evolution of rock composition and geochemical processes in the sedimentary shell of the Earth. *Sedimentology* **19**, 157–172.

Ronov, A. B. (1976). Global carbon geochemistry, volcanism, carbonate accumulation, and life. *Geokhimiya* (8) pp. 1252–1257; *Geochem. Int.* 13, No. 4, 175–196.

Ronov, A. B. (1980). *Osadochnaya Obolochka Zemli Kolichestvennyye zakonomernosti stroyeniya sostavai*, Izv. Nauka. Moscow. 80 pp; *Int. Geol. Rev.* **24**, 1313–1388 (1982).

Ronov, A. B. (1982). The Earth's sedimentary shell (quantitative patterns of its structure, compositions, and evolution). *Internat. Geol. Rev.* **24**, 1313–1388.

Ronov, A. B. and Ermishkina, A. I. (1953). Method of compiling quantitative lithological geochemical maps. *Dokl. Akad. Nauk. SSSR* **91**, 1179–1182 (in Russian).

Ronov, A. B. and Migdisov, A. A. (1971). Evolution of the chemical composition of the rocks in the shields and sediment cover of the Russian and North American Platforms. *Sedimentology* **16**, 137–185.

Ronov, A. B. and Yaroshevskiy, A. A. (1969). Chemical composition of the earth's crust. *Geophys. Monogr., Am. Geophys. Union* **13**, 37–57.

Ronov, A. B. and Yaroshevskiy, A. A. (1976). A new model for the chemical structure of the earth's crust. *Geokhimiya*, No. 12, pp. 1761–1795; *Geochem. Int.* 13 (n. 6), 89–121.

Ronov, A. B., Grinenko, V. A., Girin, Yu. P., Savina, L. I., Kasakov, G. A., and Grinenko, L. N. (1974). Effects of tectonic conditions on the concentration and isotope composition of sulfur in sediments. *Geokhimiya* **12**, 1772–1798; *Geochem. Int.* **11**, 1246–1272.

Ronov, A. B., Khain, V. E., Balukhovsky, A. N., and Seslavinsky, K. B. (1980). Quantitative analysis of Phanerozoic sedimentation. *Sediment. Geol.* **25**, 311–325.

Ronov, A. B., Khain, V. E., and Balukhovsky, A. N. (1986a). Global quantitative balance of continental and oceanic sedimentation during the last 150 million years. *Izv. Akad. Nauk SSSR* **11**, 3–11.

Ronov, A. B., Khain, V. E., and Balukhovsky, A. N. (1986b). Quantitative distribution of sediments in oceans. *Litol. Polezn. Iskop.* **2**, 3–16.

Roy, A. B. and Trüdinger, P. A. (1970). *The Biochemistry of Inorganic Compounds of Sulfur*, Cambridge Univ. Press, London, 400 pp.

Rubey, W. W. (1951). Geologic history of sea water. An attempt to state the problem. *Geol. Soc. Am. Bull.* **62**, 1111–1148.

Russell, K. L. (1970). Geochemistry and halmyrolysis of clay minerals, Rio Ameca, Mexico. *Geochim. Cosmochim. Acta* **34**, 893–907.

Rutten, M. G. (1962). *The Geological Aspects of the Origin of Life on the Earth*, Elsevier, Amsterdam, 146 pp.

Sadler, P. M. (1981). Sediment accumulation rates and the completeness of stratigraphic sections. *J. Geol.* **89**, 569–584.

Saito, K. (1978). Classification and generation of terrestrial rare gases. In *Terrestrial Rare Gases*, E. C. Alexander and M. Ozima, Eds., Center for Academic Publications, Tokyo, pp. 148–162.

Salop, L. J. (1977). *Precambrian of the Northern Hemisphere.* Elsevier, Amsterdam, 378 pp.

Saltzman, E. S., Lindh, T. B., and Holser, W. T. (1982). Secular changes in $\delta^{13}C$ and $\delta^{34}S$, global sedimentation, P_{O_2} and P_{CO_2} during the Phanerozoic. *Geol. Soc. Am. Abs.* **14**, 607.

Sandberg, P. A. (1975). New interpretation of Great Salt Lake ooids and of ancient non-skeletal carbonate mineralogy. *Sedimentology* **22**, 497–538.

Sandberg, P. A. (1983). An oscillating trend in non-skeletal carbonate mineralogy. *Nature (London)*, **305**, 19–22.

Sandberg, P. A. (1985). Aragonite cements and their occurrence in ancient limestones. *SEPM Spec. Pub.* **36**, 33–57.

Savin, S. M., Douglas, R. G., and Stehli, F. G. (1975). Tertiary marine paleotemperatures. *Geol. Soc. Am. Bull.* **86**, 1499–1510.

Sayles, F. L. (1979). The composition and diagenesis of interstitial solutions; I. Fluxes across the seawater-sediment interface in the Atlantic Ocean. *Geochim. Cosmochim. Acta* **43**, 527–546.

Sayles, F. L. and Mangelsdorf, P. C. (1977). The equilibration of clay minerals with seawater: exchange reactions. *Geochim. Cosmochim. Acta* **41**, 951–960.

Sayles, F. L. and Manglesdorf, P. C. (1979). Cation-exchange characteristics of Amazon River suspended sediment and its reaction with seawater. *Geochim. Cosmochim. Acta* **43**, 767–779.

Scarfe, C. M. and Smith, D. G. W. (1977). Secondary minerals in some basaltic rocks from DSDP leg 37. *Can. J. Earth Sci.*, **14**, 903–910.

Scharff, M. L. (1980). Die Verteilung von Stickstoff, Schwefel, Schwefel-isotopen sowie Mn, Zn, Fe, und Cu in Sedimenten des Atlantischen Ozeans (DSDP Bohrkerne). Dissertation, George-August-Universität, Göttingen, Germany.

Schidlowski, M. (1979). Antiquity and evolutionary status of bacterial sulfate reduction: Sulfur isotope evidence. *Origins Life* **9**, 299–331.

Schidlowski, M. (1982). Content and isotopic composition of reduced carbon in sediments. In *Mineral Deposits and the Evolution of the Biosphere*, H. D. Holland and M. Schidlowski, Eds., Springer-Verlag, Berlin, pp. 103–122.

Schidlowski, M. (1987). Application of stable carbon isotopes to early biochemical evolution on Earth. *Ann. Rev. Earth. Planet. Sci.* **15**, 47–72.

Schidlowski, M. and Junge, C. E. (1981). Coupling among the terrestrial sulfur, carbon and oxygen cycles: Numerical modeling based on revised Phanerozoic carbon isotope record. *Geochim. Cosmochim. Acta* **45**, 589–594.

Schidlowski, M., Eichmann, R., and Junge, C. E. (1975). Precambrian sedimentary carbonates: Carbon and oxygen isotope geochemistry and implications for the terrestrial oxygen budget. *Precambrian Res.* **2**, 1–69.

Schidlowski, M., Junge, C. E., and Pietrek, H. (1977). Sulfur isotope variations in marine sulfate evaporites and the Phanerozoic oxygen budget. *JGR, J. Geophys. Res.* **83,** 2557–2565.

Schidlowski, M., Hayes, J. M., and Kaplan, I. R. (1983). Isotopic inferences of ancient biochemistries: Carbon, sulfur, hydrogen, and nitrogen. In *Earth's Earliest Biosphere,* J. W. Schopf, Ed., Princeton Univ. Press, Princeton, NJ, pp. 149–186.

Schlanger, S. O., and Jenkyns, H. C. (1976). Cretaceous oceanic anoxic events—causes and consequences. *Geol. en Mijnbouw* **55,** 179–184.

Schlanger, S. O., Jenkyns, H. C., and Premoli-Silva, I. (1981). Volcanism and vertical tectonics in the Pacific Basin related to global Cretaceous transgressions. *Earth Planet. Sci. Lett.* **52,** 435–449.

Schlanger, S. O., Arthur, M. A., Jenkyns, H. C., and Scholle, P. A. (1987). The Cenomanian-Turonian oceanic anoxic event, I. Stratigraphy and distribution of organic carbon-rich beds and the marine $\delta^{13}C$ excursion. In *Marine Petroleum Source Rocks,* J. Brooks and A. J. Fleet, Eds., Geol. Soc. Lond. Spec. Pub. 26, pp. 371–399.

Scholl, D. M., Marlow, H. S., and Cooper, A. K. (1977). Sediment subduction and offscraping at Pacific margins. In *Island Arcs, Deep Sea Trenches and Back-Arc Basins,* M. Talwani and W. G. Pitman, III, Eds., Am. Geophys. Union, Washington, DC, pp. 199–210.

Scholle, P. A. and Arthur, M. (1980). Carbon isotopic fluctuations in Cretaceous pelagic limestones: Potential stratigraphic and petroleum exploration tool. *Am. Assoc. Pet. Geol. Bul.* **64,** 67–87.

Scholle, P. A., Bebout, D. G., and Moore, C. H. (1983). Carbonate depositional environments. *Mem.—Am. Assoc. Pet. Geol.* **33,** 1–708.

Schopf, J. W., Ed. (1983). *Earth's Earliest Biosphere,* Princeton Univ. Press, Princeton, NJ, 543 pp.

Schopf, J. W. and Walter, M. R. (1983). Archean microfossils: new evidence of ancient microbes. In J. W. Schopf, Ed., *Earth's Earliest Biosphere,* Princeton Univ. Press, Princeton, NJ, pp. 214–239.

Schuchert, C. (1931). Geochronology or the age of the earth on the basis of sediments and life. *Natl. Res. Counc. Bull.* **80,** 10–64.

Schwab, F. L. (1978). Secular trends in the composition of sedimentary rock assemblages—Archean through Phanerozoic time. *Geology* **6,** 532–536.

Schwartzman, D. W. (1978). On the ambient mantle $^4He/^{40}Ar$ ratio and the coherent model of degassing of the earth. *Adv. Earth Planet. Sci.* **3,** 185–191.

Sclater, J. G., Parsons, B., and Jaupart, C. (1981). Oceans and continents: similarities and differences in the mechanism of heat loss. *JGR, J. Geophys. Res.* **86,** 11535–11552.

Scott, M. R., Scott, R. B., Rona, P. A., Butler, L. W., and Nalvelk, A. J. (1974). Rapidly accumulating manganese deposit from the median valley of the mid-Atlantic ridge. *Geophys. Res. Lett.* **1,** 355–358.

Scott, R. B., Rona, P. A., McGregor, B. A., and Scott, M. R. (1974). The TAG hydrothermal field. *Nature (London)* **251,** 301–302.

Seckbach, J. and Kaplan, I. R. (1973). Growth pattern and $^{13}C/^{12}C$ isotope fractionation of *Cyanidium cladariumu* and hot spring algal mats. *Chem. Geol.* **12,** 161–169.

Seccombe, P. K. (1977). Sulphur isotope and trace metal composition of stratiform sulphides as an ore guide in the Canadian Shield. *J. Geochem. Explor.* **8,** 117–137.

Sedimentation Seminar (1977). Magnitude and frequency of transport of solids by streams in the Mississippi Basin. *Am. J. Sci.* **277**, 862–875.

Sellers, W. D. (1965). *Physical Climatology*. Univ. of Chicago Press, Chicago, IL, 272 pp.

Seto, Y. B. and Duce, R. A. (1972). A laboratory study of iodine enrichment on atmospheric sea salt particles produced by bubbles. *J. Geophys. Res.* **77**, 5339–5349.

Seyfried, W. E. (1976). Seawater-basalt interaction from 25°–300° and 1–500 bars: implications for the origin of submarine metal-bearing hydrothermal solutions and regulation of ocean chemistry. Ph.D. Dissertation, Univ. Southern California, Los Angeles.

Seyfried, W. E., Jr., and Bischoff, J. L. (1977). Hydrothermal transport of heavy metals by seawater: The role of seawater basalt ratio. *Earth Planet. Sci. Lett.* **34**, 71–77.

Seyfried, W. E., Jr., and Bischoff, J. G. (1979). Low temperature basalt alteration by seawater; An experimental study at 70°C and 150°C. *Geochim. Cosmochim. Acta* **43**, 1937–1947.

Seyfried, W. E., Jr. and Mottl, M. J. (1977). Origin of submarine metal-rich hydrothermal solutions: Experimental basalt-seawater interaction in a seawater-dominated system at 300°C, 500 bars. In *Water-Rock Interactions*, I.A.G.C., Strasbourg, France, pp. 173–180.

Seyfried, W. E., Jr., Janecky, D. R., and Mottl, M. J. (1984). Alteration of oceanic crust: Implications for geochemical cycles of lithium and boron. *Geochim. Cosmochim. Acta* **48**, 557–569.

Seyfried, W. E., Jr., Berndt, M. E., and Janecky, D. R. (1986). Chlorine depletions and enrichments in seafloor hydrothermal fluids: Constraints from experimental basalt alteration studies. *Geochim. Cosmochim. Acta* **50**, 469–475.

Shackleton, N. J. (1987). The carbon isotope record of the Cenozoic: history of organic carbon burial and of oxygen in the ocean and atmosphere. In *Marine Petroleum Source Rocks*, J. Brooks and A. J. Fleet, Eds., Geol. Soc. Lond. Spec. Pub. 26, pp. 423–434.

Shand, S. J. (1943). *Eruptive Rocks*, T. Murby, London, 360 pp.

Shanks, W. C., Bischoff, J. L., and Rosenbauer, R. J. (1981). Seawater sulfate reduction and sulfur isotope fractionation in basaltic systems: interaction of seawater with fayalite and magnetite at 200–300°C. *Geochim. Cosmochim. Acta* **45**, 1977–1995.

Sharma, G. D. (1979). *The Alaskan Shelf; Hydrographic, Sedimentary, and Geochemical Environment*, Springer-Verlag, New York, 487 pp.

Shelton, J. E. (1978). Sulfur and pyrites. *Min. Yearb.*, pp. 1287–1307.

Sholkovitz, E. R. and Price, N. B. (1980). The major-element chemistry of suspended matter in the Amazon estuary. *Geochim. Cosmochim. Acta* **44**, 163–172.

Sibley, D. F. and Vogel, T. A. (1976). Chemical mass balance of the earth's crust: The calcium dilemma (?) and the role of pelagic sediments. *Science* **192**, 551–553.

Sibley, D. F. and Wilband, J. T. (1977). Chemical balance of the Earth's crust. *Geochim. Cosmochim. Acta* **41**, 545–554.

Sidorenko, A. V. et al. (18 authors) (1978). *Precambrian and Problems of Earth Crust Formation*, Nauka, Moscow, 311 pp. (in Russian).

Sillen, L. G. (1961). The physical chemistry of sea water. In *Oceanography*, Publ. No. 67, M. Sears, Ed., Am. Assoc. Adv. Sci., Washington, DC, pp. 549–581.

Sirevag, R., Buchana, B. B., Berry, J. A., and Troughton, J. H. (1977). Mechanisms of CO_2 fixation in bacterial photosynthesis: Studies by the carbon isotope fractionation technique. *Arch. Microbiol.* **112**, 35–38.

Slanksy, M. (1986). *Geology of Sedimentary Phosphorites*, Elsevier, Amsterdam, 210 pp.

Sleep, N. H. (1979). Thermal history and degassing of the earth: Some simple calculations. *J. Geol.* **87**, 671–686.

Sleep, N. H. and Rosendahl, B. E. (1979). Topography and tectonics of mid-oceanic ridge axes. *JGR, J. Geophys. Res.* **85**, 6831–6839.

Sleep, N. H. and Wolery, T. J. (1978). Egress of hot water from midocean ridge hydrothermal systems: Some thermal constraints. *JGR, J. Geophys. Res.* **83**, 5913–5922.

Sleep, N. H., Morton, J. L., Burns, L. E., and Wolery, T. J. (1983). Geophysical constraints on the volume of hydrothermal flow at ridge axes. In *Hydrothermal Processes at Seafloor Spreading Centers*, P. A. Rona, K. Bostrom, L. Labier and K. L. Smith, Eds., Plenum, New York, pp. 71–82.

Slemr, F., Seiler, W., and Schuster, G. (1981). Latitudinal distribution of mercury over the Atlantic Ocean. *J. Geophys. Res.* **86**, 1159–1166.

Sloss, L. L. (1976). Areas and volumes of cratonic sediments, western North America and eastern Europe. *Geology* **4**, 272–276.

Sloss, L. L. and Speed, R. C. (1974). Relationships of cratonic and continental margin tectonic episodes. In Spec. Publ.—*Soc. Econ, Paleontol. Mineral.* **22**, 98–119.

Smith, B. N. and Epstein, S. (1971). Two categories of $^{13}C/^{12}C$ ratios for higher plants. *Plant Physiol.* **47**, 380–384.

Smith, J. W. and Croxford, N. J. W. (1973). Sulphur isotope ratios in the McArthur lead-zinc-silver deposit. *Nature Phys. Sci.* **245**, 10–12.

Smith, J. W. and Croxford, N. J. W. (1975). An isotopic investigation of the environment of deposition of the McArthur mineralization. *Min. Depos.* **10**, 269–276.

Smith, J. W. and Kaplan, I. R. (1980). Endogenous carbon in carbonaceous meteorites. *Science* **167**, 1367–1370.

Southam, J. R. and Hay, W. W. (1977). The scales and dynamic models of deep-sea sedimentation. *JGR, J. Geophys. Res.* **82**, 3825–3842.

Southam, J. R. and Hay, W. W. (1981). Global sedimentary mass balance and sea level changes. In *The Sea*, C. Emiliani, Ed., Vol. 7, Wiley (Interscience), New York, pp. 1617–1685.

Spivack, A. J. and Edmond, J. M. (1987). Boron isotope exchange between seawater and the oceanic crust. *Geochim. Cosmochim. Acta* **51**, 1033–1043.

Spooner, E. T. C., Chapman, H. J., and Smewing, J. D. (1977). Strontium isotopic contamination and oxidation during ocean floor hydrothermal metamorphism of the ophiolitic rocks of the Troodos Massif, Cyprus. *Geochim. Cosmochim. Acta* **41**, 873–890.

Sprague, D. and Pollack, H. N. (1980). Heat flow in the Mesozoic and Cenozoic. *Nature (London)* **285**, 393–395.

Stakes, D. S. (1978). Submarine hydrothermal systems: Variations in mineralogy, chemistry, temperatures, and the alteration of ocean layer II. Ph.D. dissertation, Oregon State University, Eugene.

Stille, H. (1936). The present tectonic stage of the earth. *Am. Assoc. Pet. Geol. Bull.* **20**, 849–880.

Strakhov, N. M. (1964). States and development of the external geospheres and formation of sedimentary rocks in the history of the earth. *Int. Geol. Rev.* **6**, 1466–1482.

Strakhov, N. M. (1969). *Principles of Lithogenesis*, Vol. 2, Oliver and Boyd, Edinburgh, 609 pp.

Stumm, W., Ed. (1977). *Global Chemical Cycles and Their Alterations by Man*, Dahlem Konferenzen, Berlin, 347 pp.

Styrt, M. M., Brackmann, A. J., Holland, H. D., Clark, B. C., Pisutha-Arnond, V., Edlridge, C. S., and Ohmoto, H. (1981). The mineralogy and isotopic composition of sulfur in hydrothermal sulfide/sulfate deposits on the East Pacific Rise, 21°N latitude. *Earth Planet. Sci. Lett.* **53**, 382–390.

Suess, E. (1980). Particulate organic carbon flux in the oceans—surface productivity and oxygen utilization. *(London) Nature* **288**, 260–263.

Summerhayes, C. (1981). Organic facies of Middle Cretaceous black shales in deep North Atlantic. *Am. Assoc. Pet. Geol. Bull.* **65**, 2364–2380.

Sundquist, E. T. and W. S. Broecker, Eds. (1985). *The Carbon Cycle and Atmospheric CO_2: Natural Variations, Archean to Present*, Geophys. Monogr. Ser. No. 32, Am. Geophys. Union, Washington, DC, 627 pp.

Sverdrup, H. N., Johnson, M. W., and Fleming, R. H. (1942). *The Oceans*, Prentice-Hall, Englewood Cliffs, NJ, 1087 pp.

Sweeney, R. E. and Kaplan, I. R. (1980). Isotope fractionation of sulfur during sulfate reduction in marine sediments. Unpublished manuscript.

Tanner, J. T. (1978). *Guide to the Study of Animal Populations*, Univ. of Tennessee Press, Knoxville, 186 pp.

Tappan, H. (1968). Primary production, isotopes, extinctions and the atmosphere. *Palaeogeogr., Palaeoclimatol., Palaeoecol.* **4**, 187–210.

Taylor, P. S. and Stoiber, R. E. (1973). Soluble material on ash from active Central American volcanoes. *Geol. Soc. Am. Bull.* **84**, 1031–1042.

Taylor, S. R. (1964). Abundance of chemical elements in the continental crust: A new table. *Geochim. Cosmochim. Acta* **28**, 1273–1285.

Taylor, S. R. (1979). Chemical composition and evolution of continental crust: The rare earth element evidence. In *The Earth: Its Origin, Structure and Evolution*, M. W. McElhinny, Ed., Academic Press, New York, pp. 353–376.

Taylor, S. R. and McLennan, S. M. (1985). *The Continental Crust: its Composition and Evolution*, Blackwell, Oxford, 312 pp.

Taylor, S. R., McLennan, S. M., and McCulloch, M. T. (1983). Geochemistry of loess, continental crustal composition and crustal model ages. *Geochim. Cosmochim. Acta* **47**, 1897–1905.

Thode, H. G. and Monster, J. (1965). Sulfur isotope geochemistry of petroleum, evaporites, and ancient seas. *Mem.—Am. Assoc. Pet. Geol.* **4**, 367–377.

Thode, H. G., Dunford, H. B., and Shima, M. (1962). Sulfur isotope abundances in rocks of the Sudbury District and their geological significance. *Econ. Geol.* **57**, 565–578.

Thompson, G. (1983). Basalt-seawater interaction. In *Hydrothermal Processes at Seafloor Spreading Centers*, P. A. Rona, K. Bostrom, L. Laubier, and K. L. Smith, Eds., Plenum, New York, pp. 225–278.

Thompson, G. (1983). Hydrothermal fluxes in the ocean. In *Chemical Oceanography*, J. P. Riley and R. Chester, Eds., Vol. 8, Academic Press, New York, pp. 272–338.

Thompson, S. L., and Barron, E. J. (1981). Comparison of Cretaceous and present earth albedos: Implications for the causes of paleoclimates. *J. Geol.* **89**, 143–167.

Tissot, B. (1979). La répartition mondiale des combustibles fossiles. *La Récherche* **104**, 984–991.

Tissot, B. and Pelet, R. (1981). Sources and fate of organic matter in ocean sediments. *Oceanol. Acta.* No. SP, pp. 97–103.

Tolstikhin, I. N. (1978). A review: Some recent advances in isotope geochemistry of light rare gases. *Adv. Earth Planet. Sci.* **3**, pp. 33–62.

Toth, D. J. and Lerman, A. (1977). Organic matter, reactivity and sedimentation rates in the ocean. *Am. J. Sci.* **277**, 465–485.

Trendall, A. F. and Morris, R. C. (1983). *Iron-formation, Facts and Problems*, Elsevier, Amsterdam, 558 pp.

Trudinger, P. A. (1969). Assimilatory and dissimilatory metabolism of inorganic sulfur compounds by microorganisms. *Adv. Microb. Physiol.* **3**, 111–158.

Trüper, H. G. (1982). Microbial processes in the sulfuretum through time. In *Mineral Deposits and the Evolution of the Biosphere*, H. D. Holland and M. Schidlowski, Eds., Springer-Verlag, Berlin, pp. 5–30.

Tucker, M. E. (1982). Precambrian dolomites: Petrographic and isotopic evidence that they differed from Phanerozoic dolomites. *Geology* **10**, 7–12.

Tugarinov, A. I. and Bibikova, E. V. (1976). Evolution of the chemical composition of the earth crust. *Geokhimiya* (8) 1151–1159, *Geochem. Int.* **13** (4), 114–121.

Turcotte, D. L. and Burke, K. (1978). Global sea-level changes and the thermal structure of the earth. *Earth Planet. Sci. Lett.* **41**, 341–356.

Turekian, K. (1971). Rivers, tributaries, and estuaries. In *Impingement of Man on the Oceans*, D. W. Hood, Ed., Wiley, New York, pp. 9–73.

Turekian, K. K. (1983). Geochemical mass balances and cycles of the elements. In *Hydrothermal Processes at Seafloor Spreading Centers*, P. A. Rona, K. Bostrom, L. Laubier, and K. L. Smith, Eds., Plenum, New York, pp. 318–368.

Turekian, K. K. and Wedepohl, K. H. (1961). Distribution of the elements in some major units of the earth's crust. *Geol. Soc. Am. Bull.* **72**, 175–192.

Umbgrove, J. H. F. (1947). *The Pulse of the Earth*, Martinus Nijhoff, The Hague, 358 pp.

Uyeda, S. (1978). *The New View of the Earth: Moving Continents and Moving Oceans*, Freeman, San Francisco, CA, 217 pp.

Vail, P. R., Mitchum, R. M., and Thompson, S. (1977). Seismic stratigraphy and global changes of sea level. Part 4, *Mem.—Am. Assoc. Pet. Geol.* **26**, 276 pp.

Valley, J. W. and O'Neil, J. R. (1981). $^{13}C/^{12}C$ exchange between calcite and graphite: A possible thermometer in Grenville marbles. *Geochim. Cosmochim. Acta* **45**, 411–419.

Van Keulen, H., Van Laar, H. H., Louwerse, W., and Goudriaan, J. (1980). Physiological aspects of increased CO_2 concentration. *Experientia* **36**, 786–792.

Van Moort, J. C. (1972). The K_2O, CaO, MgO and CO_2 contents of shales and related rocks and their implication for sedimentary evolution since the Proterozoic. *Int. Geol. Congr., Rep. Sect. Sess., 24th, 1972*, No. 10, pp. 427–439.

Van Moort, J. C. (1973). The magnesium and calcium contents of sediments, especially pelites, as a function of age and degree of metamorphism. *Chem. Geol.* **12**, 1–37.

Veizer, J. (1973). Sedimentation in geologic history: recycling vs. evolution or recycling with evolution. *Contrib. Mineral. Petrol.* **38**, 261–278.

Veizer, J. (1976a). Evolution of ores of sedimentary affiliation through geologic history; relations to the general tendencies in evolution of the crust, hydrosphere, atmosphere and biosphere. In *Handbook of Strata-bound and Stratiform Ore Deposits Vol. III*, K. H. Wolf, Ed., Elsevier, Amsterdam, pp. 1–41.

Veizer, J. (1976b). $^{87}Sr/^{86}Sr$ evolution of seawater during geologic history and its significance as an index of crustal evolution. In *The Early History of the Earth*, B. F. Windley, Ed., Wiley, London, pp. 569–578.

Veizer, J. (1978). Secular variations in the composition of sedimentary carbonate rocks. II. Fe, Mn, Ca, Mg, Si and minor constituents. *Precambrian Res.* **6**, 381–413.

Veizer, J. (1979). Secular variations in chemical composition of sediments: A review. In *The Origin and Evolution of the Elements* L. H. Ahrens, Ed., Pergamon, Oxford, pp. 269–278.

Veizer, J. (1983). Geological evolution of the Archean-Early Proterozoic Earth. In *The Earth's Earliest Biosphere: Its Origin and Evolution*, J. W. Scopf, Ed., Princeton Univ. Press, Princeton, NJ, pp. 240–259.

Veizer, J. (1984). Recycling on the evolving earth. *Proc. Int. Geol. Congr. 27th*, Vol. 11, pp. 325–345.

Veizer, J. (1985). Carbonates and Ancient oceans: Isotopic and chemical record on time scales of 10^7–10^9 years. *Geophys. Monogr., Am. Geophys. Union* **32**, 595–601.

Veizer, J. (1987). The earth and its life: geologic record of interactions and controls. In *Origin and Evolution of the Universe: Evidence for Design*, J. M. Robson, Ed., McGill-Queen's Univ. Press, Montreal pp. 167–194.

Veizer, J. (1988a). Continental growth: Comments on "The Archean-Proterozoic transition: Evidence from Guyana and Montana" by A. K. Gibbs, C. W. Montgomery, P. A. O'day, and E. A. Erslev, *Geochim. Cosmochim. Acta*, March issue.

Veizer, J. (1988b). Solid earth as a recycling system: temporal dimensions of global tectonics. In *Physical and Chemical Weathering in Geochemical Cycles*, A. Lerman and M. Meybeck, Eds., Reidel, Dordrecht (in press).

Veizer, J. and Compston, W. (1974). $^{87}Sr/^{86}Sr$ composition of sea water during the Phanerozoic. *Geochim. Cosmochim. Acta* **38**, 1461–1484.

Veizer, J. and Compston, W. (1976). $^{87}Sr/^{86}Sr$ in Precambrian carbonates as an index of crustal evolution. *Geochim. Cosmochim. Acta* **40**, 905–914.

Veizer, J. and Garrett, D. E. (1978). Secular variations in composition of carbonate rocks. I. Alkali metals. *Precambrian Res.* **6**, 367–380.

Veizer, J. and Hoefs, J. (1976). The nature of O^{18}/O^{16} and C^{13}/C^{12} secular trends in sedimentary carbonate rocks. *Geochim. Cosmochim. Acta* **40**, 1387–1395.

Veizer, J. and Jansen, S. L. (1979). Basement and sedimentary recycling and continental evolution. *J. Geol.* **87**, 341–370.

Veizer, J. and Jansen, S. L. (1985). Basement and sedimentary recycling-2: time dimension to global tectonics. *J. Geol.* **93,** 625–643.

Veizer, J., Holser, W. T., and Wilgus, C. K. (1980). Correlation of $^{13}C/^{12}C$ and $^{34}S/^{32}S$ secular variations. *Geochim. Cosmochim. Acta* **44,** 579–587.

Veizer, J., Compston, W., Hoefs, J., and Nielsen, H. (1982). Mantle buffering of the early oceans. *Naturwissenschaften* **69,** 173–180.

Veizer, J., Fritz, P., and Jones, B. (1986). Geochemistry of braciopods: Oxygen and carbon isotopic records of Paleozoic oceans. *Geochim. Cosmochim. Acta* **50,** 1679–1696.

Veizer, J., Laznicka, P., and Jansen, S. L. (1988). Mineralization through geologic time: recycling perspective. Submitted for publication.

Vinogradov, A. P. (1967). The formation of the oceans. *Izv. Akad. Nauk SSSR, Ser. Geol.* **4,** 3–9.

Vinogradov, A. P., Ronov, A. B., and Ratynskii, V. Y. (1952). Evolution of the chemical composition of carbonate rocks. *Izv. Akad. Nauk SSSR, Ser. Geol.* **1,** 33–60.

Vinogradov, A. P., Grinenko, V. A., and Ustinov, U. S. (1962). Izotopny sostav soedineii sery v chornomore. (Isotopic composition of sulfur compounds in the Black Sea.) *Geokhimiya* pp. 851–873; *Geochem. Int.* pp. 973–997.

Vinogradov, V. I., Reimer, T. O. and Leites, A. M. (1976). The oldest sulfates in the Archean formations of the South African and Aldan shields, and the evolution of the Earth's oxygen atmosphere. *Lithol. Miner. Resour.*, **11,** 407–420.

Vishemirskij, V. S. (1978). Stratigraphic distribution of fossil fuels, *Geol. Geofiz.* **6,** 3.

Vogel, J. C. (1961). Isotope separation factor of carbon in the equilibrium system CO_2-HCO_3^--CO_3^{2-}. In *Summer Course on Nuclear Geology, Varenna, 1960,* F. G. Houtermans, E. E. Picciotto, and E. Tongiori, Eds., Lischi, Pisa, pp. 216–221.

Volkov, V. N. (1968). Coal and bituminous shales. In *Geological Framework of the U.S.S.R.*, Vol. 4, A. I. Semenov and A. D. Scheglov, Eds., Nedra, Moscow, pp. 458–471 (in Russian).

von Bertalanffy, L. (1968). *General Systems Theory*, George Braziller, New York, 289 pp.

von Brunn, V. and Mason, T. R. (1977). Siliciclastic-carbonate tidal deposits from the 3000 m.y. Pongola Supergroup, South Africa. *Sediment. Geol.* **18,** 245–255.

Von Damm, K. L. (1983). Chemistry of submarine hydrothermal solutions at 21°N, East Pacific Rise, and Guaymas Basin, Gulf of California. Ph.D. dissertation, Massachusetts Institute of Technology, Cambridge, 239 pp.

Von Damm, K. L., Grant, B., and Edmond, J. M. (1983). Preliminary report on the chemistry of hydrothermal solutions at 21°N. East Pacific Rise. In *Hydrothermal Processes at Seafloor Spreading Centers*, P. A. Rona, K. Bostrom, L. Laubier, and K. L. Smith, Eds., Plenum, New York, pp. 369–390.

Von Damm, K. L., Edmond, J. M., Grant, B., Measures, C. I., Walden, B., and Weiss, R. F. (1985). Chemistry of submarine hydrothermal solutions at 21°N, East Pacific Rise. *Geochim. Cosmochim. Acta* **49,** 2197–2220.

Wadleigh, M. A. (1982). Marine geochemical cycle of strontium. M.Sc. thesis, University of Ottawa, 187 pp.

Wadleigh, M. A., Veizer, J., and Brooks, C. (1985). Strontium and its isotopes in Canadian rivers: Fluxes and global implications. *Geochim. Cosmochim. Acta* **49,** 1727–1736.

Walker, J. C. G. (1977). *Evolution of the Atmosphere*, Macmillan, New York, 318 pp.

Walker, J. C. G. (1978). Oxygen and hydrogen in the primitive atmosphere. *Pure Appl. Geophys.* **117**, 498–512.

Walker, J. C. G. (1980). Biogeochemical cycles: Oxygen. In *Handbook of Environmental Chemistry*, O. Hutzinger, Ed., Springer-Verlag, Heidelberg, pp. 87–104.

Walker, J. C. G. (1984). How life affects the atmosphere. *BioScience* **34**, 486–491.

Walker, J. C. G. (1986). *Earth History: The Several Ages of the Earth*, Jones and Bartlett Publishers, Boston, MA, 199 pp.

Walker, J. C. G., Hays, P. B., and Kasting, J. F. (1981). A negative feedback mechanism for the long-term stabilization of Earth's surface temperature. *JGR, J. Geophys. Res.* **86**, 9776–9782.

Walker, J. C. G., Klein, C., Schidlowski, M., Schopf, J. W., Stevenson, D. J., and Walter, M. R. (1983). Environmental evolution of the Archean-Early Proterozoic Earth. In *The Earth's Earliest Biosphere: Its Origin and Evolution*, J. W. Schopf, Ed., Princeton Univ. Press, Princeton, NJ, pp. 260–290.

Walter, M. R., Buick, R., and Dunlop, J. S. R. (1980). Stromatolites, 3,400–3,500 Myr old from the North Pole area, Western Australia. *Nature (London)* **284**, 443–445.

Ward, B. B., Kilpatrick, K. A., Novelli, P. C., and Scrantom, M. I. (1987). Methane oxidation and methane fluxes in the ocean surface layer and deep anoxic waters. *Nature (London)* **327**, 226–229.

Watson, A., Lovelock, J. E., and Margulis, L. (1978). Methanogenesis, fires and the regulation of atmospheric oxygen. *BioSystems* **10**, 293–298.

Weaver, C. E. (1967). Potassium, illite and the ocean. *Geochim. Cosmochim. Acta* **31**, 2181–2196.

Wedepohl, K. H. (1960). Spurenanalytische Untersuchungen an Tiefseetonen and dan Atlantik. *Geochim. Cosmochim. Acta* **18**, 200–231.

Wedepohl, K. H. (1969). *Handbook of Geochemistry*, Vol. 1, Springer-Verlag, Berlin.

Wegener, A. (1926). *The Origin of Continents and Oceans*, 4th ed., Dover, London (1966 English Transl.).

Welhan, J. A. and Craig, H. (1979). Methane and hydrogen in East Pacific rise hydrothermal fluids. *Geophys. Res. Lett.* **6**, 829–832.

Welhan, J. A. and Craig, H. (1983). Methane, hydrogen, and helium in hydrothermal fluids at 21°N on the East Pacific Rise. In *Hydrothermal Processes at Seafloor Spreading Centers*, P. A. Rona, K. Bostrom, L. Laubier, and K. L. Smith, Eds., Plenum, New York, pp. 391–410.

Welte, D. H., Kalkreuth, W., and Hoefs, J. (1975). Age-trend in carbon isotopic composition in Paleozoic sediments. *Naturwissenschaften* **62**, 482–483.

Wendt, I. (1968). Fractionation of carbon isotopes and its temperature dependence in the system CO_2 (gas)-CO_2 (in solution) and HCO_3-CO_2 in solution. *Earth Planet. Sci. Lett.* **4**, 64–68.

Wilde, P. and Berry, W. B. N. (1984). Destabilization of the oceanic density structure and its significance to marine "extinction" events. *Paleogeogr. Paleoclimatol. Paleoecol.* **48**, 143–162.

Wilgus, C. K. (1981). A stable isotope study of Permian and Triassic marine evaporite and carbonate rocks, Western interior, U.S.A. Ph.D. dissertation, University of Oregon, Eugene.

Wilkinson, B. H. (1979). Biomineralization, paleoceanography, and the evolution of calcareous marine organisms. *Geology* **7**, 524–527.

Wilkinson, B. H., Owen, R. M., and Carroll, A. R. (1985). Submarine hydrothermal weathering, global eustasy, and carbonate polymorphism in Phanerozoic marine oolites. *J. Sediment. Petrol.* **55**, 171–183.

Williams, C. F., Narashimhan, T. N., Anderson, R. N., Zoback, M. D., and Becker, K. (1986). Convection in the oceanic crust: Simulation of observations from Deep Sea Drilling Project Hole 504B, Costa Rica Rift. *JGR, J. Geophys. Res.* **91**, 4877–4890.

Williams, J., Ed. (1978). *Carbon Dioxide, Climate and Society*, Pergamon, New York, 332 pp.

Williams, N. (1978). Studies of the base metal sulfide deposits at McArthur River, Northern Territory, Australia. II. The sulfide-S and organic-C relationships of the concordant deposits and their significance. *Econ. Geol.* **73**, 1036–1056.

Wilson, E. O. and Bossert, W. H. (1971). *A Primer of Population Biology*, Sinauer Associates, Sunderland, MA, 192 pp.

Wilson, J. L. (1975). *Carbonate Facies in Geologic History*, Springer, New York, 471 pp.

Windley, B. F. (1984). *The Evolving Continents*, 2nd ed., Wiley, New York, 399 pp.

Winkler, H. G. F. and von Platen, H. (1961). Experimentelle Gesteinsmetamorphose, IV. Bildung anatectischer Schmelzen aus metamorphisierten Grauwacken. *Geochim. Cosmochim. Acta* **24**, 48–69.

Wise, D. U. (1972). Freeboard of continents through time. *Mem.—Geol. Soc. Am.* **132**, 87–100.

Wolery, T. J. (1978). Some chemical aspects of hydrothermal processes at mid-oceanic ridges—a theoretical study. I. Basalt-sea water reaction and chemical cycling between the oceanic crust and the oceans. II. Calculation of chemical equilibrium between aqueous solutions and minerals. Ph.D. dissertation, Northwestern University, Evanston, IL, 263 pp.

Wolery, T. J. (1980). Modeling sea water-basalt interactions in a flow-through system with a temperature gradient. In *Water-Rock Interactions*, Int. Assoc. Geochem. Cosmochem. (UNESCO) Edmonton, Alberta.

Wolery, T. J. and Sleep, N. H. (1976). Hydrothermal circulation and geochemical flux at mid-ocean ridges. *J. Geol.* **84**, 249–275.

Wolfe, R. S. (1971). Microbiol formation of methane. *Adv. in Microb. Physiol.* **6**, 107–146.

Wollast, R. (1974). The silica problem. In *The Sea*, Vol. 5, Chap, 11, E. D. Goldberg, Ed., Wiley (Interscience), New York.

Wollast, R. and DeBroeu, F. (1971). Study of the behavior of dissolved silica in the estuary of the Scheldt. *Geochim. Cosmochim. Acta* **35**, 613–620.

Wollast, R., Mackenzie, F. T., and Bricker, O. P. (1968). Experimental precipitation and genesis of sepiolite at earth-surface conditions. *Am. Mineral.* **53**, 1645–1662.

Wong, W. W. and Sackett, W. M. (1978). Fractionation of stable carbon isotopes by marine phytoplankton. *Geochim. Cosmochim. Acta* **42**, 1809–1815.

Wong, W. W., Sackett, W. M., and Benedict, C. R. (1975). Isotope fractionation in photosynthetic bacteria during carbon dioxide assimilation. *Plant Physiol.* **55**, 475–479.

Wong, W. W., Benedict, C. R., and Kohel, R. J. (1979). Isotope enzymic fractionation of the stable carbon isotopes of carbon dioxide by ribulose-1.5-bisphosphate carboxylase. *Plant Physiol.* **63,** 852–856.

Worsley, T. R., and Davis, T. A. (1979). Sea-level fluctuations and deep-sea sedimentation rates. *Science* **203,** 455–456.

Wright, J., Schrader, H., and Holser, W. T. (1987). Paleoredox variations in ancient oceans recorded by rare earth elements in fossil apatite. *Geochim. Cosmochim. Acta* **51,** 631–644.

Wyllie, P. J. (1976). *The Way the Earth Works: An Introduction to the New Global Geology and its Revolutionary Development,* Wiley, New York, 296 pp.

Ycas, M. (1972). Biological effects on the early atmosphere. *Nature (London)* **238,** 163–164.

Yeats, P. A., Sundby, B., and Bewers, J. M. (1979). Manganese recycling in coastal waters. *Mar. Chem.* **8,** 43–45.

Yeh, H. W. and Epstein, S. (1981). Hydrogen and carbon isotopes of petroleum and related organic matter. *Geochim. Cosmochim. Acta* **45,** 753–762.

Yung, Y. L. and McElroy, M. B. (1979). Fixation of nitrogen in the prebiotic atmosphere. *Science* **203,** 1002–1004.

Zahnle, K. J. (1986). Photochemistry of methane and the formation of hydrocyanic acid (HCN) in the Earth's early atmosphere. *JGR, J. Geophys. Res.* **91,** 2819–2834.

Zharkov, M. A. (1974). *Paleozoiskie Solenosnie Formatsii Mira.* Izd. "Nedra," Moskua, 391 pp.

Zharkov, M. A. (1981). *History of Paleozoic Salt Accumulation,* Springer-Verlag, Berlin, 308 pp.

Ziegler, A. M., Scotese, C. R., McKerrow, W. S., Johnson, M. E., and Bambach, R. K. (1979). Paleozoic paleogeography. *Annu. Rev. Earth Planet. Sci.* **7,** 473–502.

Ziman, K. E. (1978). Source functions for CO_2 in the atmosphere. In *Carbon Dioxide, Climate and Society,* J. Williams, Ed., Pergamon, New York, 332 pp.

AUTHOR INDEX

Abbott, D.H., 82
Adams, J.A.S., 20
Agegian, C., 161
Ajtay, G.L., 108, 113, 114
Alfvén, H., 163
Allårt, J.H., 166, 214
Allègre, C.J., 100, 193, 195, 208, 212
Alt, J.C., 79, 90, 97, 127
Anderson, A.T., 99, 115
Anderson, N.R., 63
Anderson, R.N., 77, 79, 82, 95
Andrews, A.J., 79, 90, 95, 126, 127
Anonymous, 177, 181
Archer, P., 85
Armstrong, R.L., 72, 102
Arndt, N.T., 193
Arnold, M., 97
Arrhenius, G., 163, 215
Arthur, M.A., 107, 120, 140, 144, 156, 168
Aston, S.R., 22
Averitt, P., 108
Awramik, S.M., 166, 204

Baas-Becking, L.G.M., 133
Bach, W., 63
Bachinski, D.J., 126
Badham, J.P.N., 204
Badiozamani, K., 44
Bailey, N.T.S., 177
Balashov, Yu.A., 195
Ballard, R.D., 183
Barley, M.E., 169, 214
Barnes, I., 115, 117, 153
Barrell, J., 175
Barron, E.J., 113
Barth, T.F.W., 3, 17
Basset, C.A., 9
Beaty, R.D., 22
Beauford, W., 41
Becker, G.F., 13
Behrens, E.W., 131
Belyaev, S.S., 128, 131
Benedict, C.R., 129, 130, 131
Ben Othman, D., 195

Berger, W.H., 31, 34, 113, 114, 116
Berner, R.A., 25, 31, 44, 105, 114, 120, 121, 125, 126, 135, 142, 145–149, 156–158
Berry, W.B.N., 156
Bestougeff, M.A., 204
Bibikova, E.V., 175, 187, 192
Bickle, M.J., 212
Birch, F., 212
Bischoff, J.L., 23, 80, 85, 86, 97
Bjornsson, S., 23, 41, 85
Blackburn, T.H., 59
Bois, C., 204
Bolin, B., 63, 113
Bossert, W.H., 180, 199
Böstrom, K., 20
Bottinga, Y., 132
Boucot, A.J., 113
Bowers, T.S., 82, 86, 90
Bowring, S.A., 215
Boyle, E.A., 34, 38
Bradshaw, M.J., 157
Braitsch, O., 110
Brass, G.W., 101, 142
Brevart, O., 212
Broda, E., 169, 171
Broecker, W.S., 18, 31, 32, 48, 60, 61, 63, 105, 153, 164
Brotzen, O., 95
Brown, L., 102
Brumsack, H.J., 137
Brune, J., 212
Bryan, W.B., 99
Buat-Menard, P., 42
Buckland, W., 11
Buckle, T.H., 10
Buddington, A.F., 171
Buffon, Comte D., 7
Burchfield, J.D., 17
Burke, W.H., 142, 144, 159, 212
Burke, K., 149
Burnie, S.W., 171
Burns, R.C., 59
Burns, L.E., 102
Bushinskii, G.I., 204
Byers, C.W., 52

Cadle, R.D., 115, 117, 125, 131
Calder, J.A., 128
Calvert, S.E., 29
Canuto, V.M., 68
Carstens, H., 13
Casadevall, T.J., 125
Chaloner, W.G., 73
Chambers, L.A., 133, 134, 135
Chameides, W.L., 60, 69
Chandler, G.W., 85
Chang, J., 166
Chen, J., 141
Chester, R., 22
Clarke, F.W., 13
Claypool, G.E., 45, 105, 138, 141
Clayton, R.N., 90, 175
Clemmensen, L., 141
Cloud, P.E., 201, 203, 204, 215
Coats, R.R., 77
Coleman, R.G., 78
Compston, W., 144, 210
Condie, K.C., 191, 212
Converse, D.R., 91
Conway, E.J., 18
Cook, P.J., 201, 204
Cooper, J., 102
Cope, J.J., 73
Corliss, J.B., 80, 82, 91, 92, 99
Cortecci, G., 141
Cox, A., 77
Craig, H.L., 81, 83, 91, 97, 100, 101, 163
Crocket, J.H., 22
Croll, J., 12
Croxford, N.J.W., 171
Cuong, N.B., 42
Cuvier, G., 11

Dahlkamp, F-J., 217
Daly, R.A., 175
Darwin, C., 11
Dasch, E.J., 18
Davies, S.F., 212
Davis, T.A., 31
Davis, J.B., 120, 121
DeBroeu, F., 28
Deffeyes, J.S., 18, 32
Degens, E.T., 108, 113, 114, 116, 175
Deines, P., 129, 137, 142
Delaney, J.M., 99
Delaney, J.R., 114, 115, 125
De Luc, J.A., 9
Delwiche, C.C., 113
De Maillet, B., 7, 8

DeMaster, D.J., 34
DePaolo, D.J., 102
DesMarais, D.J., 100, 115
Desmarest, N., 8
Detrick, R.S., 82
Deuser, W.G., 131
DeVooys, C.G.N., 113, 114
Dickson, F.W., 23, 85
Dietz, R.S., 77, 184
Dimroth, E., 121
Donaldson, J.A., 214
Donnelly, T.H., 171
Donnelly, T.W., 79
Dorman, J., 212
Dostalek, M., 118
Drever, J.I., 18, 28, 69, 72, 92, 214
Duce, R.A., 42, 113, 116, 124
Dunlop, J.S.R., 166, 169, 204
Dunne, T., 153
Duursma, E.K., 113, 116
Dymond, J., 100, 101

Edmond, J.M., 23, 41, 50, 80, 82, 87, 88, 90, 97, 98, 101, 127
Elderfield, H., 99, 143
El Wakeel, S.K., 108
Emery, K.O., 29
Engel, A.E.J., 175, 200, 203, 210, 217, 218
Epstein, S., 53, 131, 142, 175
Eriksson, E., 125
Eriksson, K.A., 214
Ermishkina, A.I., 122
Ewart, A., 102

Fabbi, B.P., 125, 126
Fanale, F.P., 163
Fanning, K.A., 34
Faure, G., 142, 144
Fenchel, T., 59
Ferguson, J., 171
Fischer, A.G., 151
Fisher, D.E., 83, 100
Fisher, I.St.J., 142
Fisher, O., 13, 17
Fitton, W.H., 9
Fitzgerald, W.F., 42
Folk, R., 44
Forrester, J.W., 176, 178, 179, 181, 193
Francois, L.M., 147
Frey, R.W., 52
Friend, J.P., 118, 125
Fripp, R.E.P., 171
Frishman, S.A., 131

Froelich, P.N., 82
Fry, B., 53
Fryer, B.J., 213
Fuchs, G., 130, 131

Galimov, E.M., 129, 130, 136, 142
Garcia, M.O., 125
Gardner, W.S., 116
Garrels, R.M., 14, 18, 19, 22, 23, 25, 27, 31, 33, 38, 41, 43, 44, 64, 67, 105, 107, 110, 127, 130, 137, 145, 151, 152, 153, 169, 175, 200–203, 205
Garrett, D.E., 175
Gaudette, H.E., 118
Geikie, A., 8
Gerard, J-C., 147
Gerasimovskiy, V.I., 137
Gilluly, J., 175
Gimmel'farb, G.B., 203
Gingrich, P.D., 178
Glikson, A.Y., 195
Gold, D.P., 129
Goldberg, E.D., 38, 41
Goldhaber, M.B., 118, 120, 121, 134, 135
Goldschmidt, V.M., 2, 13, 206
Goldstein, S.D., 208
Goodwin, A.M., 171
Goody, R.M., 57
Gordeyev, V.V., 38
Graedel, T.T., 67
Graham, D.W., 159
Granat, L., 118, 124, 125
Grandstaff, D.E., 214
Gray, J., 113
Gregor, C.B., 14, 17, 25, 125, 175, 187
Grinenko, V.A., 137
Grossman, L., 162
Groves, D.I., 204
Guettard, J.E., 8

Hajash, A., 85
Hale, L.D., 82
Hall, J., 8, 10
Hallam, A., 149, 157
Hamilton, E.I., 29
Hamilton, P.J., 208
Hampicke, U., 113
Handa, N., 113, 114, 116
Hardy, R.W.F., 59
Hargraves, R.B., 212
Harper, H.E., 53
Harris, D.M., 114
Harrison, A.G., 135

Hart, R., 18, 87–90, 93, 99
Hart, S.R., 79, 82
Hartmann, M., 134
Hawkesworth, C.J., 102
Hay, W.W., 96, 107, 108, 114, 117, 121, 122, 127, 137, 201, 218
Hays, J.D., 150
Heath, G.R., 156
Hedges, J.I., 116, 121
Heinrichs, T.K., 169, 204
Helgeson, H.C., 18, 86, 99
Henderson-Sellers, A., 57
Herron, T.L., 82
Hess, H.H., 77, 184
Hitchcock, D.R., 123
Hobart, M.A., 79
Hoefs, J.L., 115, 142, 144, 167, 175
Hoffman, E.J., 124
Hoffman, P.F., 215
Hogan, L., 100, 101
Holeman, J.N., 38, 125
Holland, H.D., 14, 17, 23, 25, 27, 28, 33, 44, 45, 47, 49, 50, 67, 68, 72, 85, 91, 105, 113, 114, 125, 127, 167, 169, 211, 214, 215, 217
Holmes, A., 175
Holser, W.T., 27, 51, 110, 122, 125–127, 136–138, 140, 201, 204
Honnerez, J., 79
Horn, M.K., 20
Hower, J., 43
Hsü, K.J., 33
Hubberton, H.W., 126, 137
Hudson, J.D., 142
Humphris, S.E., 23, 79, 95
Hunt, J.M., 162
Hunt, T.S., 13
Hurley, P.M., 14, 175, 187, 191
Hutcheson, I., 152
Hutton, J., 7–11

Ibach, L.E.J., 114
Irwin, W.P., 115, 117, 153
Ittekkot, V., 116
Ito, E., 99
Ivanov, M.V., 118, 136

Jackson, T.A., 142
Jacobsen, S.B., 189
Jaeschke, W., 124
Jaeschke-Boyer, H., 167
Jansen, C., 193
Jansen, S.L., 125, 175, 178, 179, 184–193, 200, 205, 208–210

Jenkins, N.J., 80, 83
Jenkyns, H.C., 156
Jessop, A.M., 212
Johnson, R.M., 102
Johnston, D.A., 125
Joly, J., 17
Jorgensen, B.B., 121, 123, 124
Junge, C.E., 125, 173

Kaleska, M., 29
Kanahira, K.L., 137
Kaplan, I.R., 27, 110, 118, 120, 121, 129, 130, 137, 138
Karhu, J., 175
Karig, D.E., 84
Kasting, J.F., 61, 68, 69, 71
Kay, R.W., 84, 102
Keays, 125
Keith, M.L., 132, 157, 175
Kellogg, W.W., 118
Kemp, A.L.W., 134
Kempe, S., 113, 114, 116
Kerridge, J.F., 97, 136
Kidd, W.S.F., 212
Kimberley, M.M., 121
Kinsman, D.J.J., 122
Kirkland, D.W., 120
Kirwan, R., 8–10
Knauth, L.P., 169, 175, 214
Knoll, A.H., 53, 69
Kohlmaier, G.H., 111, 113
Kovach, J., 142
Kroner, A., 215, 217, 218
Kroopnick, P., 124, 125
Krouse, H.R., 135
Ku, T.L., 22, 38
Kuenen, Ph.H., 203, 204
Kump, L.R., 64, 67
Kuo, H.Y., 22
Kurz, M.D., 83
Kvet, R., 118

Lambert, I.B., 169, 171
Land, L.S., 44
Langmuir, C., 91
Larimer, J.S., 162
Lasaga, A.C., 48, 64, 68, 70–72, 158
Lawrence, J.R., 23
Laznicka, P., 205
Lefond, S.R., 204
LePichon, X., 77, 184
Lerman, A., 31, 34, 107, 130, 145, 156, 180, 181
Leventhal, J.S., 121, 156

Lew, M., 137
Lewis, B.T.R., 82
Lewis, C., 126
Lewis, T., 212
Li, Y-H., 19, 24, 42, 110, 136, 163, 164, 175
Libby, W.F., 129
Lindh, T.B., 138, 139, 140, 143
Lisitsyn, A.J., 38
Livingstone, D.A., 25, 26, 38, 125
Longinelli, A., 175
Lotka, A.J., 180
Lovelock, J., 219
Lowe, D.R., 166, 169, 175, 204, 214
Lowenstam, H.A., 44
Lupton, J.E., 81, 83, 100, 101
Lutz, B., 175
Lyell, C., 11
Lyle, H., 98
Lyons, W.B., 118

MacDonald, K.C., 91
Machel, H.G., 205
Mackenzie, F.T., 14, 18, 19, 22, 23, 25, 27, 31, 33, 38, 41, 43, 44, 74, 127, 150, 151, 153, 161, 169, 175, 181, 200–205, 211
Madsen, J.A., 82
Magaritz, M.D., 101, 141, 142, 144
Makela, M., 171
Malahoff, A., 63
Mangini, A., 38
Manglesdorf, P.C., 27, 28
Manheim, F.T., 29
Manuel, O.K., 22, 100
Marowsky, G., 22, 134, 171
Martin, J.M., 24, 27, 30, 38, 40, 202
Marvin, U.B., 77
Mason, T.R., 215
Mathez, E.A., 125, 128
Maynard, J.B., 18, 29, 31–34, 72, 87, 88, 93, 135, 142
McAlester, A.L., 67
McCready, R.G.L., 135
McCulloch, M.T., 193, 208
McDuff, R.E., 97
McElhinny, M.W., 201, 204
McElroy, M.B., 69
McKenzie, D.P., 212
McLennan, S.M., 175, 193, 210, 215
McPhee, J., 9
Mead, W.J., 20
Meade, R.H., 34, 202
Meadows, A.J., 57
Mekhtieva, V.L., 136

Mel'nik, Y.P., 72
Menard, H.W., 31
Menzel, D.W., 116
Meybeck, M., 24–28, 30, 38, 40, 153, 202
Meyer, B., 108, 123, 124
Michard, A., 209
Migdisov, A.A., 137, 175, 210, 213
Milliman, J.D., 34, 202
Miyake, Y., 42
Monster, J., 136, 169, 171
Mook, W.G., 131, 132
Moorbath, S., 166, 214
Moore, J.B., 99
Moore, J.G., 100, 125, 126
Moore, P.D., 52, 53
Moore, W.S., 38
Mopper, K., 108, 113, 114
Morgan, W.J., 184
Morris, R.C., 72
Morton, J.L., 82
Mottl, M.J., 23, 85–88, 91–93
Mountjoy, E.W., 205
Moyers, J.L., 42
Mroz, E.J., 41
Muelenbachs, K., 90
Muenow, D.W., 125
Murray, J., 9

Naldrett, A.J., 126
Nanz, R.H., 175
Newmark, R.L., 90
Nicol, D.L., 171
Nielsen, H., 110, 132, 134, 138, 137, 172
Nieuwenkamp, W., 1–4, 13
Nuti, S., 175

Odum, H.T., 176, 196–199
Ohmoto, H., 132, 136
O'Leary, M.H., 129, 131
Olson, J.M., 69
O'Neil, J.R., 132, 133
O'Nions, R.K., 95, 101, 193, 208
Orr, W., 118
Ospovat, A.M., 8
Oversby, V., 102
Owen, R.M., 70

Palmer, M.R., 143
Pardue, J.W., 131
Parker, P.L., 116, 121, 128
Patchett, P.J., 193, 208
Peck, H.D., 171
Pelet, R., 113, 114, 116
Peng, T-H., 63

Perry, E.C., 171, 175, 212
Perry, E.J., 105, 110, 130, 137, 145, 151, 153
Petrenchuk, O.P., 125
Pflug, H.D., 167
Pielou, E.C., 180
Pigott, J.D., 44, 74, 150, 151, 161
Pilson, M.E.Q., 34
Piper, D.Z., 22
Pitman, W.C., 115, 150, 159
Playfair, J., 9
Pollack, H.N., 184
Popp, B.N., 142, 175
Porrenga, D.H., 29
Postgate, J.R., 133
Pressler, J.W., 125
Price, N.B., 34
Puchelt, H., 126, 137
Pytkowicz, R.M., 18, 48

Rahn, K.S., 42
Raiswell, R., 120, 121, 145–149, 156, 157
Rand, J.R., 14, 175, 187, 191
Randle, K., 22
Rea, D.K., 70
Reade, T.M., 12
Redfield, A.C., 60
Rees, C.E., 126, 133, 134
Reibach, P.H., 130
Reimer, T.O., 169, 204
Renard, M., 159
Richardson, S.M., 71
Riley, J.R., 108
Ringwood, A.E., 95
Ripley, E.M., 171
Rittenberg, S.C., 134, 136
Robertson, D.S., 217
Rodgers, C.D., 57
Rona, P.A., 79, 81
Ronov, A., 14, 19, 23, 24, 84, 95, 105–110, 121, 122, 125–127, 130, 137, 138, 175, 187, 191, 199–204, 206, 210, 211, 213, 215
Rosenbauer, R.J., 80
Rosendahl, B.E., 82
Rousseau, D., 208
Roy, A.B., 133
Rubey, W.W., 206, 211
Russell, K.L., 18, 32
Rutten, M.G., 217
Rye, R.O., 132, 136

Sackett, W.M., 131
Sadler, P.M., 178
Saito, K., 100

AUTHOR INDEX

Salop, L.J., 175
Saltzman, E.S., 138, 139, 143
Sandberg, P.A., 44, 74, 161
Savin, S.M., 158
Sayles, F.L., 27, 28, 29
Scarfe, C.M., 79
Scharff, M.L., 137, 142
Schidlowski, M., 72, 110, 131, 132, 137, 142, 162–169, 173, 175
Schlanger, S.O., 150, 156
Scholl, D.M., 84
Scholle, P.A., 156, 205, 218
Schopf, J.W., 43, 72, 205
Schuchert, C., 175
Schwab, F.L., 175, 210
Schwartzman, D.W., 100
Sclater, J.G., 184, 187, 192
Scott, R.B., 82, 125
Seccombe, P.K., 171
Seckbach, J., 131
Seilacher, A., 52
Sellers, W.D., 55, 56, 58
Seto, Y.B., 42
Seyfried, W.E., 23, 82, 85, 86, 88, 91, 97, 99, 101
Shackleton, N.J., 160
Shand, S.J., 13
Shanks, W.C., 136
Sharma, G.D., 29
Sharman, G.F., 84
Shelton, J.E., 125
Shepard, S.M.F., 97
Sholkovitz, E.R., 34
Sibley, D.F., 24, 95, 96, 100, 117, 206
Sidorenko, A.V., 203
Sillén, L.G., 18
Sirevåg, R., 130, 131
Slansky, M., 204
Sleep, N.H., 23, 79, 82, 85, 87, 88, 93, 95, 98–100, 114, 125, 127, 206
Slemr, F., 42
Sloss, L.L., 151–155, 187
Smith, B.N., 53, 131
Smith, D.G.W., 79
Smith, J.W., 129, 171
Smith, S.M., 31
Southam, J.R., 96, 107, 114, 117, 121, 127, 137, 201, 218
Speed, R.C., 151–155
Spivack, A.J., 82, 90, 91, 92
Spooner, E.T.C., 101
Sprague, D., 184
Stakes, D.S., 79

Stanworth, C.W., 204
Staudigel, H., 79, 82
Stille, H., 175
Stoiber, R.E., 125, 128
Strakhov, N.M., 175, 200, 201, 203, 204
Stumm, W., 55
Styrt, M.M., 97
Suess, E., 114
Summerhayes, C., 114, 116
Sundquist, E.T., 61, 105
Sverdrup, H.N., 19
Sweeney, R.E., 121, 134

Tanner, J.T., 180
Tappan, H., 67
Taylor, H.P., 82, 86, 90
Taylor, P.S., 125, 128
Taylor, S.R., 20, 175, 193, 208, 210, 215
Thode, H.G., 134, 136, 171
Thompson, G., 23, 79, 87–89, 95
Thompson, S.L., 113
Tissot, B., 113, 114, 116, 120, 204
Tolstikhin, I.N., 101
Toth, D.J., 156
Trendall, A.F., 72, 105
Trudinger, P.A., 133, 134, 135
Trüper, H.G., 133, 171
Tsunogai, S., 42
Tucker, M.E., 205
Tugarinov, A.I., 175, 187, 192
Turcotte, D.L., 149
Turekian, K., 20, 22, 25, 81, 91

Umbgrove, J.H.F., 175
Uyeda, S., 77

Vail, P.R., 127, 144, 149
Valley, J.W., 132, 133
Van Keulen, H., 111
Van Moort, J.C., 175
Veizer, J., 11, 45, 113, 115, 125, 139, 142, 144, 173, 175, 178, 179, 184–189, 192–195, 199–218
Vening Meinesz, F.A., 2
Vinogradov, A.P., 134, 171, 175, 206
Vishmirskij, V.S., 204
Vogel, T.A., 24, 95, 117
Vogel, J.C., 131
Volkov, V.N., 204
von Bartalanffy, L., 176, 179, 193
von Brunn, V., 215
Von Damm, K.L., 23, 50, 80, 87, 88, 97
von Platen, H., 13

AUTHOR INDEX

Wadleigh, M.A., 212
Walker, J.C.G., 57–62, 65–74, 113, 162, 168, 203, 204, 211, 217
Walshaw, C.D., 57
Walter, M.R., 166
Ward, B.B., 67
Wasserburg, G.J., 102, 189, 192, 208
Watson, A., 73
Weaver, C.E., 43
Weber, J.N., 132, 175
Wedepohl, K.H., 20, 22, 59, 110
Wegener, A., 77
Weiss, N.O, 212
Welhan, J.A., 81, 91, 97
Welte, D.H., 142
Wendt, I., 131
Werner, A.G., 8
Wilband, J.T., 95, 96, 100, 206
Wilde, P., 156
Wilgus, C.K., 141, 144
Wilkinson, B.H., 73, 74, 161
Williams, C.F., 82
Williams, J., 63
Williams, N., 121
Wilson, E.O., 180, 199
Wilson, J.L., 205

Windley, B.F., 214
Winkler, H.G.F., 13
Wise, D.U., 99
Wolery, T.J., 23, 79, 85–88, 92–100, 114, 125, 127, 206
Wolfe, R.S., 67
Wollast, R., 18, 28
Wong, W.W., 129, 131
Worsley, T.R., 31
Wright, J., 142
Wyllie, P.J., 77, 99

Yarbrough, H.D., 121
Yaroshevskiy, A.A., 19, 23, 24, 84, 95, 105
Ycas, M., 67
Yeats, P.A., 41
Yeh, H.W., 142
Yung, Y.L., 69

Zahnle, K.J., 68
Zharkov, M.A., 122, 126, 204
Ziegler, A.M., 151
Ziman, K.E., 108
Zoback, M.D., 82
Zoller, W.H., 41

SUBJECT INDEX

Adiabatic decompression, 80
Aerobic respiration, 63
Aerosols, 41, 42
Age distribution models, 179–183, 187, 188, 193
Age of earth, 13, 17
Agricola, 6
Air temperature, 56, 57
Algae, 131. *See also* Cyanobacteria
Alkalinity, 18, 75
Alteration of basalt:
 depth, 91
 at high T, 79–84
 at low T, 89–102
Ammonia, 59, 70
Anaerobic decarboxylation, 62
Angiosperms, 130
Anhydrite, 32, 204. *See also* Sulfate
 in MOR basalts, 79, 97
Anoxic events, 50, 51, 127, 140, 156
Antimony, 41, 42
Apatite, 142
Archean, 138, 169, 171, 172, 192, 194, 203, 204, 207–215
Archean geothermal gradients, 212
Argillite, 203, 205, 210. *See also* Shale
Argon isotopes, 95, 100
Arkose, 201, 203, 205, 215. *See also* Sandstone
Arsenic, 40, 97
Assimilatory sulfate reduction, 132, 135
Authigenic minerals, 28
Auvergne, 8
Average rock:
 igneous, 20, 122
 sedimentary, 20, 24
Axial vents, 80–83. *See also* Off-axis vents
 ages of, 91

Back-arc spreading, 82
Bacon, Francis, 9, 10
Bacteria, 118, 121, 131, 133. *See also* Sulfate, reduction
 photosynthetic, 168
Bacterial sulfate reduction, antiquity of, 169

Baltic Sea, 134
Banded iron-formation, *see* Iron-formation
Barants shelf, 29
Barite, 169, 171, 204, 213
Barium, 41
Basalt:
 alteration at high T, 79–84, 91
 alteration at low T, 89–102
 reaction with seawater, 23, 41, 50, 77–103, 117, 127
Basaltic crust, *see* Oceanic crust
Basins:
 active margins, 184, 187–190, 199, 201–202, 218
 passive margins, 185, 187–190, 199, 201–202, 218
Bearing shelf, 29
Becquerl, Henri, 12
Beryllium isotopes, 102
Big bang hypothesis, 6
Biogenicity, 172
Biosphere, 176, 218, 219
Bioturbation, 51, 120
Birch–Uchi greenstone belt, 171
Black, J., 10
Black Sea, 119, 134
Boltwood, B.B., 12
Boron, 82, 91, 92
Boron isotopes, 90
Bristol Channel, 29
Bromine, 42
Brucite, 45
Bruno, Giordano, 6
Buzzard's Bay, 29

Cadmium, 38, 41, 42
Cahill Formation, 171
Calcium:
 in basalt–seawater reaction, 87, 94–97
 excess in rock cycle, 22–24, 94–97, 117
 ion-exchange and, 27
Calcium carbonate, 31. *See also* Carbonate minerals
Cambrian:
 isotope geochemistry, 138, 142

S/C ratios, 121, 157
  sea level, 149, 151, 157
CAM plants, 130, 131
Carbonate compensation depth, 116
Carbonate ions, in oceans, 74–76
Carbonate minerals:
  deposition, 43, 112
  equilibrium with seawater, 31, 116
  in pelagic sediments, 22–24, 94–97, 117
  in skeletons, 52
Carbonate rocks:
  geologic record 201–204, 210, 213, 215, 216
  proportion, 19, 23
Carbonatite, 129
Carbon, residence time of, 63, 117
Carbon budget, 62, 111–117
Carbon cycle coupling to other cycles, 64, 143–161
Carbon dioxide:
  atmospheric, 61, 111
  and carbonate deposition, 32–44, 74–76, 152
  geologic record of, 70–72, 161, 162
  release by metamorphism, 71, 113, 117, 152, 167
  and temperature, 61, 74–76, 158
  and weathering, 49, 65
Carbon fluxes, 111–117
Carboniferous, 139, 149–151, 157
Carbon isotopes:
  abiologic fractionations, 131, 132, 164
  biologic fractionations, 129–131
  geologic record of, 136–145, 163–168
  in MOR basalts, 100, 137
Carbon monoxide, 67, 162
Carbon/phosphorous ratio, 167
Carbon reservoirs, 64, 106–111, 163, 167
Carbon/sulfur ratio, *see* Sulfur/carbon ratio
Cariaco Trench, 119
Catastrophism, 8, 11
Cation exchange, 27, 59
Cavendish, Henry, 7
Celadonite, 92
Cerium, 142
Charcoal, and atmospheric oxygen, 73
Charles II (of England), 9
Chert, 45, 203, 204
Chloride, 33
Chlorine, 99
Christian apologists, 6
Classical magmatism, 2, 11, 14
Clastics, geologic record, 201–203, 215, 216
Clausius, R., 12

Clay content, of altered basalt, 90
Clays, reaction with seawater, 18, 27, 59
Coal, 157, 204, 218. *See also* Fossil fuel(s)
Cobalt, 41
Computer simulation of basalt–seawater reaction, 86
Conduction of heat, 81
Conglomerate, 203, 214, 215
Continental crust, 182, 187–190, 192–198, 203, 206, 207, 209, 211, 214
Continental drift, 5
Continents, growth, 13, 192, 197, 198, 215
Continuous rock series, 2
Cook, Captain James, 7
Copper, 41, 42, 97
Cosmology, 5
Costa Rica Rift, 90
C3 plants, 130, 131, 166
C4 plants, 53, 130, 131
Cratonization, 175, 191, 194, 203, 207, 215, 216, 218
Cratons, 185, 194, 198, 203, 215
Cretaceous:
  C isotopes, 140, 144, 156
  CO_2 level, 72
  sea level, 150, 157, 160
  Sr isotopes, 142
  temperature record, 158
Cyanobacteria, 131
Cycle:
  exogenic, 175, 201, 217, 219
  magmatic, 214
  orogenic, 196, 198
Cyclic aspect of geology, 5
Cyclic salts, 25, 41

D'Alembert, Jean Le Rond, 7
da Vinci, Leonardo, 12
Deductive philosophy, 10
Deep-sea sediments, *see* Pelagic sediment
Deforestation, 63
Degassing of mantle, 100, 162–164
Deism, 10
Denitrification, 59, 69
Desulfovibrio, 118
Deuterium, 101
Devonian, 157
Diamond, 129
Diatoms, 34, 52
Diderot, Denis, 7
Diffusion, 28, 85, 120
Dimethyl sulfide, 123
Dissimilatory sulfate reduction, 132, 169
Dissolved load composition, 24–27, 202

Dissolved organics, *see* Organic matter, degree of use by bacteria
Dolomite, 31, 44
Dolostone, 201, 203, 205

Earliest fossils, 166, 167
Earth, age of, 13, 17
East Pacific Rise, 80–82
Ecclesiastes, Book of, 6
École normale (Paris), 9
Enthalpy, 81
Equilibrium, in hydrothermal systems, 85
Equilibrium models, 18
Erosion of crystalline rocks, 14
Eskola, P., 2
Eukaryotes, 51, 131, 205
Euxinic basins, 135, 142, 156, 157
Evaporation, 56
Evaporite events, 50
Evaporites:
 deposition, 32, 45
 geologic record, 124–127, 204, 213, 216
 proportion, 19, 122
 S isotope age curves, 137–145
 S isotope fractionation, 136
 timing, 37, 51
Evolution effect on rock record, 51–53, 164–171

Fermentation, 121
Ferric/ferrous ratio of MOR basalts, 98, 127, 162
Fig Tree Group, 171
Flint nodules, 10
Flow rates, in hydrothermal systems, 93
Foraminifera, 52, 95
Fortescue Group, 171
Fossil fuel(s):
 combustion, 25, 61, 63, 74–76, 111
 geologic record, 204, 217, 218
 mass, 108
French Revolution, 10
Frood Series, 171

Gaia hypothesis, 219
Galapagos Ridge, 80–82, 97, 127
Galilei, Galileo, 6
Genesis, Book of, 6
Geochemical Society, 14
Geognosy, 8
Geological Society, 10
Geosynclinal sediments, 108, 109, 123
Germanium, 82
Germany, classical magmatism in, 11

Glass, basaltic, 83
Glaucophane schist, 187
Glen Tilt, 8, 10
Global tectonic theory, 16
Godavari Delta, 29
Goethite, 120
Gold, 41, 42, 97
Gold conglomerates, 72
Graphite, 132, 133
Gravitational accretion, 12
Graywacke, 201–205, 214, 218. *See also* Sandstone
Greenhouse effect:
 from CO_2, 61, 74–76
 from methane, 68
 from water vapor, 57, 58
Greenstone, 93, 192, 198, 203, 212, 214–216
Growth of continents, 13, 192, 197, 198, 215
Growth of sedimentary mass, 13–15, 199, 207, 209, 215, 216
Guaymas Basin, 81
Gulf of Alaska, 29
Gulf of California, 134
Gulf of Paria, 29
Gypsum, 32, 45, 46, 204

Halite, 32, 46, 126, 204
Halley, Sir Edmund, 12
Hamersley Basin, 171
Heat budget:
 of atmosphere, 57–61, 68, 74–76
 of mid-ocean ridges, 81
Heat flow, 79–84, 212, 219
Helium, 83
Helium isotopes, 80, 81, 101
Hematite, 120
Hydration, of basalt, 99
Hydrogen:
 in MOR hydrothermal system, 97
 in primitive atmosphere, 65, 68, 162
Hydrothermal circulation depth, 81
Hydrothermal vents and heat flow, 79–84. *See also* Axial vents; Off-axis vents

Ichor, 13
Iengra Series, 171
Inquisition, The, 6
Iodine, 42
Ion-exchange, 27
Iron, *See also* Ferric/ferrous ratio of MOR basalts
 in basalt–seawater reaction, 85
 in pyrite formation, 120
 in ridge–crest sediment, 98, 99

Iron-formation:
  Algoma type, 213
  and atmospheric oxygen, 69, 72, 167
  Superior type, 215
Island arcs, 84, 99–102, 198, 207, 213
Isotope age curves, 138–145
Isotopic mass balance, 137, 166
Isua, Greenland, 165, 171

Juan de Fuca Ridge, 82
Jurassic, 142, 151, 203

Kant, Immanuel, 7
Kelvin, Lord (William Thomson), 11, 12
Kerogen, 163, 165
Kimberlite, 129
Kinetic isotope effect, 134
Kinetic models, 18
Koheleth the Sage, 6
Krantz leaf anatomy, 130
Krayenhoof, C.R.T., 3
Krypton, 100
Kupferschiefer, 134, 171

Land plants, 52, 73, 121, 142, 157, 205, 217
Laplace, Pierre de, 7
Latent heat, 80
Lau Basin, 82
Lavoisier, Antoine, 7
Lead, 41, 42
Lead isotopes, 102
Lightning and atmospheric N, 60
Limestone, *see* Carbonate rocks
Linnaeus, Carolus, 7
Lithium, 40, 82, 91
Lithosphere, 77
Load:
  dissolved in rivers, 24–27, 202
  suspended in rivers, 27, 30, 302
Locke, John, 10
Luther, Martin, 6

Magma chambers, 82
Magnesium:
  in basalt–seawater reaction, 28, 88, 92–94
  ratio to calcium, 44
Magnetic anomalies of ocean floor, 14
Manganese:
  and atmospheric oxygen, 69
  in basalt–seawater reaction, 85
  in ridge-crest sediment, 98, 99
Mantle, 182, 183, 192–198, 207, 208, 212–219

Mercury, 41
Metal-rich sediments, 98–99
Metamorphic rocks, classical magmatism and, 11
Metamorphism:
  and carbon isotopes, 132, 133, 166
  and CO_2 release, 71, 113, 117, 152, 167
Metazoa, 51, 217
Meteorites, 129, 162
Methane, 67, 97
Methanogenesis, 63, 165
Michipicoten iron-formation, 171
Mid-Atlantic Ridge, 81, 82
Mid-ocean ridges, 182, 197, 212
  and CO_2, 70
  heat flow from, 80, 81
  reactions at, 23, 41, 50, 79–84, 117, 127, 136
Miller–Urey process, 166
Mineral deposits, 192, 205, 213–218
Models of C–S system, 145–148
Molybdenum, 41
Mosaic time scale, 7, 8
Mudstone, 203. *See also* Shale

Napoleon, 10
Neodymium isotopes, 102
Neon, 100
Neptunism, 7, 8
Nickel, 41
Niger Delta, 29
Nile Delta, 7
Nitrate, 60
Nitrogen, 59–61, 69–70, 162
Nitrogen fixation, 59, 70
Nonesuch Shale, 171

Oceanic crust, 182–184, 187–190, 195, 196, 206–218
  age of, 78
  thickness of alteration, 94
Oceanic mixing state, 157, 212
Oceanic sediment, subduction of, 84, 99–102
Off-axis vents, 82, 83, 91. *See also* Axial vents
Oil, *see* Fossil fuel(s)
Old Red Sandstone, 7, 10
Onverwacht Group, 171
Onwatin Slate, 171
Ooids, aragonite *vs.* calcite, 44, 74, 161
Open system sulfate reduction, 135, 142
Ophiolite, 182, 187
Ordovician, 121, 126, 149
Oregon, shelf, 29

Organic matter, degree of use by bacteria, 116, 117, 120, 121. *See also* Carbon, residence time of
Orinoco Delta, 29
Orogenic belts, 185, 187–190, 196, 201, 202, 218
Orthogneiss, 11
Oxford University, 11
Oxidation state of mantle, 162
Oxygen:
  in atmosphere, 65–67
  in basalt–seawater reaction, 97–99
  biogeochemical cycles of, 66
  geologic record of, 64, 68, 69
  isotopes:
    and basalt–seawater reaction, 82, 90
    in carbonates, 44, 142, 158, 160
    in evaporites, 142
  minimum layer, 214
  and organic matter, 66, 121

Palagonite, 94. *See also* Basalt, alteration
Pangea, 150, 161
Paragneiss, 11
Paris Basin, 11
Particulate organics, *see* Organic matter, degree of use by bacteria
Patterson, Claire, 13
Pelagic sediment, 19, 108, 109, 117
  subduction, 95, 96, 102, 112
Permeability, of MOR basalts, 90
Permian, 138–141, 144, 161
Persedimentary theory, 2
Philipsite, 92
pH of oceans CO_2 and, 74–76
Phosphate, ratio to N, 60
Phosphorites, 201, 202, 204, 205, 216, 217
Phosphorous, 41, 42, 153, 157, 167
Photochemical reactions, 60
Photolysis of water, 68
Photosynthesis, 61, 65, 112, 129, 166
  oxygen-evolving, 68, 168
Pine Creek geosyncline, 171
Plants, *see* Land plants; Algae
Plate tectonics, 77–84, 184, 192, 197, 198, 217, 218
Platform sediments, 108, 109, 185–190, 199–202
Platinum, 97
Plutonism, 7, 8, 9
Pollution, 25
Population dynamics, 176, 180, 187, 197, 218, 219
Pore water, 27, 28

Porosity of MOR basalts, 82
Potassium, in basalt–seawater reaction, 87, 90–92
Potassium isotopes, 100
Prebiologic atmosphere, 68
Precambrian evaporites, 169
Precambrian oxygen, 69, 72, 73, 169
Precipitation, atmospheric, 56, 58, 123
Primitive atmosphere, *see individual compounds*
Prokaryotes, 130, 169, 204
Proterozoic, 138, 171, 192, 194, 199, 203–205, 209, 215–218
Protestant Church, 6
Pyrite, 120, 121. *See also* Sulfur
  S isotopes of, 135

Quartz, in basalt–seawater reaction, 86
Quartzite, 203, 215. *See also* Sandstone
Quartz–pebble conglomerate, 73

Radioactivity, 12
Radiolaria, 52
Raguin, E., 3
Rainwater, 25
Rare earths, 38
Rate of deposition:
  of evaporites, 126
  of sulfides, 33, 127, 135, 148
Rate of sedimentation, 120, 142, 156
Rationalism, 7, 9
Rayleigh, Lord (R.J. Strutt), 12
Rayleigh distillation, 135
Reconstitution, *see* Silicate reconstitution
Red beds, 72, 203, 217
Redox reactions, 62
Reformation, the, 7
Reformed Church, 6
Relative humidity, 56
Residence time, 179, 180, 194, 212
  of C, 63, 117
  of S, 123
  of water vapor, 56
Respiration, 62, 112
Reverse weathering, 23, 64, 153
Reykjanes brines, 85
Ridge, *see* Mid-ocean ridges
Ridge volume, 150, 158, 159
River water, 24, 116. *See also* Load
Rock cycle, 5, 14, 15
Rock/water ratios, 85, 91, 93
Rosenbusch, Harry, 11
Rot chemical event, 140, 144
Roth, Justus, 11

Royal Society, 9
Royal Society of Edinburgh, 7, 9, 10
Rubidium, 91
Rubidium isotopes, 101
Rutherford, Ernest, 12

Salt content of oceans, 12, 13, 17
Samarium, 95, 194
Sandstone, 19, 201, 203, 205, 215. *See also* Graywacke; Arkose
Sarawak Shelf, 29
Saturated vapor pressure, 56, 58
Sea-floor spreading, rate, 49, 71, 83
Sea level:
  and isotope curves, 149–151
  and spreading rate, 71, 150, 158, 159
  and tectonics, 50
Seawater at high T, 80, 85
Sebakwian Group, 171
Sederholm, J.J., 2
Sediment:
  composition, 29, 206, 210
  reaction with seawater, 28, 82
Sedimentary mass growth, 13–15, 199, 207, 209, 215, 216
Sedimentary rocks:
  geologic record of, 43, 151, 192, 199–205
  proportions of, 19, 107–109, 162
Sedimentation rate, *see* Rate of sedimentation
Sediments, subduction of, 95, 96, 102, 112
Selenium, 41, 97
Sepiolite, 45
Septarian concretions, 10
Shale, 19, 201–203, 205
  C content, 121, 204
Silica, biogenic, 33
Silicate reconstitution, 23, 64, 153
Silicon, in basalt–seawater reaction, 88
Silurian, 205
Silver, 38, 41, 42, 97
Smectite, 92
Smith, William, 10
Sodium:
  in basalt–seawater reaction, 87
  cyclic behavior, 2, 12–14
  ion-exchange and, 27
Solar luminosity, 57
Soviet Academy of Sciences, 14
Spencer Gulf, 134
Steady-state, 17
Strontium, 159, 160, 212
Strontium isotopes, 101, 102, 142–144, 158–160, 211, 212

Subduction:
  of hydrated basalt, 99
  of sediments, 95, 96, 102, 112, 117, 167
Subduction zones, 84
Sudbury District, 171
Sulfate:
  build-up in oceans, 25
  concentration, effect on reduction rate, 135, 148
  reduction:
    abiogenic, 63
    bacterial, 118–121, 133, 169, 170
    in basalt–seawater reaction, 86, 97
    kinetics of, 86
Sulfide deposition, 33, 126, 127. *See also* Rate of deposition
Sulfide minerals, in MOR hydrothermal system, 97
Sulfite, 168
Sulfur:
  atmospheric, 123
  in basalt–seawater reaction, 95, 97, 127
  budget, 118–129
  dioxide, 123, 168
  fluxes, 123–129
  isotopes:
    abiogenic fractionations, 136
    biologic fractionations, 132–136
    geologic record of, 136–145, 168–173
    in MOR basalts, 137
    in river water, 27
  reservoirs, 121–123
Sulfur/carbon ratio, 121, 127, 145, 149, 156–158
Suspended load, 27, 30, 202

Tectonic realms, 184–190, 199–202, 218
Tectonics:
  and CO_2 levels, 70
  effect on seawater, 49, 77–84
  and isotope curves, 144, 148–161
Terrestrial vegetation, *see* Land plants
Tertiary, 204, 213, 217
  temperature record, 72, 158
Thorium, 38
Tin, 41
Triassic, 140–144, 150
Troposphere, 56
Tungsten, 41

Unconformity, 7, 10, 11
Uniformitarianism, 8, 11
Uranium, 38
  and atmospheric oxygen, 72

Ussher, Archbishop, 6
Utrecht University, 2

Van Marum, Martin, 3
Vents, see Axial vents; Off-axis vents
Vesicles, 83
Voight, J.K.W., 7
Volatiles, in basalt–seawater reaction, 99
Volcanism, and sulfur dioxide, 124
Voltaire, 7
Von Buch, L., 3

Warrawoona Group, 171
Water content, of basalt, 90, 99–101
Water vapor, 55, 68

Weathering, 49, 65
 of basalt, see Alteration of basalt, at low T
 and pyrite S measurement, 122
Wegmann, C.E., 3
Wernerian Society, 2, 8
Whitehead, A.N., 16
White Pine mine, 171
Wilbrink, Anna, 1
Woman River iron-formation, 171

Xenon, 100

Yilgarn Block, 171

Zinc, 41, 42, 85, 97